Transiting Exoplanets

The methods used in the detection and characterisation of exoplanets are presented through the study of transiting systems in this unique textbook for advanced undergraduates. From determining the atmospheric properties of transiting exoplanets to measuring the planetary orbit's alignment with the stellar spin, students will discover what these measurements imply for reinvigorated theories of planet formation and evolution. Produced by academics drawing on decades of Open University experience in supported open learning, the book is completely self-contained with numerous worked examples and exercises (with full solutions provided), and illustrated in full-colour throughout. Designed to be worked through sequentially by a self-guided student, it also includes clearly identified key facts and equations as well as informative chapter summaries and an Appendix of useful data.

Carole A. Haswell is a Senior Lecturer in Physics and Astronomy in the Department of Physics and Astronomy, The Open University. She worked on accretion flows around black hole binary star systems until becoming fascinated by the field of exoplanets. Her research work now focuses on observations of transiting exoplanets.

Cover image: An artist's impression of the 'hot Jupiter' exoplanet HD 189733b transiting across the face of its star. (ESA, C. Carreau)

Transiting Exoplanets

Author:

Carole A. Haswell

CAMBRIDGE UNIVERSITY PRESS

Cambridge, New York, Melbourne, Madrid, Cape Town, Singapore, São Paulo, Delhi, Dubai, Tokyo

Cambridge University Press
The Edinburgh Building, Cambridge CB2 8RU, UK

In association with THE OPEN UNIVERSITY

The Open University, Walton Hall, Milton Keynes MK7 6AA, UK

Published in the United States of America by Cambridge University Press, New York.

www.cambridge.org
Information on this title: www.cambridge.org/9780521133203

First published 2010.

Edited and designed by The Open University.

Typeset by The Open University.

Printed and bound in the United Kingdom by Latimer Trend and Company Ltd, Plymouth.

This book forms part of an Open University course S382 *Astrophysics*. Details of this and other Open University courses can be obtained from the Student Registration and Enquiry Service, The Open University, PO Box 197, Milton Keynes MK7 6BJ, United Kingdom: tel. +44 (0)845 300 60 90, email general-enquiries@open.ac.uk

http://www.open.ac.uk

British Library Cataloguing in Publication Data available on request.

Library of Congress Cataloguing in Publication Data available on request.

ISBN 978-0-521-19183-8 Hardback
ISBN 978-0-521-13938-0 Paperback

Additional resources for this publication at www.cambridge.org/9780521139380

1.1

TRANSITING EXOPLANETS

Introduction

One thing is certain about this book: by the time you read it, parts of it will be out of date. The study of exoplanets, planets orbiting around stars other than the Sun, is a new and fast-moving field. Important new discoveries are announced on a weekly basis. This is arguably the most exciting and fastest-growing field in astrophysics. Teams of astronomers are competing to be the first to find habitable planets like our own Earth, and are constantly discovering a host of unexpected and amazingly detailed characteristics of the new worlds. Since 1995, when the first exoplanet was discovered orbiting a Sun-like star, over 400 of them have been identified. A comprehensive review of the field of exoplanets is beyond the scope of this book, so we have chosen to focus on the subset of exoplanets that are observed to transit their host star (Figure 1).

Figure 1 An artist's impression of the transit of HD 209458 b across its star.

These transiting planets are of paramount importance to our understanding of the formation and evolution of planets. During a transit, the apparent brightness of the host star drops by a fraction that is proportional to the area of the planet: thus we can measure the sizes of transiting planets, even though we cannot see the planets themselves. Indeed, the transiting exoplanets are the only planets outside our own Solar System with known sizes. Knowing a planet's size allows its density to be deduced and its bulk composition to be inferred. Furthermore, by performing precise spectroscopic measurements during and out of transit, the atmospheric composition of the planet can be detected. Spectroscopic measurements during transit also reveal information about the orientation of the planet's orbit with respect to the stellar spin. In some cases, light from the transiting planet itself can be detected; since the size of the planet is known, this can be interpreted in terms of an empirical effective temperature for the planet. That we can learn so much about planets that we can't directly see is a triumph of twenty-first century science.

This book is divided into eight chapters, the first of which sets the scene by examining our own solar neighbourhood, and discusses the various methods by which exoplanets are detected using the Solar System planets as test cases. Chapter 2 describes how transiting exoplanets are discovered, and develops the related mathematics. In Chapter 3 we see how the transit light curve is used to derive precise values for the radius of the transiting planet and its orbital inclination. Chapter 4 examines the known exoplanet population in the context of the selection effects inherent in the detection methods used to find them. Chapter 5 discusses the information on the planetary atmosphere and on the stellar spin that can be deduced from spectroscopic studies of exoplanet transits. In Chapter 6 the light from the planet itself is discussed, while in Chapter 7 the dynamics of transiting exoplanets are analyzed. Finally, in Chapter 8 we briefly discuss the prospects for further research in this area, including the prospects for discovering habitable worlds.

The book is designed to be worked through in sequence; some aspects of later chapters build on the knowledge gained in earlier chapters. So, while you could dip in at any point, you will find if you do so that you are often referred back to concepts developed elsewhere in the book. If the book is studied sequentially it provides a self-contained, self-study course in the astrophysics of transiting exoplanets.

A special comment should be made about the exercises in this book. You may be tempted to regard them as optional extras to help you to revise. *Do not fall into this trap!* The exercises are not part of the *revision*, they are part of the *learning*. Several important concepts are developed through the exercises and nowhere else. Therefore you should attempt each of them when you come to it. You will find full solutions for all exercises at the end of this book, but do try to complete an exercise yourself first before looking at the answer. An Appendix containing physical constants is included at the end of the book; use these values as appropriate in your calculations.

For most calculations presented here, use of a scientific calculator is essential. In some cases, you will be able to work out order of magnitude estimates without the use of a calculator, and such estimates are invariably useful to check whether an expression is correct. In some calculations you may find that use of a computer spreadsheet, or graphing calculator, provides a convenient means of visualizing a particular function. If you have access to such tools, please feel free to use them.

Chapter 1 Our Solar System from afar

Introduction

In 1972 a NASA spacecraft called Pioneer 10 was launched. It is the fastest-moving human artefact to have left the Earth: 11 hours after launch it was further away than the Moon. In late 1973 it passed Jupiter, taking the first close-up images of the largest planet in our Solar System. For 25 years Pioneer 10 transmitted observations of the far reaches of our Solar System back to Earth, passing Pluto in 1983 (Figure 1.1).

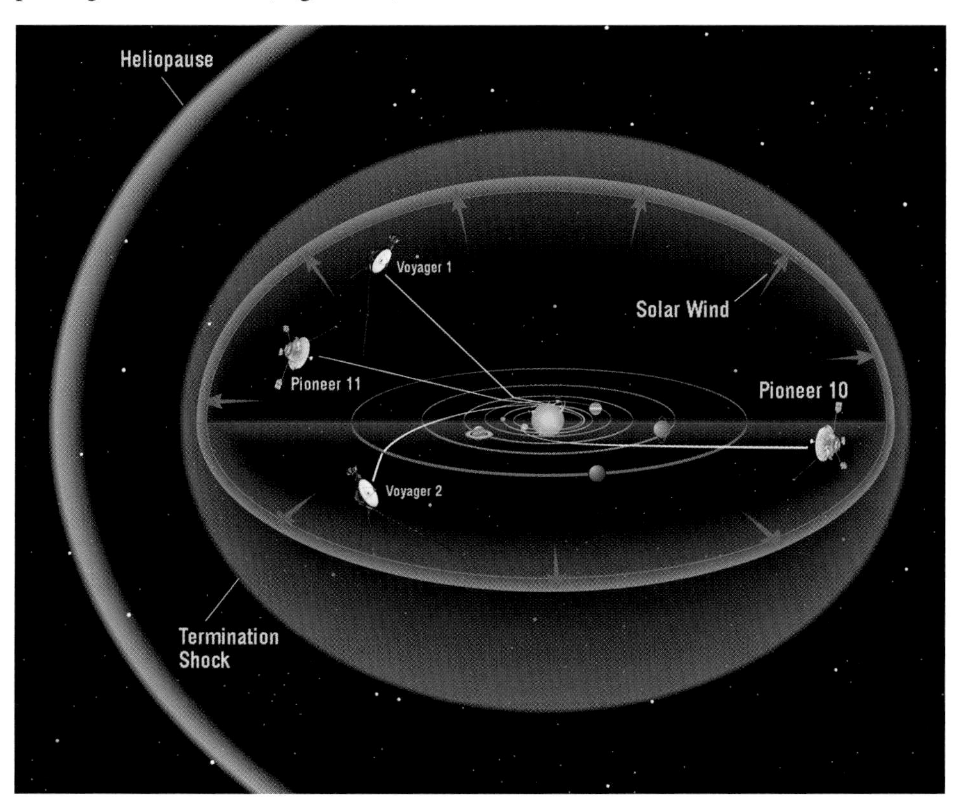

Figure 1.1 The trajectories of Pioneer 10 and other space probes. The heliopause is the boundary between the heliosphere, which is dominated by the solar wind, and interstellar space. The solar wind is initially supersonic and becomes subsonic at the termination shock. Note: all objects are *much* smaller than they have been drawn!

At final radio contact, it was 82 AU (or equivalently 1.2×10^{13} m) from the Sun. It continues to coast silently towards the red star Aldebaran in the constellation Taurus. It is 68 light-years to Aldebaran, and it will take Pioneer 10 two million years to cover this distance. During this time, of course, Aldebaran will have moved.

Any single star may have several different names. This is especially true of bright stars, e.g. Aldebaran is also known as α Tau. We have generally tried to adopt the names most frequently used in the exoplanet literature.

Figures 1.2 and 1.3 show the Sun's place in our Galaxy. The 10-light-year scale-bar in Figure 1.2a is 2000 times the radius of Neptune's orbit: Neptune is the furthest planet from the Sun shown in Figure 1.1. The structure of our own Solar System shown in Figure 1.1 is invisibly tiny on the scale of Figure 1.2a. To render them visible, all the stars in Figure 1.2a are shown larger than they really are. Interstellar space is sparsely populated by stars. The successive parts of Figures 1.2 and 1.3 zoom out from the view of our immediate neighbourhood. In Figure 1.2b we see a more or less random pattern of stars, which includes the Hyades cluster and the bright stars in Ursa Major. All the stars shown in Figure 1.2b belong to our local spiral arm, which is called the Orion Arm.

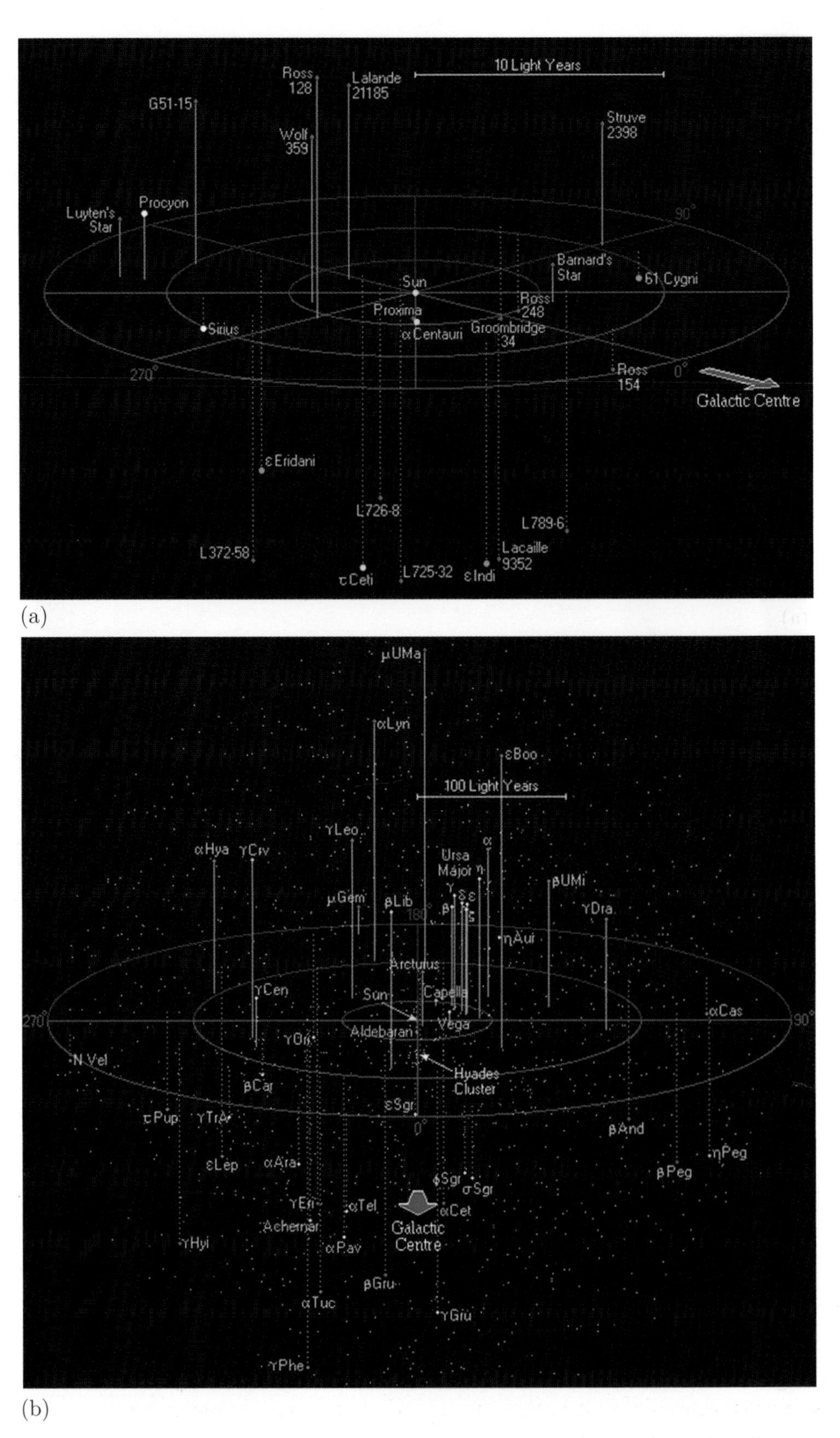

Figure 1.2 (a) The stars within 12.5 light-years of the Sun. (b) The solar neighbourhood within 250 light-years.

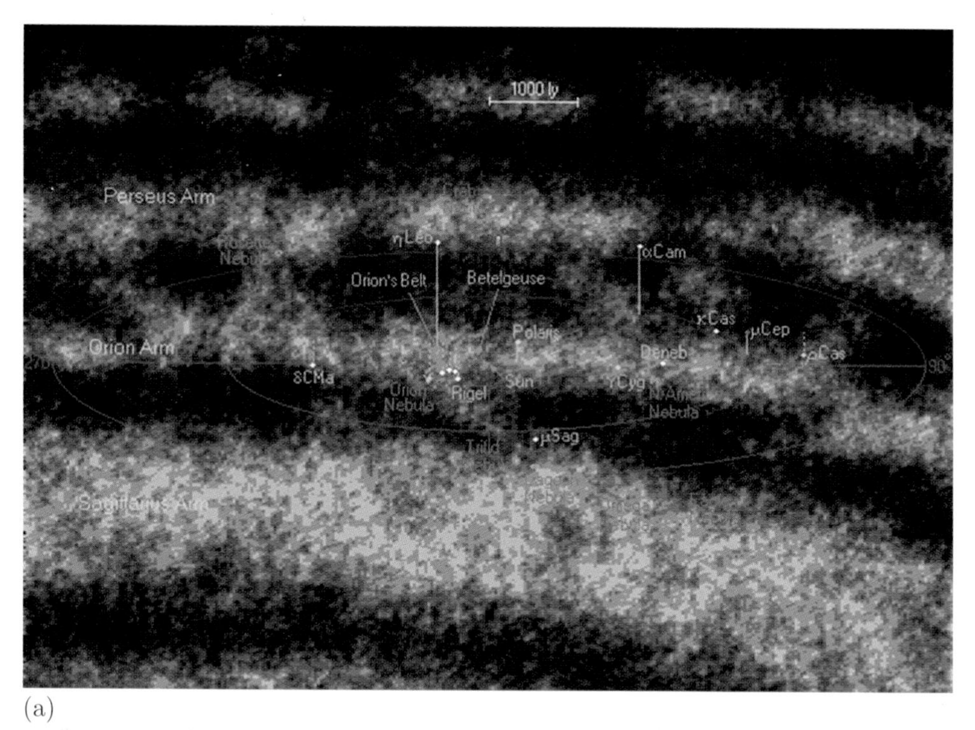

(a)

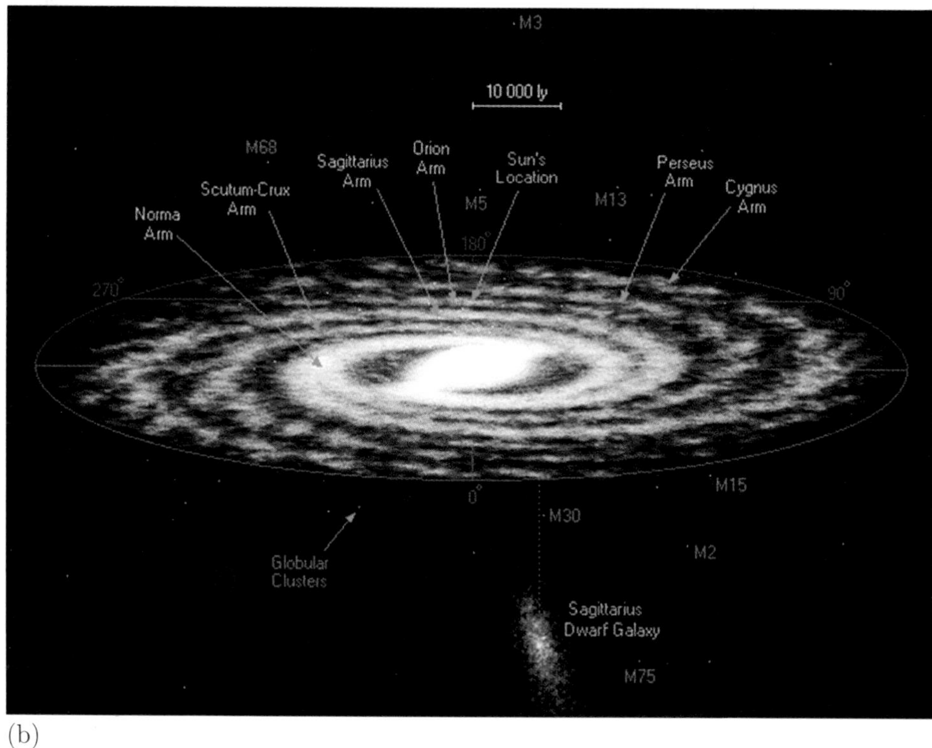

(b)

Figure 1.3 (a) Our part of the Orion Arm and the neighbouring arms. (b) Our Milky Way Galaxy.

Figure 1.3a shows our sector of the Galaxy: a view centred on the Sun, which is a nondescript star lost in the host of stars comprising the Orion Arm. The bright stars of Orion, familiar to many people from the naked-eye night sky, are prominent in this view: they give our local spiral arm its name. Finally, Figure 1.3b shows our entire Galaxy.

Exercise 1.1 Pioneer 10 is coasting through interstellar space at a speed of approximately $12\,\mathrm{km\,s^{-1}}$. Confirm the time that it will take to cover the 68 light-years to Aldebaran's present position. *Hint*: The Appendix gives conversion factors between units. ■

The nearest stars and planets

Planets have been detected around 6 of the 100 nearest stars (October 2009), including the Sun. Throughout this book, where new results are frequent, we indicate in parentheses the date at which a statement was made, as we did in the previous sentence. Table 1.1 lists the seven known planetary systems within 8 pc.

Table 1.1 Stars within 8 pc with known planets. (In accordance with the IAU we list 8 planets in the Solar System; we will not comment further on the status of Pluto and similar dwarf planets.)

Name (alias)	Parallax, π/arcsec	Spectral type	V	M_V	Mass ($\mathrm{M_\odot}$)	Known planets
Sun	—	G2V	−26.72	4.85	1.00	8
ε Eri (GJ 144)	0.309 99 ±0.00079	K2V	3.73	6.19	0.85	1
GJ 674	0.220 25 ±0.00159	M3.0V	9.38	11.09	0.36	1
GJ 876 (IL Aqr)	0.212 59 ±0.00196	M3.5V	10.17	11.81	0.27	3
GJ 832 (HD 204961)	0.202 52 ±0.00196	M1.5V	8.66	10.19	0.45	1
GJ 581 (HO Lib)	0.159 29 ±0.00210	M2.5V	10.56	11.57	0.30	3
Fomalhaut (αPsA/GJ881)	0.130 08 ±0.00092	A3V	1.16	1.73	2.1	1

Confusingly, the standard symbol for astronomical parallax is π, which is more frequently used as the symbol for the constant relating the diameter and circumference of a circle. Usually it is easy to work out which meaning is intended.

These data were taken from the information on the 100 nearest stars on the Research Consortium on Nearby Stars (RECONS) website, which is updated annually. Fomalhaut, which is nearby, but not one of the 100 nearest stars, was added because the planet Fomalhaut b was recently discovered, as we will see in Subsection 1.1.1. Table 1.1 gives the values of the parallax, π, the spectral type, the apparent V band magnitude, V, and the absolute V band magnitude, M_V, as well as the stellar mass and the number of known planets. It is very noticeable that over half of these nearby planet host stars are M dwarfs, the lowest-mass, dimmest and slowest-evolving main sequence stars. By the time you read this, more nearby planets will probably have been discovered.

The nearest stars have distances directly determined by geometry using trigonometric parallax, and the RECONS sample (Figure 1.4) includes 249 systems within 10 pc, or equivalently 32.6 light-years, all with distances known to 10% or better. 72% of the stars are M dwarfs. M dwarfs are the faintest main sequence stars, so although they predominate in the solar neighbourhood, they are more difficult to detect than other more luminous stars, particularly at large distances.

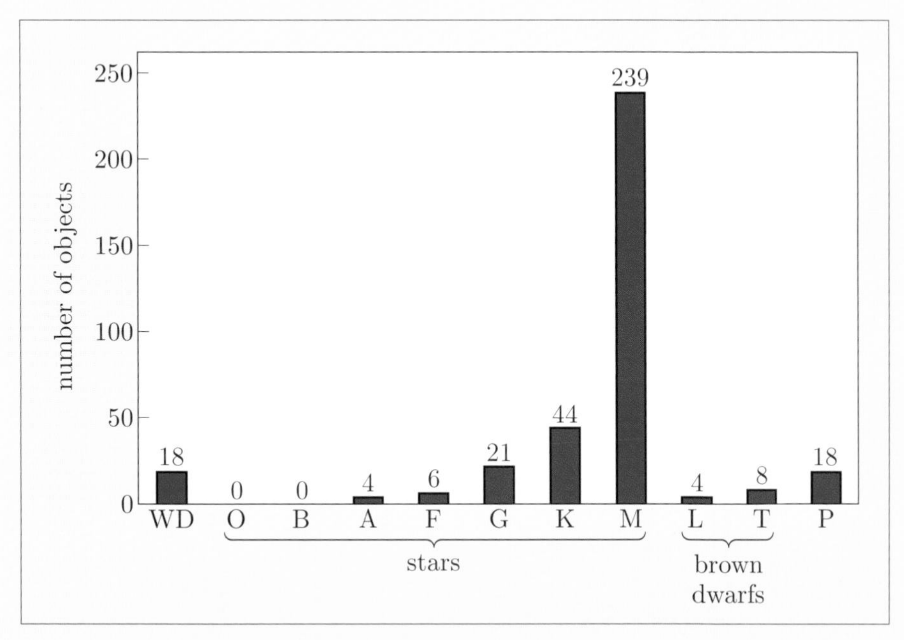

Figure 1.4 Known objects within 10 parsecs (RECONS 2008). The leftmost column shows white dwarfs, columns 2–8 show stars subdivided by spectral type, columns 9 and 10 show L and T class brown dwarfs, and the last column shows known planets, including eight in our own Solar System. The stars in the census are overwhelmingly M dwarf stars, i.e. low-mass main sequence stars.

The census illustrated in Figure 1.4 is likely to be more representative of stellar demographics than a sample limited by apparent brightness would be. The numbers in this census include the Sun and its eight planets. The 362 objects represented in Figure 1.4 are comprised of 170 systems containing a single object, 58 double systems (either binary stars or star plus single planet systems), and 21 multiple systems, including our own Solar System. New objects are being discovered continually, and this census grew by 20% between January 2000 and January 2008.

While not quite close enough to be included in Table 1.1, or in the census illustrated in Figure 1.4, HD 160691 at a distance of 15.3 pc deserves a mention: it has 4 known planets. The recently discovered planet GJ 832 b is a giant planet with an orbital period between 9 and 10 years: i.e. it has pronounced similarities to the giant planets in our own Solar System. GJ 832 b has mass $M_\mathrm{P} = 0.64 \pm 0.06\,\mathrm{M_J}$, where $\mathrm{M_J}$ is the mass of Jupiter, i.e. $1898.13 \pm 0.19 \times 10^{24}$ kg. The properties of its host star, GJ 832, are given in Table 1.1.

- With reference to Table 1.1, what is the distance to the nearest known exoplanetary system?

- ε Eri (GJ 144) has the largest parallax of all known exoplanetary systems, with $\pi = 0.309\,99 \pm 0.000\,79$ arcsec. The distance, d, in parsecs is given by

$$\frac{d}{\mathrm{pc}} = \frac{1}{\pi/\mathrm{arcsec}},$$

so the distance to ε Eri is $0.309\,99^{-1}$ pc, or 3.226 pc. The uncertainty on the parallax measurement is 0.25%, so the uncertainty in the distance determination is approximately 0.25% of 3.226 pc. Hence, to four significant figures, the distance to the nearest known exoplanetary system is 3.226 ± 0.008 pc. Converting this to SI units, 1 pc is 3.086×10^{16} m, so the distance to GJ 144 is $(9.96 \pm 0.02) \times 10^{16}$ m.

As you can see in Figure 1.2b, Aldebaran is actually an extremely nearby star when we consider our place in the Milky Way Galaxy, so the example of Pioneer 10's journey shows just how distant the stars are, compared to the interplanetary distances in our own Solar System. Until 1995, we knew of no planets orbiting around other stars similar to our own Sun. Given the distances involved, and the extreme dimness of planets compared to main sequence stars, this is hardly surprising.

Stars, planets and brown dwarfs

This book will primarily discuss giant planets and their host stars. Objects with masses too small to ignite hydrogen burning in their cores, yet massive enough to have fused deuterium, are known as **brown dwarfs**. This implies that brown dwarfs have masses less than about 80 M_J. While there is no universally agreed definition, giant planets are often defined as objects that are less massive than brown dwarfs and have failed to ignite deuterium burning. Adopting this definition, planets have masses less than about 13 M_J. Stars, brown dwarfs and giant planets form a sequence of declining mass.

Figure 1.5a compares two examples of main sequence stars with two brown dwarfs, one old and one young, and giant planets with masses of 1 M_J and 10 M_J. It is clear that the brown dwarfs and giant planets are all roughly the same size, while the Sun and the MV star Gliese 229A differ radically in size and effective temperature, despite having masses of the same order of magnitude. These most basic characteristics arise because of the physics operating to support these various objects against self-gravity, some of which we will explore in Chapter 4 of this book. When discussing exoplanets, astronomers use the mass of Jupiter, M_J, or the mass of the Earth, $M_\oplus$, as a convenient mass unit, just as the mass of the Sun is used as a convenient unit in discussing stars. The values of these units are given in the Appendix. (Table 4.1 in Chapter 4 gives conversion factors between the units.)

Figure 1.5b shows part of the Sun and the Solar System planets, immediately revealing the huge difference in size between the giant planets and the terrestrial planets. It is because giant planets are so much bigger than terrestrial planets that almost all the transiting exoplanets discovered so far (Dec.2009) are giant planets. As we will see in Chapter 4, giant planets are

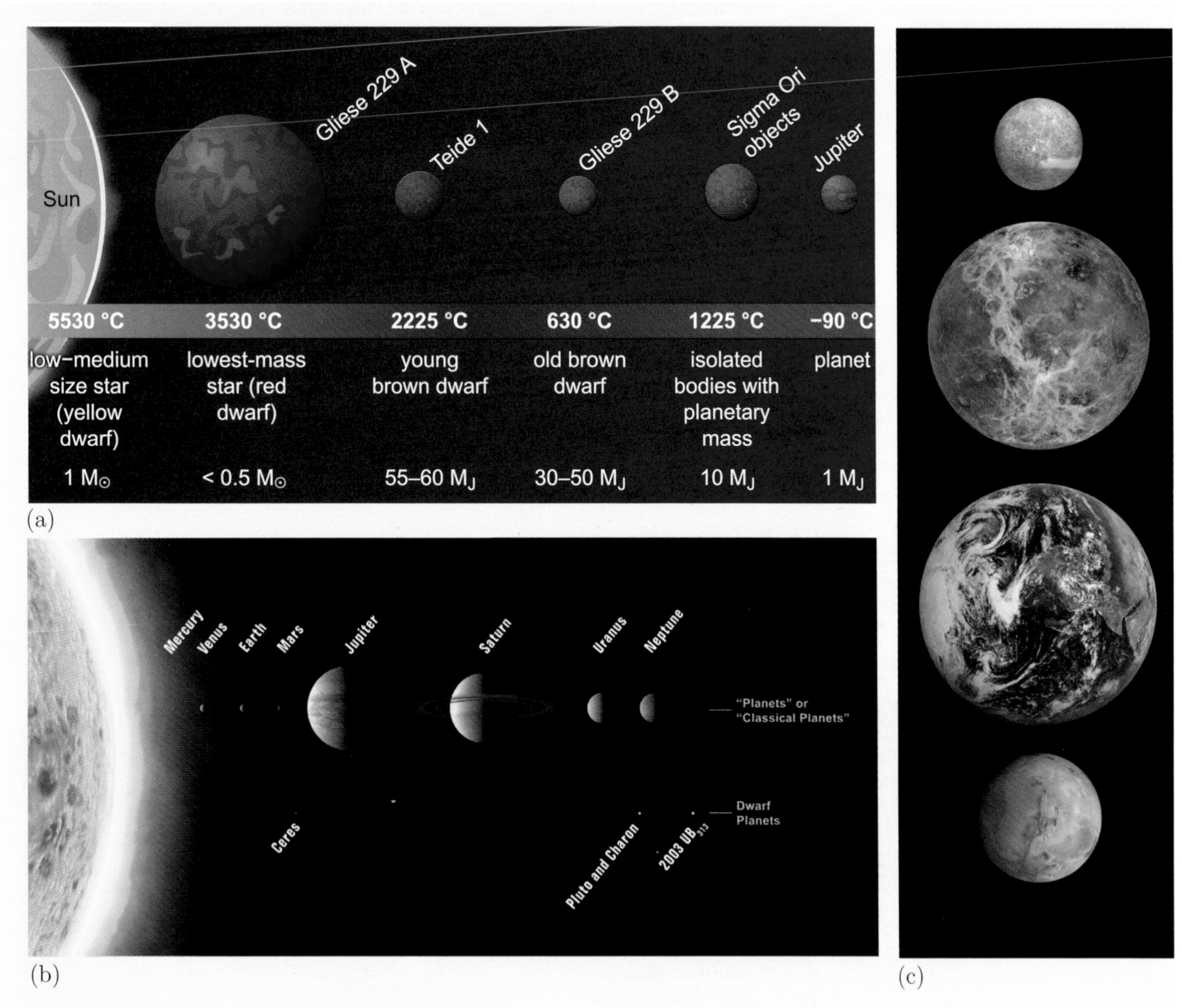

Figure 1.5 Comparisons of stars, brown dwarfs and planets. (a) Two main sequence stars, two brown dwarfs, and two objects of planetary mass, arranged in a sequence of declining mass. Neither the radius nor the surface temperature follows the mass sequence exactly. (b) The Sun and its planets. The giant planets are much larger than the terrestrial planets. (c) The Solar System's terrestrial planets in order of distance from the Sun: Mercury, Venus, Earth and Mars, from top to bottom.

formed predominantly of gas or ice, i.e. low-density material. The terrestrial planets, as we know from our intimate experience of the Earth, have solid surfaces composed of rock, and are denser than giant planets by about a factor of three. Jupiter is over 300 times more massive than the Earth, and has a radius about 10 times bigger, i.e. a volume 1000 times bigger. Generally, terrestrial planets are thought to have lower mass than giant planets: this is certainly true in the Solar System. There may, however, be **mini-Neptune** ice giant exoplanets with lower masses than **super-Earth** terrestrial exoplanets. Since few transiting exoplanets with masses below $10\,M_\oplus$ are yet known (Dec. 2009), this remains an open question, though one that we will explore later.

Could extraterrestrial astronomers detect the Solar System planets?

So far, in this Introduction and in the boxes entitled 'The nearest stars and planets' and 'Stars, planets and brown dwarfs', we have set our own Solar System in the context of its place in our Milky Way Galaxy, and briefly compared **main sequence stars** with brown dwarfs and planets. The subject of the *book* is transiting exoplanets, but to interpret our findings about exoplanets, we need first to understand the various methods that have been or could be used to find them. To this end, in the remainder of this chapter we will use the familiar Solar System planets to illustrate the various exoplanet detection methods. We will examine the question: *If the Galaxy harbours intelligent life outside our own Solar System, could these hypothetical extraterrestrials detect the Solar System planets?* Each of the exoplanet detection methods will be applied to the Solar System, and through this the limitations for each method will be explored.

1.1 Direct imaging

The Solar System planets were all, of course, detected from Earth using direct imaging (Figure 1.6). Obviously, direct imaging works best for nearby objects, and becomes more difficult as the distance, d, to an object increases, as Figure 1.7 demonstrates.

(a)

(b)

(c)

Figure 1.6 NASA images of Saturn and Jupiter. (a) The back-lit rings of Saturn observed by Cassini. (b) Saturn's moon Tethys and a close-up of the rings. (c) Jupiter, the largest planet in the Solar System.

This is particularly true if the object to be imaged is close to a brighter object. Since the optical light from the Solar System's planets is overwhelmingly reflected sunlight, this is clearly the case for them. The biggest and most luminous planet in the Solar System, Jupiter, reflects about 70% of the sunlight that it intercepts, and consequently has a luminosity of $\sim 10^{-9}\,L_{\odot}$.

- Compare the position of Voyager 1 when the images in Figure 1.7 were taken, with the distance to the nearest known exoplanetary system. How many times further away is the nearest known exoplanet?

- ❍ Voyager 1 was 6.4×10^{12} m from the Earth, while the distance to the nearest known exoplanetary system, ε Eri, is $9.96 \pm 0.02 \times 10^{16}$ m. The ratio of these two distances is 15 560: the nearest known exoplanetary system is over 15 000 times further away.

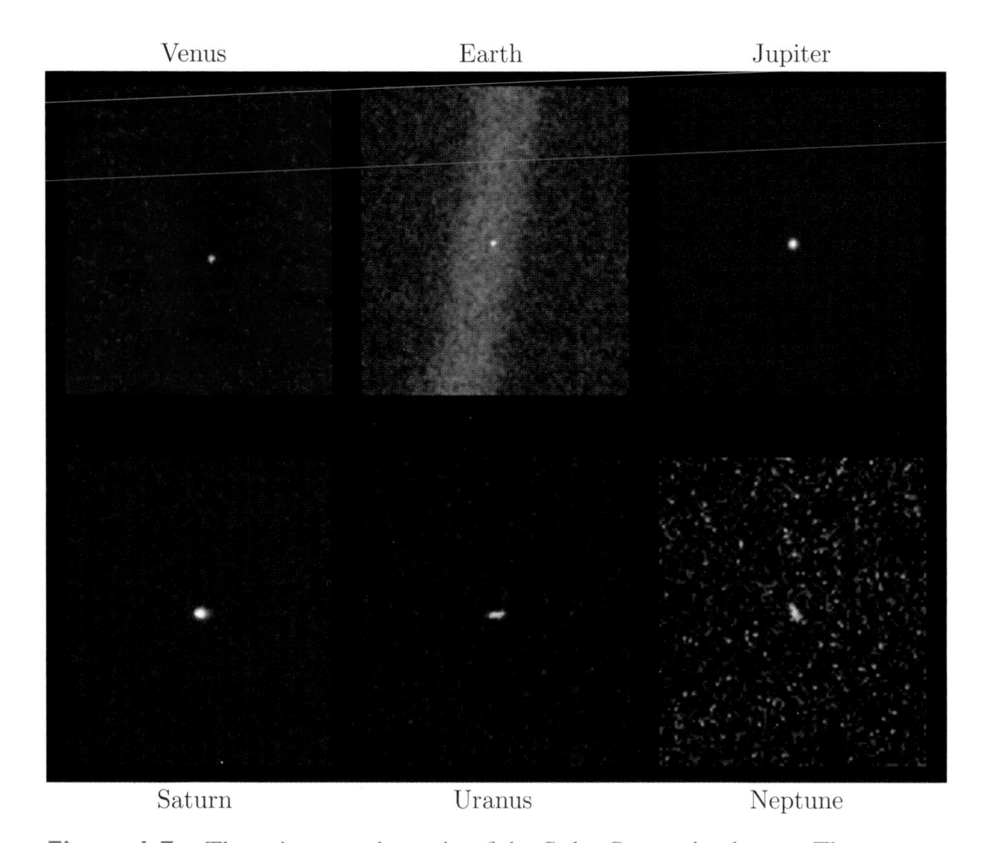

Figure 1.7 These images show six of the Solar System's planets. They were taken by Voyager 1 from a position similar to where it is shown in Figure 1.1, more than 6.4×10^{12} m from Earth and about 32 degrees above the **ecliptic plane**. Mercury was too close to the Sun to be seen; Mars was not detectable by the Voyager cameras due to scattered sunlight in the optics. Top row, left to right, are Venus, Earth and Jupiter; the bottom row shows Saturn, Uranus and Neptune. The background features in the images are artefacts resulting from the magnification. Jupiter and Saturn were resolved by the camera, but Uranus and Neptune appear larger than they should because of spacecraft motion during the 15 s exposure. Earth appears in the centre of the scattered light rays resulting from the Sun. Earth was a crescent only 0.12 pixels in size. Venus was 0.11 pixels in diameter.

Jupiter is a prominent object in our own night sky, but viewed from ε Eri it is very close to the Sun, which outshines it by a factor of 10^9. Jupiter's **orbital semi-major axis**, a_J, is 7.784×10^{11} m. Using the **small angle formula**, the **angular separation** between the Sun and Jupiter as viewed at the distance of ε Eri is

$$\begin{aligned}\theta &= \frac{a_J}{d_{\varepsilon\,\mathrm{Eri}}}\ \text{radians}\\ &= \frac{7.784 \times 10^{11}\ \mathrm{m}}{9.96 \times 10^{16}\ \mathrm{m}}\ \mathrm{rad}\\ &= 7.815 \times 10^{-6}\ \mathrm{rad}\\ &= 7.815 \times 10^{-6} \times \frac{360 \times 60 \times 60}{2\pi}\ \mathrm{arcsec}\\ &= 1.61\ \mathrm{arcsec}.\end{aligned}$$

- Does the angular separation of Jupiter and the Sun at the distance of ε Eri depend on the direction from which the Solar System is viewed?
- Yes. The angular separation calculated above is the maximum value, which would be attained only if the system were viewed from above so that the plane of Jupiter's orbit coincides with the **plane of the sky**. Even for edge-on systems, though, the full separation can be observed at two positions on the orbit.

Thus detecting Jupiter from the distance of ε Eri by direct imaging would be extremely challenging, as there is an object $\sim 10^9$ times brighter within 2 arcsec. If Jupiter were brighter and the Sun were fainter this would help, and one way to achieve this is to observe in the infrared. At these wavelengths Jupiter's thermal emission peaks but the Sun's emission peaks at much shorter wavelengths, so the contrast ratio between the Sun and Jupiter is less extreme. Figure 1.8 shows the **spectral energy distributions** of the Sun and four of the Solar System planets as they would be observed at a distance of 10 pc. This figure shows that the contrast ratio between the Sun and the planets becomes generally more favourable at longer wavelengths.

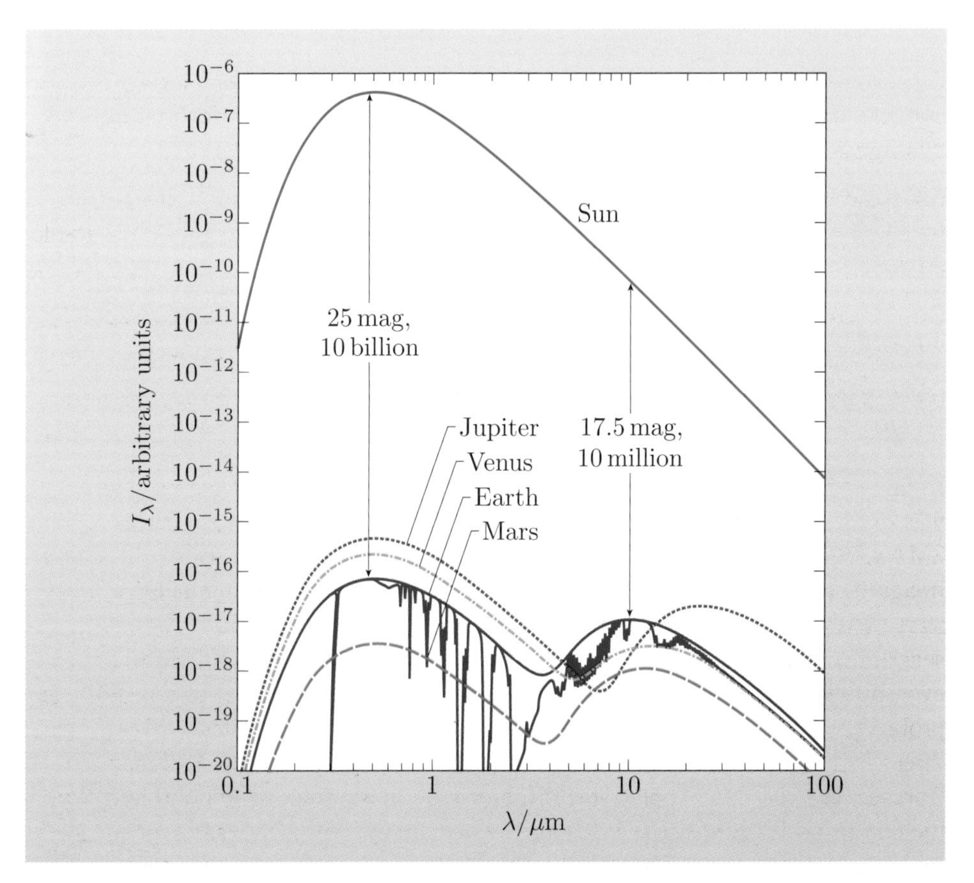

Figure 1.8 The spectral energy distributions of the Sun, Jupiter, Mars, Earth and Venus. The curve for the Earth is far more detailed than the curves for the other objects because empirical data have been included. For each of the planets, the spectral energy distribution is composed of two broad components: peaking at around $0.5\,\mu$m is the reflected solar spectrum, while the thermal emission of the planet itself peaks at 9–20 μm.

- Jupiter is the biggest planet in the Solar System, yet at wavelengths around $10\,\mu$m the Earth emits more light. Why is this?

- The Earth is closer to the Sun, and consequently it receives more **insolation**, i.e. more sunlight falls per unit surface area. Thus the Earth is heated to a higher temperature than Jupiter, and the Earth's thermal emission peaks at around $10\,\mu$m. The amount of light emitted is the surface area multiplied by the flux per unit area. The latter depends strongly on temperature, thus the Earth is brighter at $10\,\mu$m despite having approximately 100 times less emitting area.

Planets can have internal sources of energy in addition to the insolation that they receive; for example, radioactive decay of unstable elements generates energy within the Earth. Radioactive decay is generally thought to be insignificant in giant planets as their composition is predominantly H and He. More important for the planets that we will discuss in this book is internal heating generated by gravitational contraction, which is known as **Kelvin–Helmholtz contraction**. Jupiter emits almost twice as much heat as it absorbs from the Sun, and all four giant planets in the Solar System radiate some power generated by Kelvin–Helmholtz contraction.

Exercise 1.2 Figure 1.8 shows the spectral energy distributions for the Sun and four of our Solar System's planets. Consider the following with reference to this figure.

(a) State what wavelength gives the most favourable contrast ratio with the Sun for the detection of Jupiter. What is the approximate value of this most favourable contrast ratio?

(b) At what wavelength does the spectral energy distribution of Jupiter peak? Is this the same wavelength as you stated in part (a)? Explain your answer.

(c) What advantages are there to using a wavelength of around $20\,\mu$m for direct imaging observations of Jupiter from interstellar distances? ■

In fact, infrared imaging has detected an exoplanet directly. A giant exoplanet was discovered at an angular distance of 0.78 arcsec from the brown dwarf 2MASSWJ 1207334-393254 by infrared imaging, as shown in Figure 1.9. Direct imaging is most effective for bright planets in distant orbits around nearby faint stars. These requirements mean that direct imaging has limited applicability in the search for exoplanets. Returning to our hypothetical extraterrestrial astronomers, even if they were located on one of the nearest known exoplanets, they would face problems in detecting even the most favourable of the Solar System's planets by direct imaging. The angular separation of Jupiter from the Sun is less than 2 arcsec, and the Sun is one of the brighter stars in the solar neighbourhood, with the contrast ratio between the Sun and Jupiter being over 10^4 even at the most favourable wavelength. Thus it seems unlikely that hypothetical extraterrestrials would first detect the Solar System planets through direct imaging.

Despite the difficulties, astronomers are designing instruments specifically to detect exoplanets by direct imaging. These instruments employ sophisticated techniques to overcome the overwhelming light from the planets' host stars. The Gemini Planet Imager (GPI) is designed to detect planets with contrast ratios as

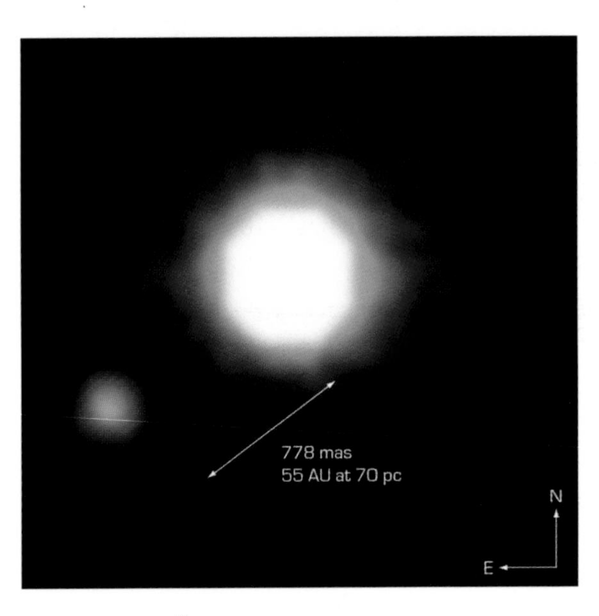

Figure 1.9 False-colour image of the brown dwarf 2MASSWJ 1207334-393254 using H, K and L band infrared filters. Blue indicates pixels brightest in H, while red indicates pixels brightest in L. The companion exoplanet is relatively bright in the L band and thus appears red. The exoplanet has an effective temperature of 1250 ± 200 K.

extreme as 10^8 using a device called a **coronagraph**. A spectacular recent result using this method is summarized in Subsection 1.1.1 below. As Figure 1.8 shows, this technology could in principle render the Earth detectable. In practice, to detect the Earth, the extraterrestrials' telescope would need to be implausibly close to the Solar System.

- If the contrast ratio of the Earth and the Sun is less than the threshold for planet detection with the GPI, what limits the detectability of the Earth with such an instrument at large distances?

- There are two limiting factors. First, the angular separation of the Earth and the Sun decreases with distance; at large distances the two objects are not spatially resolved. The angular separation diminishes as $1/d$. Second, the telescope needs to have sufficient Earthlight falling on it to permit detection of the Earth. The number of photons reaching the telescope diminishes as $1/d^2$.

The SPHERE instrument is scheduled to be deployed on the Very Large Telescope (VLT) in 2010, and aims to detect some nearby young giant exoplanets at large separations from their host star. Giant planets are brightest early in their life, when they are contracting relatively quickly. The Kelvin–Helmholtz powered luminosity diminishes as a giant planet contracts and cools with age, as we will see in Subsection 4.4.3.

1.1.1 Coronagraphy

As the name implies, coronagraphs were invented to study the solar corona by blocking the light from the disc (i.e. the photosphere) of the Sun. An occulting element is included within the instrument, and is used to block the bright object, allowing fainter features to be detected. If an exoplanetary system is close enough for the star and exoplanet to be resolved, then a coronagraph may permit detection of an exoplanet where without it the detector would be flooded with light from the central star.

Fomalhaut (α PsA) is one of the 20 brightest stars in the sky; some of its characteristics were listed in Table 1.1. It is surrounded by a dust disc that appears elliptical and has a sharp inner edge. This led to speculation that there might be a planet orbiting Fomalhaut just inside the dust ring, and creating the sharp inner edge of the ring rather as the moons of Saturn shepherd its rings (cf. Figure 1.6). Figure 1.10 was created using the Hubble Space Telescope's coronagraph, combining two images taken in 2004 and 2006. In this figure the orbital motion of the planet Fomalhaut b is clearly revealed. Fomalhaut b is $\sim 10^9$ times fainter than Fomalhaut, and is ~ 100 AU from the central star.

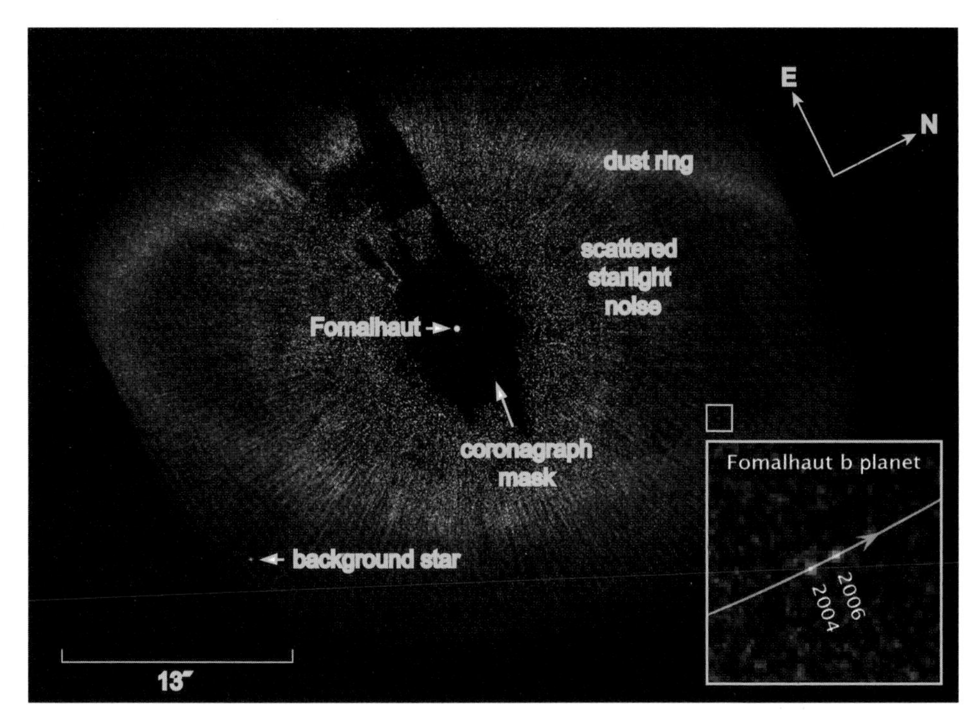

Figure 1.10 False-colour image showing the region surrounding Fomalhaut. The black shape protruding downwards from 11 o'clock is the coronagraph mask, which has been used to block the light from the star located as indicated at the centre of the image. The square box is the region surrounding the planet, which has been blown up and inset. The motion of the planet, called Fomalhaut b, along its elliptical orbit between 2004 and 2006 is clearly seen. At the distance of Fomalhaut $13''$ corresponds to 100 AU.

There are designs (though not necessarily the funding) for a space telescope with a coronagraph optimized to detect terrestrial exoplanets: Fomalhaut b may be the first of a significant number of exoplanets detected in this way. Our hypothetical extraterrestrial astronomers could possibly detect the Solar System's planets using coronagraphy, provided that they are close enough to the Sun to be able to block its light without also obscuring the locations of the Sun's planets.

1.1.2 Angular difference imaging

Another spectacular recent result, shown in Figure 1.11, uses some other clever techniques to overcome the adverse planet to star contrast ratio. Figure 1.11 reveals three giant planets around the A5V star HR 8799. The technique employed is called angular difference imaging, an optimal way of combining many short exposures to reveal faint features. HR 8799's age is between 30 Myr and 160 Myr, i.e. it is a young star. Because this planetary system is young, the giant planets are relatively bright, making their detection by direct imaging possible. The three

Myr indicates 10^6 yr, or a megayear.

planets have angular separations of 0.63, 0.95 and 1.73 arcsec from the central star, corresponding to projected separations of 24, 38 and 68 AU. The orbital motion, counter-clockwise in Figure 1.11, was detected by comparing images taken in 2004, 2007 and 2008. The masses of these three bodies are estimated to be between 5 M_J and 13 M_J.

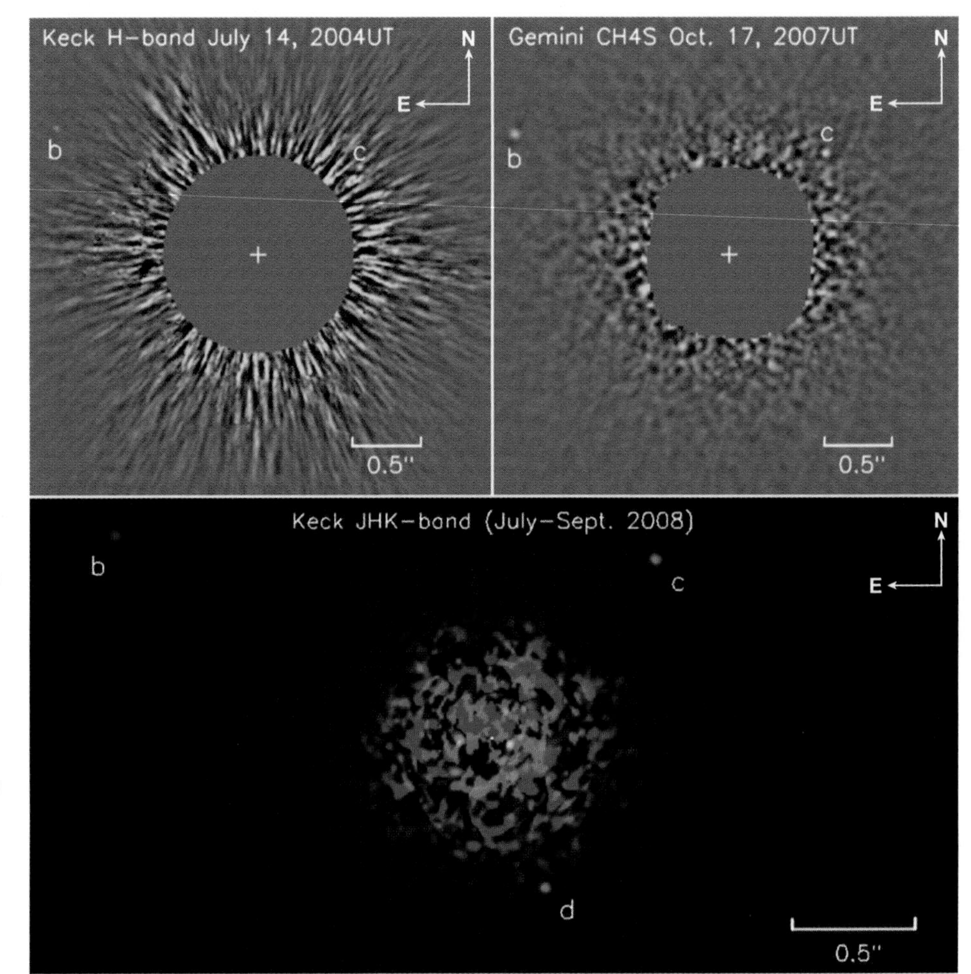

Figure 1.11 The three planets around HR 8799. The central star is overwhelmingly bright and has been digitally removed from the images, causing the featureless or mottled round central region in each panel. The three planets, whose names are HR 8799 b, c and d, are labelled b, c and d. At the distance of HR 8799, 0.5″ corresponds to 20 AU.

1.2 Astrometry

Hipparcos was an ESA satellite that operated between 1989 and 1993.

Astrometry is the science of accurately measuring the positions of stars. The Hipparcos satellite measured the positions of over 100 000 stars to a precision of 1 milliarcsec, and the Gaia satellite, due for launch by the European Space Agency (ESA) in spring 2012, will measure positions to a precision of 10 microarcsec (μarcsec). Astrometry offers a way to indirectly detect the presence of planets. Though we often casually refer to planets orbiting around stars, in fact planets and stars both possess mass, and consequently all the bodies in a planetary system orbit around the **barycentre**, or common centre of mass of the system. Figure 1.12 illustrates this.

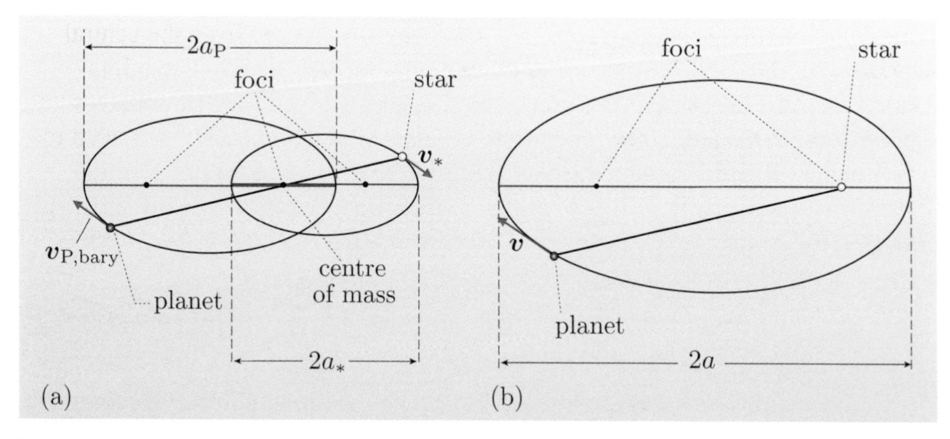

Figure 1.12 (a) The elliptical orbits of a planet and its host star about the barycentre (or centre of mass) of the system. (b) In astrocentric coordinates, i.e. coordinates centred on the star, the planet executes an elliptical orbit with the star at one focus. The semi-major axis of this astrocentric orbit, a, is the sum of the semi-major axes of the star's orbit and the planet's orbit in the barycentric coordinates shown here.

In general, orbits under the inverse square force law are conic sections, the loci obtained by intersecting a cone with a plane. For planets, the orbits must be closed, i.e. they are either circular or, more generally, elliptical, with the eccentricity of the ellipse increasing as the misalignment of the normal to the plane with the axis of the cone increases. Of course, a circle is simply an ellipse with an eccentricity, e, of 0.

There is no physical significance to the cone or the plane; it simply happens that there is a mathematical coincidence.

The orbital period, P, the semi-major axis, a, and the star and planet masses are related by (the generalization of) **Kepler's third law**:

$$\frac{a^3}{P^2} = \frac{G(M_* + M_{\rm P})}{4\pi^2}, \qquad (1.1)$$

where $a = a_* + a_{\rm P}$ as indicated in Figure 1.12.

The Sun constitutes over 99.8% of the Solar System's mass, so the barycentre of the Solar System is close to, but not exactly at, the Sun's own centre of mass. As the planets orbit around the barycentre, the Sun executes a smaller **reflex orbit**, keeping the centre of mass of the system fixed at the barycentre.

In general, the motion of a star in a reflex orbit has a semi-major axis, a_*, that can, in principle, be detected as an angular displacement, β, when the star is viewed from distance d. This angle of the astrometric wobble, β, is proportional to the semi-major axis of the star's reflex orbit:

$$\beta = \frac{a_*}{d}.$$

However, since

$$a_* = \frac{M_{\rm P}}{M_*} a_{\rm P},$$

we have

$$\beta = \frac{M_{\rm P} a_{\rm P}}{M_* d}, \qquad (1.2)$$

and we see from Equation 1.2 that the astrometric wobble method is most effective for finding high-mass planets in wide orbits around nearby relatively low-mass stars.

Worked Example 1.1

(a) Calculate the orbital semi-major axis, $a_\odot$, of the Sun's reflex orbit in response to Jupiter's orbital motion. The mass of Jupiter, M_J, is 1.90×10^{27} kg, the mass of the Sun is 1.99×10^{30} kg, and Jupiter's orbital semi-major axis, a_J, is 7.784×10^{11} m. Jupiter constitutes about 70% of the mass in the Solar System apart from the Sun, so you may ignore the other lesser bodies in the system.

(b) Hence calculate the distance, d_{Gaia}, at which the astrometric wobble angle, β, due to the Sun's reflex orbit is greater than $\beta_{\text{Gaia}} = 10\,\mu$arcsec.

(c) State whether a Gaia satellite positioned anywhere within the sphere of radius d_{Gaia} would be able to detect the astrometric wobble due to the Sun's reflex orbit, explaining your reasoning.

(d) Comment on the fraction of the Galaxy's volume over which a hypothetical extraterrestrial Gaia could detect Jupiter's presence by the astrometry method.

Solution

(a) The barycentre of the Sun–Jupiter system is such that

$$M_J a_J = a_\odot M_\odot, \tag{1.3}$$

so

$$a_\odot = \frac{M_J}{M_\odot} a_J = \frac{1.90 \times 10^{27}}{1.99 \times 10^{30}} \times 7.784 \times 10^{11}\ \text{m}$$
$$= 7.432 \times 10^{8}\ \text{m}.$$

The semi-major axis of the Sun's reflex orbit is 7.43×10^8 m.

(b) The distance d_{Gaia} is given by the small angle formula. To use this formula we must first convert the angle β_{Gaia} to radians:

$$1 \times 10^{-5}\ \text{arcsec} = 1 \times 10^{-5} \times \frac{2 \times \pi}{360 \times 60 \times 60}\ \text{rad}$$
$$= 4.85 \times 10^{-11}\ \text{rad}.$$

Here we have multiplied by the 2π radians in a full circle and divided by the $360 \times 60 \times 60$ arcseconds in a full circle, thus converting our angular units. Here π is *not* the parallax.

So $\beta_{\text{Gaia}} = 10\,\mu\text{arcsec} = 4.85 \times 10^{-11}$ rad. The astrometric wobble angle of the Sun's reflex orbit is given by

$$\beta = \frac{a_\odot}{d}, \tag{1.4}$$

so d_{Gaia} is given by

$$d_{\text{Gaia}} = \frac{a_\odot}{\beta_{\text{Gaia}}}. \tag{1.5}$$

Substituting values,

$$d_{\text{Gaia}} = \frac{a_\odot}{4.85 \times 10^{-11}} = \frac{7.43 \times 10^{8}}{4.85 \times 10^{-11}}\ \text{m} = 1.532 \times 10^{19}\ \text{m}.$$

Clearly the SI unit of metres is not a good choice for expressing interstellar distances, so we will express this distance in parsecs:

$$d_{\text{Gaia}} = \frac{1.532 \times 10^{19}\,\text{m}}{3.086 \times 10^{16}\,\text{m}\,\text{pc}^{-1}} = 4.96 \times 10^{2}\,\text{pc} = 496\,\text{pc}.$$

Hence the limiting distance is around 500 pc, to the one significant figure of the stated angular precision (10 μarcsec).

(c) If the extraterrestrials' Gaia satellite is ideally positioned so that Jupiter's orbit is viewed face on, then the satellite will see the true elliptical shape of the Sun's reflex orbit. At any other orientation, the satellite will see a foreshortened shape; but even in the least favourable orientation, with **orbital inclination** $i = 90°$, as in Figure 1.13a, the satellite will see the Sun move back and forth along a line whose length is given by the size of the orbit. Only in the extremely unlucky case of a highly eccentric orbit, viewed from the least favourable inclination *and* azimuth, will the angular deviation be substantially less than that corresponding to the semi-major axis. Since Jupiter has a low orbital eccentricity, the extraterrestrial Gaia could detect the Sun's reflex wobble from anywhere within the sphere of radius d_{Gaia}.

The azimuth is the angle around the orbit: for a highly eccentric orbit, the projection on the sky will vary with azimuth, being smallest when the orbit is viewed from an extension of the semi-major axis.

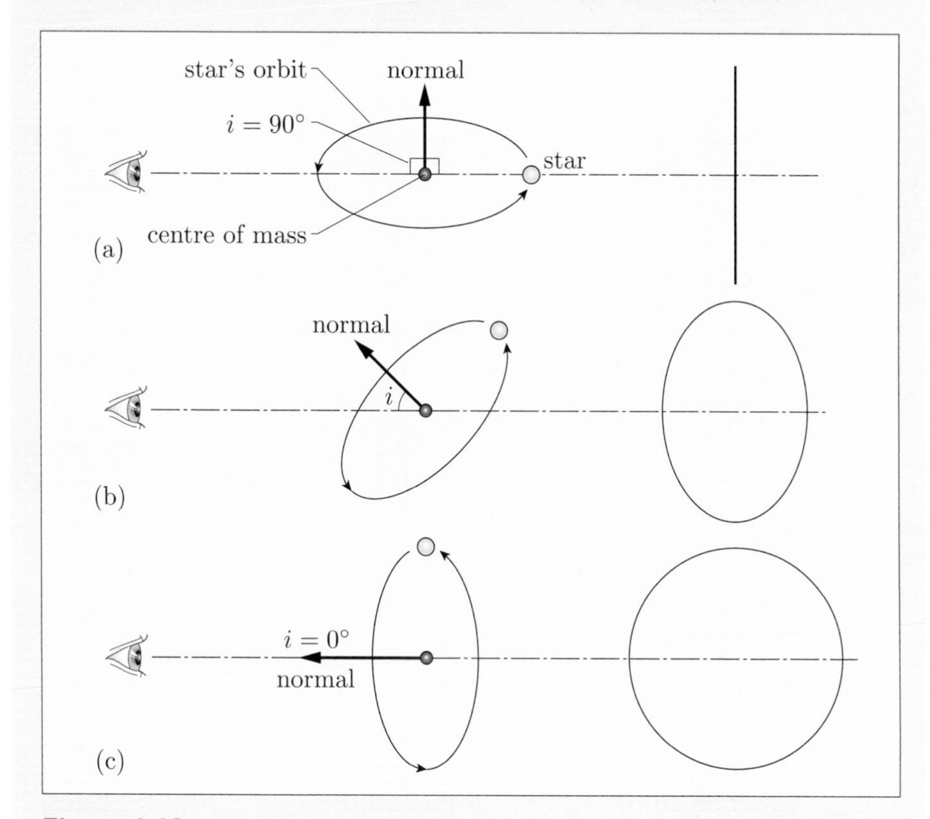

Figure 1.13 The shape of the orbit that the Gaia satellite sees is a function of the orbital inclination, i. In panel (a), $i = 90°$ and the orbit appears as a line traced on the sky. Panel (b) shows an intermediate value of i, and the orbit appears elliptical, with the true orbital shape being foreshortened. Panel (c) shows the situation where $i = 0°$; only in this case does the observer see the true shape of the orbit.

(d) The distance d_{Gaia} is about 500 pc, i.e. about 1500 ly. Comparing this with Figure 1.3a, we can see that this length scale is comparable to the width of the spiral arm structure in the Galaxy. The astrometric wobble of the Sun due to Jupiter's pull could be detected by advanced, spacecraft-building extraterrestrial civilizations in the same region of the Orion Arm as the Sun. The astrometric wobble would be too small to be detected by a Gaia-like satellite built by more remote extraterrestrials.

● How does the semi-major axis of the Sun's reflex orbit in response to Jupiter, $a_{\odot}$, compare with the solar radius, $R_{\odot}$? What does this imply for the position of the barycentre of the Solar System?

❍ The value of $R_{\odot}$ is 6.96×10^8 m, which is only just smaller than the value that we calculated for $a_{\odot}$, namely 7.43×10^8 m. This means that the barycentre of the Solar System is just outside the surface of the Sun.

● Will the barycentre of the Solar System remain at the same distance from the centre of the Sun as the eight planets in the Solar System all orbit around the barycentre? Explain your answer.

❍ No, the eight planets all have different orbital periods, so sometimes they will all be located on the same side of the Sun. When this happens, the barycentre will be pulled further from the centre of the Sun than the value that we calculated using Jupiter alone. At some other times most of the seven lesser planets will be on the opposite side of the Sun from Jupiter, in which case they will pull the barycentre closer to the centre of the Sun than implied by our calculations using Jupiter alone.

● For astrometric discovery prospects, is there any disadvantage inherent in a planet having a wide orbit?

❍ While the size of the astrometric wobble increases as the planet's semi-major axis increases, so too does the orbital period (i.e. the planet's 'year'). Kepler's third law tells us that $P \propto a^{3/2}$, so as the orbital size increases, the length of time required to measure a whole orbit increases even faster.

Our calculations show that our hypothetical extraterrestrial astronomers would be able to detect Jupiter's presence so long as they were within about 500 pc of the Sun and were able to construct an instrument with Gaia's capabilities. We do not yet know how common planetary systems like our own are, but once Gaia is launched we will begin to find out!

1.3 Radial velocity measurements

Like the astrometric method, the radial velocity method of planet detection relies on detecting the host star's reflex orbit. In this case, rather than detecting the change in position of the host star as it progresses around its orbit, we detect the change in velocity of the host star. The radial velocity, i.e. the velocity of an object directly towards or away from the observer, can be sensitively measured using the Doppler shift of the emitted light. This has been a powerful technique in many areas of astrophysics and was adopted in the 1980s by astronomers searching for Jupiter-like planets around nearby Sun-like stars.

1.3.1 The stellar reflex orbit with a single planet

The motion of a planet is most simply analyzed in the rest frame of the host star, i.e. the **astrocentric frame**. In this frame the planet executes an orbit with semi-major axis a, period P and eccentricity e, as shown in Figures 1.12b and 1.14.

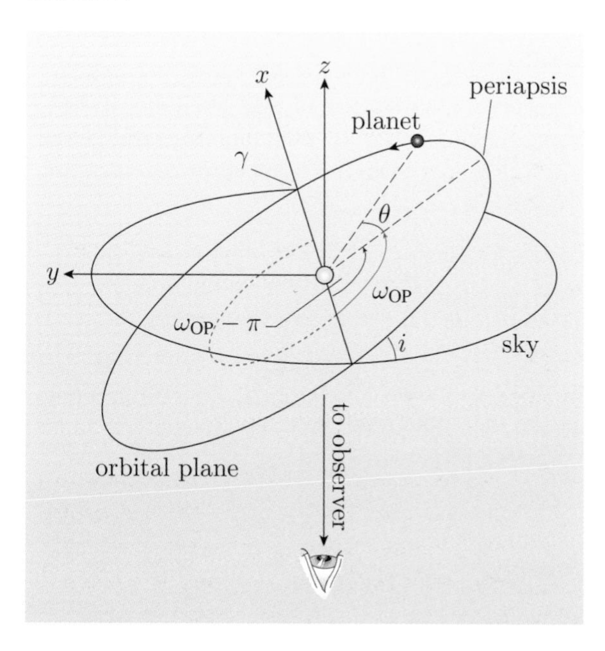

Figure 1.14 The astrocentric orbit of a planet. The observer is positioned in the direction of the bottom of the page, and is viewing the system along the direction of the z-axis. The x-axis is in the plane of the sky and is oriented so that it intersects the planet's orbit at the point where the z-component of the planet's velocity is towards the observer. Finally, the y-axis is the third direction making up a right-handed Cartesian coordinate system. The planet is closest to the star at the pericentre, whose position is defined by the angle $\omega_{\rm OP}$.

The point at which the planet is closest to the star is called the **pericentre**, and the star is positioned at one focus of the elliptical orbit. We will adopt a Cartesian coordinate system, centred on the host star, oriented as shown in Figure 1.14. The angle between the plane of the sky and the plane of the orbit is the orbital inclination, i. The x-axis is defined by the intersection of the orbit with the plane of the sky as seen by the observer. Positive x is on the side of the orbit where the planet moves towards the observer. The point labelled γ is the intersection of the orbit with the positive x-axis. The angle θ is the **true anomaly**, which measures how far around the orbit from the pericentre the planet has travelled. This angle is measured at the position of the host star. The angle $\omega_{\rm OP}$ measures the orientation of the pericentre with respect to γ. The velocity, $\boldsymbol{v}$, has components in the x-, y- and z-directions that are given by

$$
\begin{aligned}
v_x &= -\frac{2\pi a}{P\sqrt{1-e^2}}\left(\sin(\theta+\omega_{\rm OP}) + e\sin\omega_{\rm OP}\right),\\
v_y &= -\frac{2\pi a\cos i}{P\sqrt{1-e^2}}\left(\cos(\theta+\omega_{\rm OP}) + e\cos\omega_{\rm OP}\right),\\
v_z &= -\frac{2\pi a\sin i}{P\sqrt{1-e^2}}\left(\cos(\theta+\omega_{\rm OP}) + e\cos\omega_{\rm OP}\right).
\end{aligned}
\tag{1.6}
$$

What astronomers can actually observe is the star, not the planet, so we need the analogous equations for the reflex orbit of the star. So far we've considered the orbit of the planet around the star, i.e. in an 'astrocentric' frame. The star is actually moving too, executing its reflex orbit around the barycentre, as illustrated in Figure 1.12a.

The velocity, $\boldsymbol{v}$, of the planet in the astrocentric frame (Figure 1.12b) is given by simple vector subtraction:

$$\boldsymbol{v} = \boldsymbol{v}_{\mathrm{P,bary}} - \boldsymbol{v}_{*}, \tag{1.7}$$

where $\boldsymbol{v}$ is the astrocentric velocity given in Equation 1.6, and $\boldsymbol{v}_*$ is the velocity of the star in the barycentric frame, i.e. the velocity of the star as it performs its reflex orbit about the barycentre as shown in Figure 1.12a. Equation 1.3 gave us the ratio of the semi-major axes of the orbit of a planet and the reflex orbit of its star. The relationship is much more general than our use of it in Section 1.2: for any planetary system, the barycentre remains fixed in its own inertial frame, so at all times the distances of the star and the planet from the barycentre will vary proportionately:

$$M_* \boldsymbol{r}_* = -M_{\mathrm{P}} \boldsymbol{r}_{\mathrm{P}}. \tag{1.8}$$

Here we have used vector notation, and since the planet and the star are in opposite directions from the barycentre, this equation introduces a minus sign into Equation 1.3, which used the (scalar) distance rather than the vector displacement. Differentiating with respect to time, we then obtain the general relationship, in the barycentric frame, between the velocities of the planet and the star's reflex orbit:

$$M_* \boldsymbol{v}_* = -M_{\mathrm{P}}\, \boldsymbol{v}_{\mathrm{P,bary}}. \tag{1.9}$$

- What do Equations 1.8 and 1.9 imply for the shapes of the two ellipses in Figure 1.12a?

- The ellipses are the same shape, and differ only in size.

Consequently, using Equation 1.9 to substitute for $\boldsymbol{v}_{\mathrm{P,bary}}$ in Equation 1.7, and making the barycentric velocity of the star the subject of the resulting equation, we obtain

The fraction multiplying $\boldsymbol{v}$ is known as the **reduced mass** of the planet, and is analogous to that quantity in simple two-body problems in other areas of physics.

$$\boldsymbol{v}_* = -\frac{M_{\mathrm{P}}}{M_{\mathrm{P}} + M_*}\boldsymbol{v}, \tag{1.10}$$

where $\boldsymbol{v}$ is the astrocentric orbital velocity of the planet, as given in Equations 1.6. Finally, we should note that in general the barycentre of the planetary system being observed will have a non-zero velocity, $\boldsymbol{V}_0$, with respect to the observer. $\boldsymbol{V}_0$ changes on the timescale of the star's orbit around the centre of the Galaxy, i.e. hundreds of Myr. Thus we can regard it as a constant as this timescale is much longer than the timescale for the orbits of planets around stars. From the point of view of an observer in an inertial frame, therefore, the reflex velocity of the star is $\boldsymbol{V} = \boldsymbol{v}_* + \boldsymbol{V}_0$. So we now have a completely general expression, in the frame of an inertial observer, for the reflex velocity of a star in response to the motion of a planet around it:

$$\begin{aligned}
V_x &= V_{0,x} + \frac{2\pi a M_{\mathrm{P}}}{(M_{\mathrm{P}} + M_*)P\sqrt{1-e^2}}\left(\sin(\theta + \omega_{\mathrm{OP}}) + e\sin\omega_{\mathrm{OP}}\right), \\
V_y &= V_{0,y} + \frac{2\pi a M_{\mathrm{P}}\cos i}{(M_{\mathrm{P}} + M_*)P\sqrt{1-e^2}}\left(\cos(\theta + \omega_{\mathrm{OP}}) + e\cos\omega_{\mathrm{OP}}\right), \\
V_z &= V_{0,z} + \frac{2\pi a M_{\mathrm{P}}\sin i}{(M_{\mathrm{P}} + M_*)P\sqrt{1-e^2}}\left(\cos(\theta + \omega_{\mathrm{OP}}) + e\cos\omega_{\mathrm{OP}}\right).
\end{aligned} \tag{1.11}$$

- How do the three components of $\mathbf{V}$ contribute to the observed stellar radial velocity?

❍ The radial velocity is the motion directly towards or away from the observer. The coordinate system that we adopted has the z-axis directed away from the observer, while the x- and y-axes are in the plane of the sky. The radial velocity is given by V_z. The other two components do not contribute to the radial velocity.

- Which of the terms in the expression for the stellar radial velocity are time-dependent? Describe in simple physical terms how the time-dependence arises.

❍ The only time-dependent term in the expression for V_z is the true anomaly, i.e. the angle $\theta(t)$, which changes continuously as the planet and star proceed along their orbits around the barycentre.

Exercise 1.3 Use Kepler's third law to estimate the time for the Sun to orbit around the Galaxy. You may assume the mass of the Galaxy is $10^{12}\,\mathrm{M}_\odot$ and may be approximated as a point mass at the centre of the Galaxy, and that the distance of the Sun from the centre of the Galaxy (known as the **Galactocentric distance**) is $8\,\mathrm{kpc}$. ■

The radial velocity, which we will henceforth simply call V, of a star executing a reflex motion as a result of a single planet in an elliptical orbit is given by

$$V(t) = V_{0,z} + \frac{2\pi a M_\mathrm{P} \sin i}{(M_\mathrm{P} + M_*)P\sqrt{1-e^2}}\left(\cos(\theta(t) + \omega_\mathrm{OP}) + e\cos\omega_\mathrm{OP}\right). \quad (1.12)$$

$\theta(t)$ completes a full cycle of 360° (or 2π radians) each orbit. Unless the orbit is circular, θ does not change linearly with time; instead, it obeys Kepler's second law, or equivalently the **law of conservation of angular momentum**.

Exercise 1.4 Use Equation 1.12 for the radial velocity to answer the following.

(a) What is the dependence of the observed radial velocity variation on the orbital inclination, i? State this dependence as a proportionality, and draw diagrams illustrating the values of i for the maximum, the minimum and an intermediate value of the observed radial velocity. You may assume that all the characteristics of the observed planetary system, except for its orientation, remain constant.

(b) For fixed values of all parameters except the orbital eccentricity, e, how does the amplitude of the observed radial velocity variation change as the eccentricity varies? (By definition, the eccentricity of an ellipse lies in the range $0 \leq e \leq 1$.) Explain your answer fully, with reference to the behaviour of the relevant terms in Equation 1.12.

(c) Adopting the approximation that Jupiter is the only planet in the Solar System, calculate the observed radial velocity amplitude, A_RV, for the Sun's reflex orbit. Express your answer as a function of the unknown orbital inclination, i. Use the following (approximate) values of constants to evaluate your answer: $\mathrm{M_J} = 2\times10^{27}\,\mathrm{kg}$, $\mathrm{M}_\odot = 2\times10^{30}\,\mathrm{kg}$, $\mathrm{a_J} = 8\times10^{11}\,\mathrm{m}$; Jupiter's orbital period is $\mathrm{P_J} = 12$ years, and its orbital eccentricity is $\mathrm{e_J} = 0.05$. ■

An important result from Exercise 1.4 is that the amplitude of the reflex radial velocity variations predicted by Equation 1.12 is

$$A_{\mathrm{RV}} = \frac{2\pi a M_{\mathrm{P}} \sin i}{(M_{\mathrm{P}} + M_*) P \sqrt{1 - e^2}}. \qquad (1.13)$$

Applying this to the Sun's reflex orbit due to Jupiter, we find a radial velocity amplitude of $A_{\mathrm{RV}} \leq 13\,\mathrm{m\,s^{-1}}$.

Astronomers measure radial velocities using the Doppler shift of the features in the stellar spectrum. The wavelength change, $\Delta\lambda$, due to the Doppler shift is given by

$$\frac{\Delta\lambda}{\lambda} = \frac{V}{c}, \qquad (1.14)$$

so to detect the reflex orbit of the Sun due to Jupiter's presence, our hypothetical extraterrestrial astronomers would need to measure the wavelength shifts of features in the solar spectrum to a precision of

$$\frac{\Delta\lambda}{\lambda} = \frac{13\,\mathrm{m\,s^{-1}}}{3.0 \times 10^8\,\mathrm{m\,s^{-1}}} = 4.2 \times 10^{-8},$$

where we have substituted the value for the speed of light, c, in SI units. This sounds challenging, but it is possible with current technology; terrestrial astronomers have aspired to this since the 1980s. For bright stars with prominent spectral features it is now possible to measure radial velocities to precisions of better than $1\,\mathrm{m\,s^{-1}}$, i.e. motions slower than walking speed can be detected! Figure 1.15 shows the radial velocity measurements used to infer the presence of the planet around GJ 832 (also known as HD 204691); this planet has a lower mass than Jupiter and has $P = 9.4$ years. GJ 832's reflex radial velocity amplitude slightly exceeds that of Jupiter because, as we saw in Table 1.1, the star has mass less than half that of the Sun, and the planet has a shorter orbital period than Jupiter. Our Sun happens to have a spectral type (G2V) that makes it particularly suitable for radial velocity measurements: it has a host of sharp, well-defined photospheric absorption lines. By measuring the shifts in the observed wavelengths of these lines, our hypothetical extraterrestrial astronomers would be able to detect the presence of Jupiter unless they happen to be unlucky and view the Sun–Jupiter system from an orbital inclination, i, close to 0°. The other limiting factor, assuming that they have technology comparable to ours, is their distance from the Sun. To make radial velocity measurements of the required precision, astronomers need to collect a lot of light, and as noted previously the flux of photons from the Sun drops off as $1/d^2$.

GJ 832 is one of the 100 nearest known stars listed Table 1.1.

Exoplanet naming convention

The planet discovered around the star GJ 832 is called GJ 832 b. This is the general convention: the first exoplanet discovered around any star is given its host star's name with 'b' appended. The second planet discovered in the system is labelled 'c', and so on. This distinguishes planets from stellar companions, which have capital letters appended: for example, the stars α Cen A and α Cen B. The planets in a multiple planet system are always labelled b, c, d, ... *in the order of discovery*; the distances of these planets from the host star can consequently be in any order.

- If a planet were discovered orbiting around the star α Cen A, what would it be called?

- Assuming that it was the first planet discovered orbiting around the star, it would have the star's name with 'b' appended: α Cen A b.

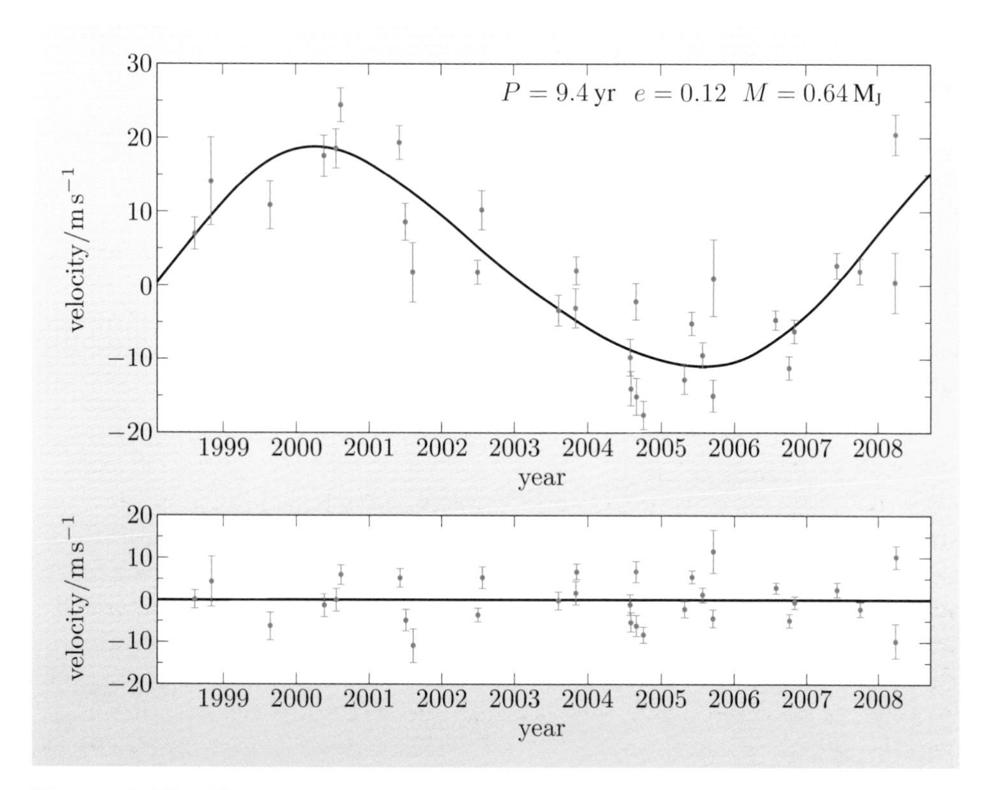

Figure 1.15 Upper panel: radial velocity measurements revealing the presence of the planet around GJ 832, with the best-fitting solution of the form of Equation 1.12 overplotted as a dashed line. The parameters corresponding to this fit for the planet GJ 832 b are indicated. This planet's orbit has similarities to that of Jupiter. Lower panel: the deviations of the measured data points from the best-fitting solution.

Figure 1.15 shows radial velocity measurements as a function of time. The best-fitting model solution, which is overplotted, was generated using Equation 1.12. If we examine Equation 1.12, we see that it predicts the same value for $V(t)$ every time $\theta(t)$ completes a full cycle, i.e. every time the star returns to the same position on its orbit. In the case of GJ 832 b, the orbital period is long, and it is easy to make observations sampling each part of the orbit. For shorter orbital periods, sometimes only a single measurement is made during a particular orbit; to make a graph like Figure 1.15, many orbits would need to be plotted on the horizontal axis, and the plotted points would be very thinly spread out. A more efficient way of presenting such data is to **phase-fold**, so that instead of plotting time on the horizontal axis, one plots **orbital phase**, ϕ:

The timetable giving orbital phase for any value of time, t, is known as an **ephemeris**.

$$\phi = \frac{t - T_0}{P} - N_{\rm orb}, \tag{1.15}$$

where t is time; T_0 is a **fiducial** time, for example, T_0 could be fixed at the first observed time at which the star is farthest from the observer; P is the orbital period; and N_{orb} is an integer such that $0 \leq \phi \leq 1$. As the star proceeds around a full orbit, t increases by an amount equal to the orbital period, P, and ϕ covers the full range of values $0 \leq \phi \leq 1$. Once per orbit, when the star crosses the fiducial point corresponding to T_0, the integer N_{orb} increments its count: N_{orb} is known as the **orbit number**.

At first glance the curve shown in Figure 1.15 looks very much like the familiar sinusoidal curve that one expects to see whenever there is motion in a circle. If you look carefully, however, you should see that the curve does not have the perfect symmetries of a sine curve: it corresponds to an orbit with finite eccentricity, $e = 0.12$. For more extreme values of the eccentricity, the deviations from a sine curve become more pronounced, as shown in Figure 1.16. The eccentricities of the solutions shown in Figure 1.16 range from $e = 0.17$ for HD 6434 to $e = 0.41$ for HD 65216.

Exercise 1.5 In Exercise 1.4 we found that the amplitude of the stellar reflex orbit's radial velocity is

$$A_{\mathrm{RV}} = \frac{2\pi a M_{\mathrm{P}} \sin i}{(M_{\mathrm{P}} + M_*) P \sqrt{1 - e^2}}. \qquad \text{(Eqn 1.13)}$$

(a) This equation has the orbital semi-major axis in the numerator, so at first glance it appears that the radial velocity amplitude increases as the planet's orbital semi-major axis increases. Does A_{RV} in fact increase as a planet's orbital semi-major axis increases? Explain your reasoning.

(b) State, with reasoning, how the radial velocity amplitude depends on planet mass, star mass, semi-major axis and eccentricity. ■

1.3.2 Reflex radial velocity for many non-interacting planets

So far we have only considered a single planet. For more than one planet, so long as the planets do not significantly perturb each other's elliptical orbits, we can simply combine all of the elliptical reflex orbits about the barycentre. The observed radial velocity will be

$$V = V_{0,\hat{z}} + \sum_{k=1}^{n} A_k \left(\cos(\theta_k + \omega_{\mathrm{OP},k}) + e_k \cos \omega_{\mathrm{OP},k} \right), \qquad (1.16)$$

where there are n planets, each with their own mass, M_k, and **instantaneous orbital parameters** a_k, e_k, θ_k and $\omega_{\mathrm{OP},k}$, and their own instantaneous value of A_k given by

$$A_k = \frac{2\pi a_k M_k \sin i}{M_{\mathrm{total}} P_k \sqrt{1 - e_k^2}}, \qquad (1.17)$$

where M_{total} is the sum of the masses of the n planets and the star. Generally, Equation 1.16 is an approximation. The instantaneous orbital parameters will evolve over several cycles of the longest planet's period because of the gravitational interactions between the planets.

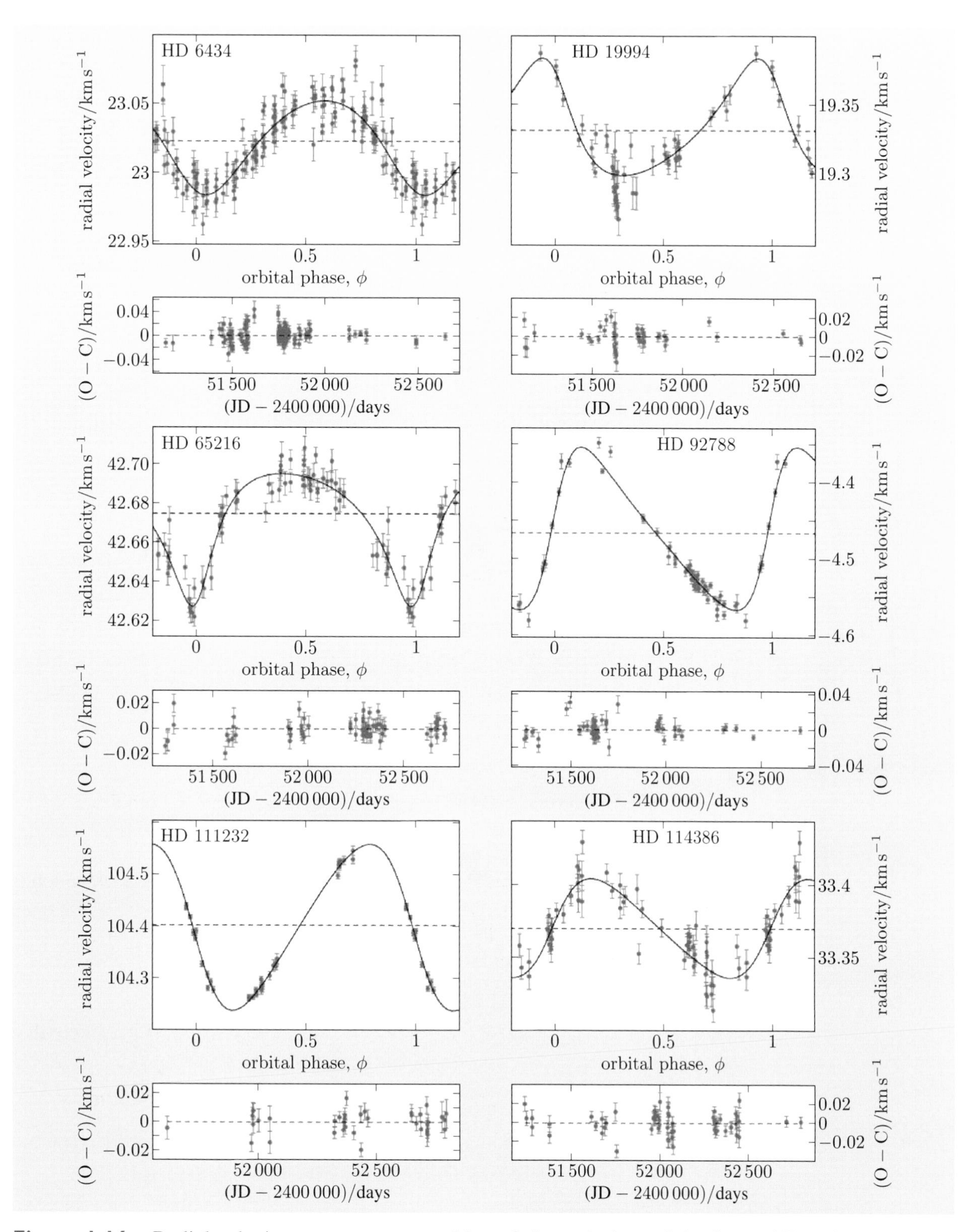

Figure 1.16 Radial velocity measurements and best-fitting solutions of the form of Equation 1.12. Upper panels: radial velocity measurements and solutions folded and plotted as a function of orbital phase. Lower panels: the deviations of the measured data points from the best-fitting solution, plotted as a function of time. Data are shown for six different stars, illustrating some of the many diverse curves that are described by Equation 1.12. '(O – C)' denotes 'observed' minus 'calculated' and indicates how well the model fits the measurements.

- What conditions are required for Equation 1.16 to be valid?

- The planets' mutual gravitational attraction at all times must be negligible compared to the gravitational force exerted by the central star on each planet. This requires the planets to all be of small mass compared to the star, and to have large orbital separations. If this is not the case, the simple two-body solution for the astrocentric orbit of the planet will not be valid.

Figure 1.17 shows two examples of two-planet fits using Equation 1.16. In the case of HD 82943, the planet inducing the larger-amplitude reflex motion has approximately twice the orbital period of the second planet, so a pattern of alternating extreme and less extreme negative radial velocity variations is produced. In the case of HD 169830, the shorter-period planet completes about seven orbits for one orbit of the longer-period planet. This figure demonstrates how continued monitoring of the radial velocities of known planet hosts can reveal the existence of additional planets.

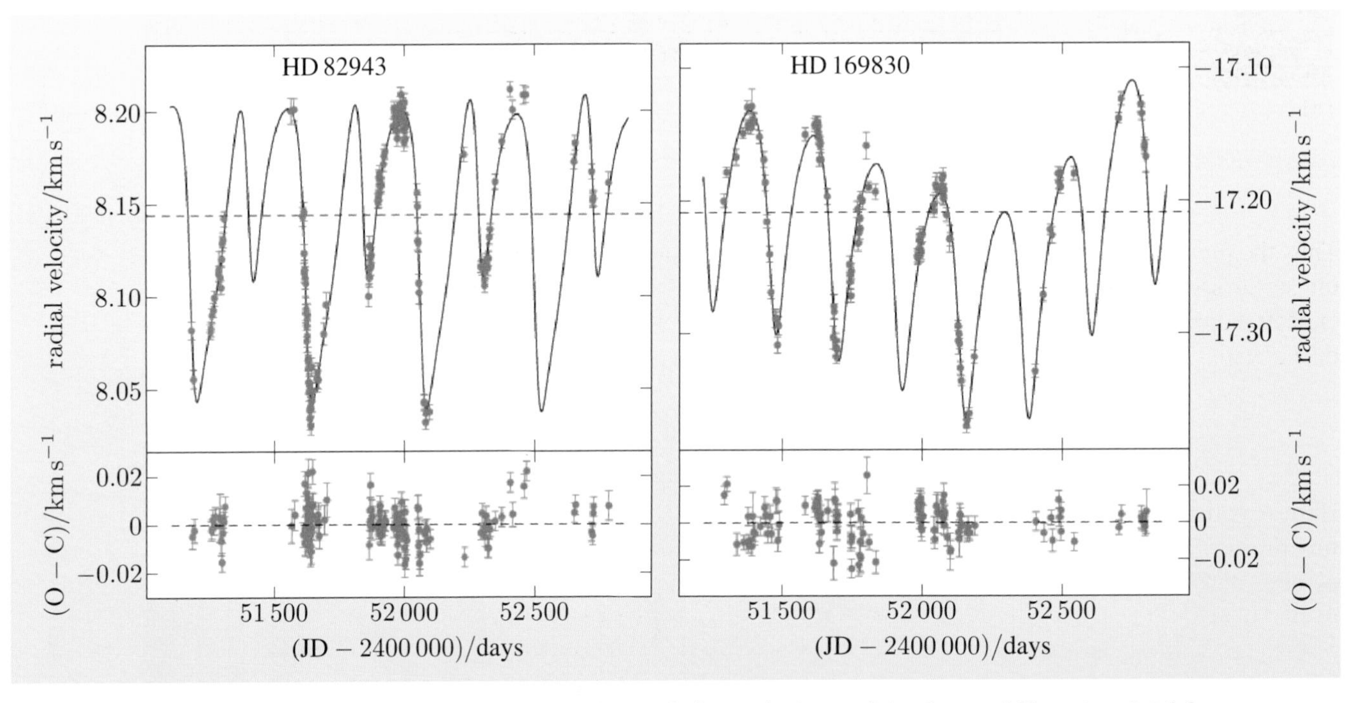

Figure 1.17 Radial velocity measurements and best-fitting solutions of the form of Equation 1.16 for two planets overplotted. Upper panels: radial velocity measurements and solutions plotted as a function of time. Lower panels: the deviations of the measured data points from the best-fitting solution, plotted as a function of time. Data are shown for two different stars, illustrating two of the many diverse curves that are described by Equation 1.16.

The radial velocity technique has been extremely successful in detecting planets around nearby stars. The majority of known exoplanets (October 2009) were detected using it. The technique is limited in its applicability in two significant ways: it can measure the radial velocity precisely only if the stellar spectrum contains suitable features, and only if the star appears bright enough. Even for the brightest of planet host stars, current telescopes can only make precise enough velocity measurements for stars within ∼2000 pc (or ∼6000 light-years). Finally, the results from radial velocity measurements all carry the unknown factor $\sin i$, unless the orbital inclination, i, can be determined using some other method.

1.4 Transits

The transit technique is second only to direct imaging in its simplicity. Figure 1.18 shows transits of Venus and Mercury: when a planet passes in front of the disc of a star, it blocks some of the light that the observer would normally receive.

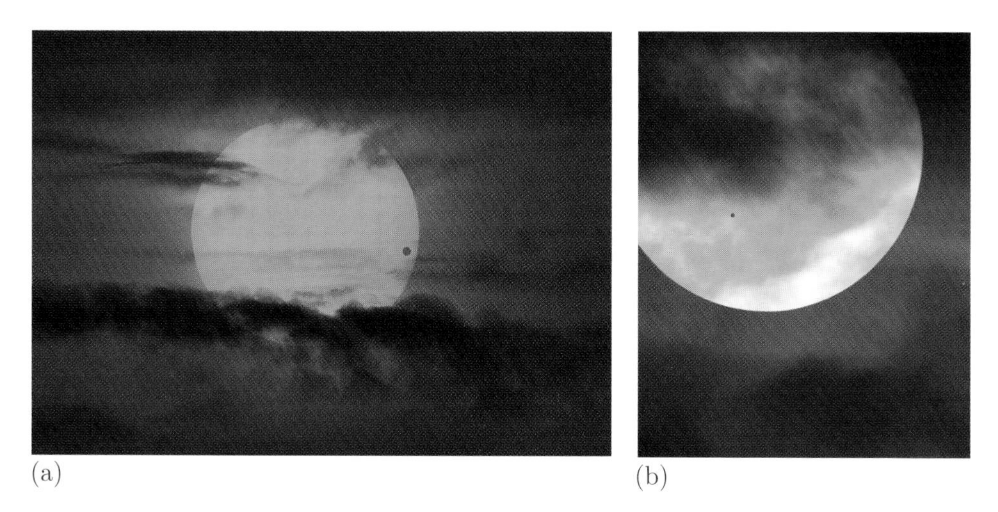

(a) (b)

Figure 1.18 Photographs of transits of (a) Venus and (b) Mercury as observed from Earth on partially cloudy days.

Thus the presence of an opaque object orbiting around a star may be inferred if the star is seen to dip in brightness periodically. The size of the dip in brightness expected during the transit can be estimated simply from the fraction of the stellar disc covered by the planet:

$$\frac{\Delta F}{F} = \frac{R_{\mathrm{P}}^2}{R_*^2}. \tag{1.18}$$

Here F is the flux measured from the star, and ΔF is the observed change in this flux during the transit. The right-hand side is simply the ratio of the areas of the planet's and star's discs. Equation 1.18 gives us a straightforward and joyous result: if a planet transits its host star, we immediately have an estimate of the size of the planet, in terms of the size of its host star. Equation 1.18 applies when the host star is viewed from an interstellar distance. The geometry for transits of Solar System planets viewed from Earth is slightly more complicated.

- Generally, the fraction of a background object obscured by a smaller foreground object depends on the relative distances of the two objects from the observer; for example, it is easy to obscure the Moon with your fist. Why does Equation 1.18 not need to account for the relative distances of the star and planet from the observer?

- As we noted in the Introduction, the distances between stars are very much larger than the typical sizes of planetary orbits. Thus the distance between any observer and an exoplanet is identical to high precision to the distance between the same observer and the host star. Expressed geometrically, the rays of light reaching the observer from the exoplanet host star are parallel, so Equation 1.18 holds.

1.4.1 Transit depth for terrestrial and giant planets

Equation 1.18 allows us to calculate the depth of the transit light curve that would be observed by our hypothetical extraterrestrial astronomers for any of the Solar System planets. For example, the Earth transiting the Sun would cause a dip

$$\frac{\Delta F_{\mathrm{E}}}{F} = \frac{\mathrm{R}_{\oplus}^2}{\mathrm{R}_{\odot}^2} = \left(\frac{6.4 \times 10^3\,\mathrm{km}}{7.0 \times 10^5\,\mathrm{km}}\right)^2 = 8 \times 10^{-5},$$

where we have substituted in the values of the radius of the Earth, $\mathrm{R}_{\oplus}$, and the radius of the Sun, $\mathrm{R}_{\odot}$, to two significant figures, and reported the result to one significant figure. Similarly, a transit by Jupiter would cause a dip of

$$\frac{\Delta F_{\mathrm{J}}}{F} = \frac{\mathrm{R}_{\mathrm{J}}^2}{\mathrm{R}_{\odot}^2} = \left(\frac{7.0 \times 10^4\,\mathrm{km}}{7.0 \times 10^5\,\mathrm{km}}\right)^2 = 1 \times 10^{-2}.$$

As a 'rule of thumb', a giant planet transit will cause a dip of $\sim 1\%$ in the light curve of the host star, while a terrestrial planet transit will cause a dip of $\sim 10^{-2}\%$. The first of these figures is easily within the precision of ground-based photometric instruments. In fact, in the 1950s Otto Struve predicted that transits of Jupiter-like exoplanets could be detected if such planets were orbiting at favourable orbital inclinations. The transit depth for a terrestrial planet is 100 times smaller, and requires a photometric precision of better than 10^{-4}. This is impossible to obtain reliably from Earth-bound telescopes because of the constantly changing transparency of the Earth's atmosphere. For this reason, while there are scores of known transiting giant exoplanets, we know of only one transiting terrestrial planet, CoRoT-7 b, though we believe that the discovery of another is about to be announced (December 2009). Further discoveries are anticipated imminently as the French/ESA satellite CoRoT continues its mission, the NASA Kepler satellite begins announcing results, and the first transit surveys optimized for M dwarf stars produce results. Further space missions are being designed specifically to find terrestrial planets; we will discuss them in Chapter 8.

- If an astronomer detects a regular 1% dip in the light from a star, can they immediately conclude that they have detected a transiting Jupiter-sized planet orbiting around that star?

- No. The conclusion can only be that there is a Jupiter-sized opaque body orbiting the star. To prove that this body is a planet, rather than, for example, a brown dwarf star, the mass of the transiting body needs to be ascertained.

- How can the astronomer ascertain the mass of the object that is transiting the star?

- By using the radial velocity technique described in Section 1.3.

- Why are M dwarf stars particularly promising for the discovery of terrestrial planet transits?

- M dwarfs are the smallest stars, so a small planet produces a relatively large transit signal in an M dwarf, as Equation 1.18 shows.

The transit technique is extremely powerful; however, radial velocity confirmation of planet status is vital for transit candidates. The transiting extrasolar planets are the *only* planets outside our own Solar System with directly measured sizes. Fortunately, the transit technique and the radial velocity technique are beautifully complementary: radial velocity measurements allow the mass of the transiting body to be deduced, while the presence of transits immediately constrains the value of the orbital inclination, i, thus removing the major uncertainty in the interpretation of radial velocity measurements. For the transiting extrasolar planets, therefore, it is possible to deduce accurate and precise masses, radii and a whole host of other quantities. This wealth of empirical information makes the transiting planets invaluable, and underpins our choice of subjects for this book.

1.4.2 Geometric probability of a transit

How likely are our hypothetical extraterrestrial astronomers to discover transits of the Solar System planets? A transit will be seen if the orbital plane is sufficiently close to the observer's line of sight. Figure 1.19 illustrates the geometry of this situation; for the purpose of this discussion we will assume a circular orbit.

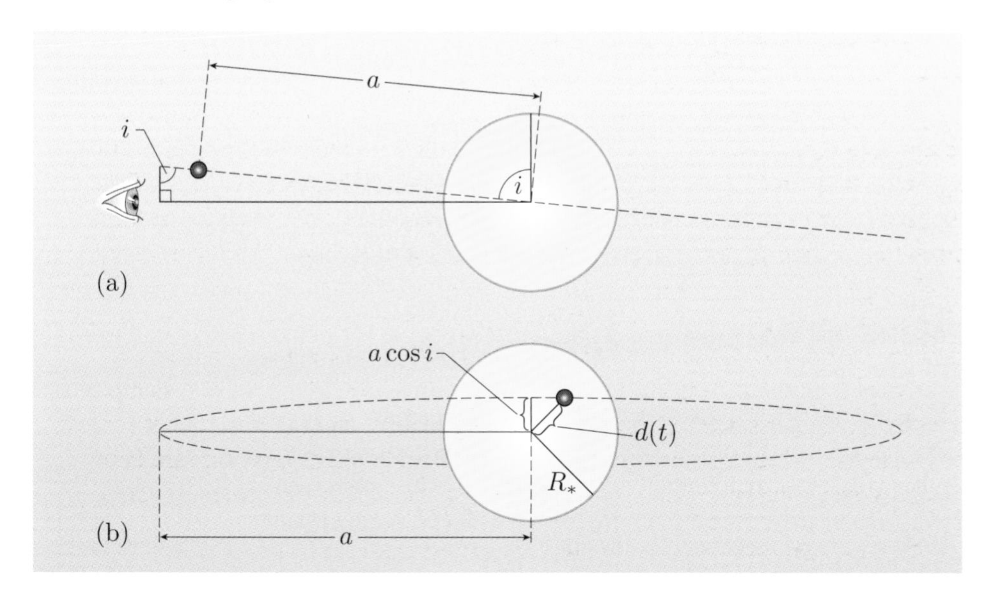

Figure 1.19 (a) The geometry of a transit as viewed from the side. The distant observer (not seen) views the system with orbital inclination i. (b) The geometry of the system from the observer's viewing direction. Note that the observer does not 'see' this geometry: the star is an unresolved point source.

For a transit to be seen, the disc of the planet must pass across the disc of the star. Referring to Figure 1.19b, the closest approach of the centre of the planet's disc to the centre of the star's disc occurs at inferior conjunction, when the planet is closest to the observer. At this orbital phase, by convention referred to as phase $\phi = 0.0$, the distance between the centres of the two discs is

$$d(\phi = 0.0) = a \cos i. \qquad (1.19)$$

Note: the right-hand side of Equation 1.19 is *not* the arccosine function (which is sometimes typeset as acos).

Note that here and throughout this book, a is the semi-major axis of the orbit; in this case we are assuming a circular orbit, so a is simply the radius of the transiting planet's orbit.

For the planet's disc to occult the star's disc, therefore, the orbital inclination, i, must satisfy

$$a \cos i \leq R_* + R_\mathrm{P}. \tag{1.20}$$

- What happens if $R_* - R_\mathrm{P} < a\cos i \leq R_* + R_\mathrm{P}$? Would the transit depth be given by Equation 1.18 in this case?

- For $R_* - R_\mathrm{P} \leq a\cos i \leq R_* + R_\mathrm{P}$, we have a grazing transit: the disc of the planet only partially falls in front of the disc of the star, so a transit is observed, but its depth is less than that give by Equation 1.18 because a smaller area of the stellar disc is occulted.

The projection of the unit vector normal to the orbital plane onto the plane of the sky is $\cos i$, and is equally likely to take on any random value between 0 and 1. To simplify notation below, we temporarily replace the variable $\cos i$ with x. For the purposes of our discussion we will assume that our extraterrestrial astronomers have the necessary technology and are located in a random direction. Thus the probability of our extraterrestrials detecting a transit of a particular Solar System planet is the probability that the random inclination satisfies Equation 1.20:

$$\begin{aligned}\text{geometric transit probability} &= \frac{\text{number of orbits transiting}}{\text{all orbits}}\\ &= \frac{\int_0^{(R_*+R_\mathrm{P})/a}\,\mathrm{d}x}{\int_0^1\,\mathrm{d}x},\end{aligned}$$

so

$$\text{geometric transit probability} = \frac{R_* + R_\mathrm{P}}{a} \approx \frac{R_*}{a}. \tag{1.21}$$

Equation 1.21 shows that transits are most probable for planets with small orbits and large parent stars.

The probability for transits being observable is small; as Figure 1.20 shows, it is less than 1% for all of the Solar System's planets except Mercury. It is rather unlikely that our hypothetical extraterrestrial astronomers would be fortunate enough to observe transits for any of the Solar System's planets.

1.5 Microlensing

Gravitational microlensing exploits the lensing effect of the general relativistic curvature of spacetime to detect planets. For the effect to occur, a chance alignment of stars from the point of view of the observer is required. These alignments do occur from time to time, and are most frequent if one looks at regions of the Galaxy that are densely populated with stars. Terrestrial astronomers working on microlensing observe in the directions of regions that are densely populated with stars. The most obvious and best-studied of these is the **Galactic bulge**: the dense region of stars around the centre of our Galaxy. The bulge of the nearest external spiral galaxy, M31, has also been targeted, as have the Magellanic Clouds. Over 2000 microlensing events have been observed in the Galactic bulge studies, and in a handful of these events, planets have been detected around the foreground, lensing star.

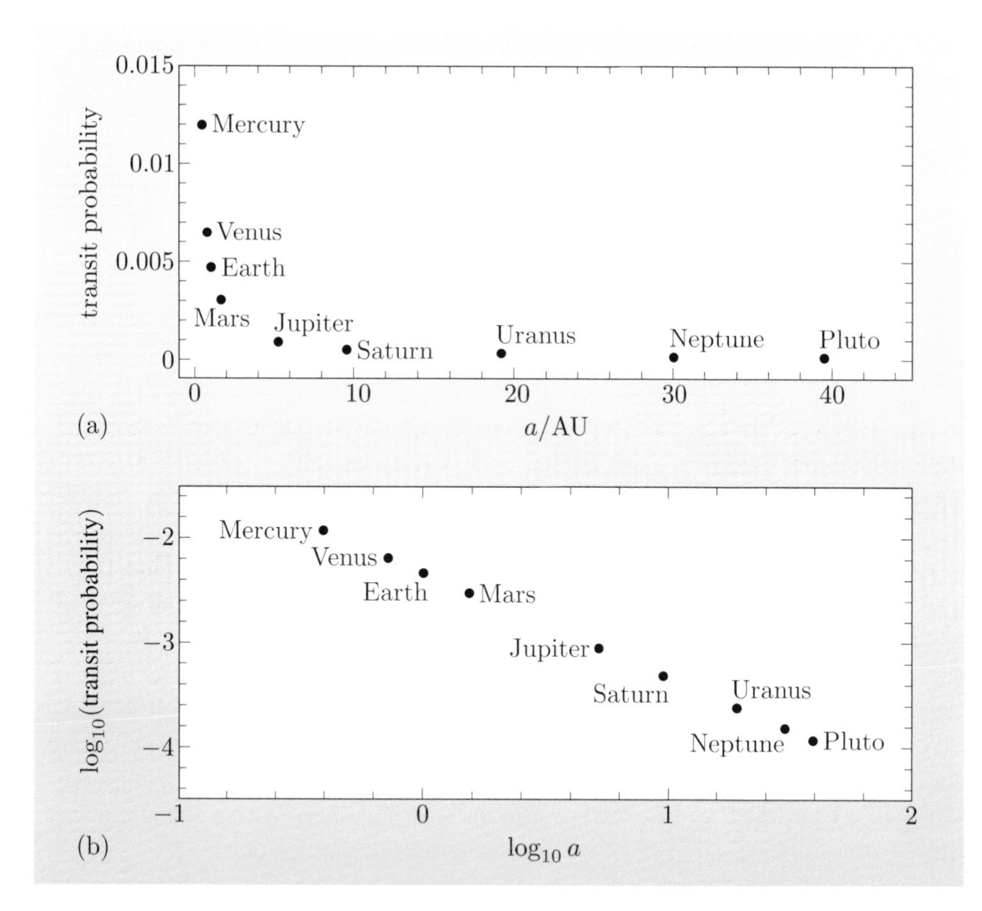

Figure 1.20 The probability of transits of each of the Solar System planets being observable for randomly positioned extraterrestrials. (a) The probability drops off as a^{-1}, and even for Mercury is only just over 1%. (b) Plotted on a log(probability) versus log(a) graph, the relationship is a straight line of gradient −1.

Figure 1.21 shows our position relative to the Galactic bulge. The planets detected by microlensing are positioned just on our side of the Galactic bulge. Since the Sun is positioned towards the edge of the Milky Way Galaxy, as shown in Figure 1.21, to view it against the dense stellar background of the Galactic bulge, an observer would need to be situated in the sparsely populated regions at the edge of the Galaxy.

Figure 1.21 A side-on view of our Galaxy produced from a composite of 2MASS photometry, our Sun's distance from the centre is indicated. The dots schematically indicate the locations of exoplanets discovered by three primary methods of exoplanet discovery: yellow dots, radial velocity method; red dots, transits; blue dots, microlensing. The planets discovered by direct imaging are all very close to our Sun.

Since these regions have low abundances of elements heavier than He, and the density of stars there is low, this part of the Galaxy is *a priori* not the most likely to contain technological extraterrestrial civilizations. In addition, the probability of any given star becoming aligned with a background star and acting as a microlens is vanishingly small; this is a consequence of the very small size of stars compared to interstellar distances. Einstein realized this, and predicted in 1936 that microlensing would occur, though noting that 'there is no great chance of observing this phenomenon'. The reason why we have detected microlensing events is that modern technology permits continuous monitoring of millions of stars, to outweigh the vanishingly small probability ($\sim 10^{-6}$) of any individual star participating in a microlensing alignment at any particular time. Even Einstein could not have begun to anticipate the powerful technologies that have revolutionized astrophysics within the last half century. Putting the two factors together, i.e. the improbability of the Sun participating in a microlensing alignment and the expected dearth of technological extraterrestrials on the edges of the Milky Way, we can conclude that it is highly unlikely that Solar System planets would be detected by hypothetical extraterrestrial astronomers via the microlensing method.

Though it has detected only ten planets (December 2009), gravitational microlensing is important because it is the method that has found the majority of terrestrial exoplanets; recently, however, terrestrial mass planets have been discovered by both transit and radial velocity techniques (December 2009). Notably, gravitational microlensing may also have detected a planet in the galaxy M31; it is probably the only method that could conceivably detect planets outside the Milky Way. To offset these advantages, microlensing has the serious disadvantage that the lensing event is a one-off occurrence: there is no opportunity for confirming and refining the observations. This means that the parameters are rather ill-constrained, and can be meaningfully discussed only in a statistical sense. The statistical analysis of microlensing results implies that no more than a third of solar-type stars host giant planets. The topic of exoplanet detection by microlensing could easily fill an entire book; since we are focusing on transiting exoplanets, we exhort the interested reader to look elsewhere for these details.

Summary of Chapter 1

1. The Sun is one of the more luminous stars in the solar neighbourhood. The majority of objects within 10 parsecs are isolated M dwarf stars. Six of the nearest 100 stars (including the Sun) harbour known planets (October 2009).
2. The exoplanets in a multiple planet system are labelled b, c, d, ... *in the order of discovery*.
3. Direct imaging as a method for exoplanet detection is limited to bright planets in distant orbits around nearby faint stars. It is most effective in the infrared where the contrast ratio is favourable. A giant exoplanet has been discovered orbiting a brown dwarf using the direct imaging method.
4. A planet's light is composed of reflected starlight, which has a spectral energy distribution close to that of the host star, and a thermal emission component peaking at longer wavelengths.
5. The thermal emission from giant planets is partially powered by gravitational (Kelvin–Helmholtz) contraction. This emission is brightest for young giant planets.
6. A young planet has been detected around Fomalhaut using coronagraphy, and three young planets have been discovered around HD 8799. Exoplanet

imaging instruments like SPHERE and GPI should make further discoveries soon (October 2009).

7. The orbits of planets are elliptical. In astrocentric coordinates, the host star is at one focus of the ellipse. The orbital period, the semi-major axis, and the masses are related by Kepler's third law:

$$\frac{a^3}{P^2} = \frac{G(M_* + M_P)}{4\pi^2}. \qquad \text{(Eqn 1.1)}$$

The pericentre is the point on the planet's orbit that is closest to the star. The motion of the planet around the elliptical orbit is measured by the true anomaly, $\theta(t)$. The true anomaly is measured at the star, and is zero when the planet crosses the pericentre.

8. Both the host star and its planet(s) move about the barycentre (the centre of mass) of the system. For a single-planet system,

$$M_* \boldsymbol{r}_* = -M_P \boldsymbol{r}_P. \qquad \text{(Eqn 1.8)}$$

The reflex orbit of the star is, therefore, a scaled-down ellipse of the same eccentricity as the planet's orbit.

9. The reflex motion of the stellar orbit can be detected by astrometry. The Gaia satellite could detect the Sun's reflex orbital motion from a distance of about 500 pc, which is roughly the width of a spiral arm. Astrometry is most effective for massive planets in wide orbits around low-mass stars, but long-term monitoring is required to detect planets with large semi-major axes.

10. The orbital positions of the star and planet are uniquely defined by the orbital phase

$$\phi = \frac{t - T_0}{P} - N_{\text{orb}}, \qquad \text{(Eqn 1.15)}$$

where t is time, T_0 is a fiducial time, and N_{orb} is the orbit number.

11. For a single planet, the amplitude of the stellar reflex radial velocity variation is

$$A_{\text{RV}} = \frac{2\pi a M_P \sin i}{(M_P + M_*) P \sqrt{1 - e^2}}. \qquad \text{(Eqn 1.13)}$$

The largest radial velocity amplitudes are exhibited by low-mass host stars with massive close-in planets in eccentric orbits.

12. The radial velocity, V, of a star is given by

$$V = V_{0,z} + \sum_{k=1}^{n} A_k \left(\cos(\theta_k + \omega_{\text{OP},k}) + e_k \cos \omega_{\text{OP},k}\right), \qquad \text{(Eqn 1.16)}$$

where there are n planets, each with their own instantaneous parameters M_k, a_k, e_k, θ_k, $\omega_{\text{OP},k}$, and A_k given by

$$A_k = \frac{2\pi a_k M_k \sin i}{M_{\text{total}} P_k \sqrt{1 - e_k^2}}, \qquad \text{(Eqn 1.17)}$$

where M_{total} is the sum of the masses of the n planets and the star.

13. Radial velocity measurements using the Doppler shift of features in stellar spectra can be made to a precision less than $1\,\mathrm{m\,s^{-1}}$. The majority of known exoplanets (December 2009) were detected by the radial velocity method. This is the most likely method by which hypothetical extraterrestrials might detect the existence of the Sun's planets.
14. The depth of an exoplanet transit is

$$\frac{\Delta F}{F} = \frac{R_{\mathrm{P}}^2}{R_*^2}. \qquad \text{(Eqn 1.18)}$$

For Jupiter-sized and Earth-sized planets around a solar-type star, this depth is $\sim 1\%$ and $\sim 10^{-2}\%$, respectively.
15. The geometric transit probability is given by

$$\text{geometric transit probability} \approx \frac{R_*}{a}. \qquad \text{(Eqn 1.21)}$$

Transits are most likely for large planets in close-in orbits. If the Solar System were viewed from a random orientation, only Mercury, with $a = 0.4\,\mathrm{AU}$, has a transit probability exceeding 1%.
16. Terrestrial exoplanets have so far been detected primarily via the microlensing method, though they are now being discovered by the transit and radial velocity techniques (December 2009).

Chapter 2 Exoplanet discoveries by the transit method

Introduction

The very first exoplanet discovery, 51 Pegasi b, happened in 1995 using the radial velocity method. The properties of this planet came as a huge surprise: it is a Jupiter-mass planet orbiting so close to its host star that it completes an orbit once every 4.2 days. The semi-major axis of its orbit is a mere 0.052 AU; Mercury's orbit is over 7 times larger. 51 Pegasi b's orbit is 100 times smaller than that of the innermost giant planet in the Solar System. It is generally accepted that giant planets are formed early in the life of the host star, from ices that are found only in the cool regions far from the star itself. Consequently, the organization of our own Solar System, with the giant planets remote from the star, was expected to be invariably the case for exoplanetary systems. 51 Pegasi b, and the dozens of other giant exoplanets with even shorter orbital periods, were not expected to exist. Once they were found, a scientific frontier opened: discoveries came thick and fast.

Planet-mass objects around pulsars were discovered earlier.

Migration of planets, so that giant planets can move inwards after they form, had been suggested as a theoretical possibility before 51 Pegasi b's discovery, but had been largely overlooked by the astronomical community.

As we saw in Chapter 1, the radial velocity technique is most sensitive to massive planets close to their host star, so once astronomers knew to search for giant planets with such small orbits, the radial velocity discoveries flowed rapidly. Happily, as we saw in Subsection 1.4.2, the probability for a transit is highest for such tiny orbits. For the parameters of 51 Pegasi b, the probability of a randomly oriented orbit producing a transit is over 10%, and for closer planets around smaller stars the probability is even higher. As early as 1952 it was realized that transits of giant planets around their host stars would produce detectable photometric signals. For exoplanetary systems organized like our own Solar System, the small probability of any given star showing a transit, and the long intervals (12 years for an exact analogue of Jupiter) between the expected transits discouraged the search for them. The discovery of 51 Pegasi b and the other **hot Jupiters** changed that.

In this chapter we will examine how astronomers search for transiting exoplanets. We begin by discussing the first known transiting exoplanet. Following this we will consider the various factors that determine how many transiting planets can be discovered by a particular survey. These factors influence the design of the telescopes adopted by the various transit search teams. Having understood the underlying design considerations, we then examine how transit search programmes operate in practice, drawing examples from the SuperWASP survey.

2.1 The hot Jupiters, STARE, and HD 209458 b, the first transiting exoplanet

In 1999 the STARE telescope (Figure 2.1) observed the first continuous light curve showing the ingress, transit floor, and egress due to exoplanet transit. Once hot Jupiters had been discovered, they were observed to catch the dimming of the star due to a transit, should the planet's orbit be favourably aligned.

Figure 2.1 The STARE telescope, which observed the first complete exoplanet transit. The telescope has a 10 cm aperture.

The equipment required to detect the transit of a hot Jupiter is cheaper, simpler and more readily available than that required to make a radial velocity detection of the same planet. The Elodie spectrograph and the 1.93 m telescope used to make the radial velocity detection of 51 Pegasi b are shown in Figure 2.2. Even a cursory comparison of Figures 2.1 and 2.2 reveals the relative sophistication of the equipment needed for the radial velocity measurements.

(a)

(b)

Figure 2.2 (a) The Elodie spectrograph. (b) The 1.93 m telescope at Observatoire de Haute Provence. These were used to make the first detection of an exoplanet using the radial velocity technique.

Worked Example 2.1

Figure 2.3 shows the reflex radial velocity curve for HD 209458. Assume that the orbit is circular (as in the orbital fit shown in the figure), and note that the convention used throughout astronomy is that objects moving away from us have positive radial velocity.

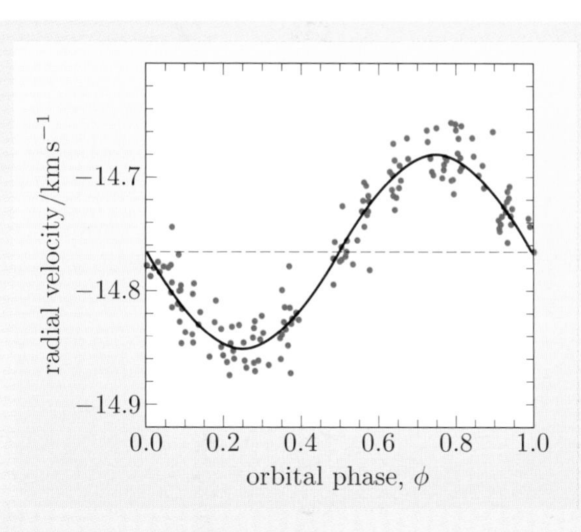

Figure 2.3 The reflex radial velocity curve of HD 209458. The solid line is the best-fitting sinusoid, as expected for a circular orbit, corresponding to Equation 1.12 with zero eccentricity. Note that by convention the radial velocity is defined to be positive for objects moving away from the observer.

(a) Why are all the values on the velocity axis of Figure 2.3 negative?

(b) At which orbital phases is the star moving fastest towards and away from the Earth-bound observer?

(c) At which orbital phases is the star closest and farthest from the observer?

(d) At which orbital phases is the planet closest and farthest from the observer?

(e) At what orbital phase will the planet transit the star, assuming that the orbit is favourably aligned?

Solution

(a) The centre of mass of the HD 209458 system is moving towards the Solar System with a velocity that exceeds the reflex radial velocity amplitude.

(b) Objects moving away from us have positive radial velocity, so the fastest motion away from the Earth is the largest positive value of the radial velocity. We could look at the individual data points in Figure 2.3 and pick out the point with the largest velocity, and give its phase as our answer here. There is a substantial scatter among the measurements though, and this indicates that the uncertainty in the individual measurements is something like $30\,\mathrm{m\,s^{-1}}$. We are told to assume a circular orbit, and the curve plotted gives us the radial velocity variation of the best-fitting circular orbit. The curve uses all of the data points to inform its value at each orbital phase. Therefore our best answer will come from using the curve rather than the data points. Motion in a circle will lead to a sinusoidal (i.e. shaped like the sine function) radial velocity curve. In Equation 1.12 it is the $\cos(\theta(t) + \omega_{\mathrm{OP}})$ term that gives the sinusoidal shape, a cosine being the same shape as a sine, but with a phase offset. We can see that the fitted curve in Figure 2.3 is phased to have its mean value at orbital phases 0.0 and 0.5, therefore we can deduce that its maximum value occurs at phase 0.75. (We could have read this maximum value directly from the graph, but that would involve getting a ruler out; we probably got a quicker, better answer using our knowledge of the sine function and the easy-to-measure axis-crossing.)

This is an example of the power of fitting a model to noisy data. Whenever we know the form of a relationship (e.g. the radial velocity variations are described by Equation 1.12), we can fit a curve of the appropriate form to the noisy data. Then we can estimate the fundamental physical parameters, in this case the orbital period of the planet, P, the amplitude of the star's reflex radial velocity variations, $A_{\rm RV}$, and the phasing of the planet's motion. This general procedure is ubiquitous in astrophysics, and indeed in all quantitative empirical sciences.

The star moves fastest away from the Earth-bound observer at phase 0.75, and moves fastest towards us at phase 0.25.

(c) A sinusoidal radial velocity curve arises from motion in a circle. If the star is moving fastest away from us at phase 0.75, then it will be farthest from us a quarter of a circle later, i.e. at phase 1.0 (which is equivalent, of course, to phase 0.0). The star will be closest to us at the opposite side of the circle, i.e. at phase 0.5.

(d) As Figure 1.12a shows, the planet and the star will always be on opposite sides of their common centre of mass, and a straight line joining their positions will always pass through the centre of mass. Therefore as each proceeds around its orbit, the planet will be closest to us when the star is farthest from us, and vice versa.

The planet is closest to us at phase 0.0 and farthest from us at phase 0.5.

(e) If the planet's orbit is aligned favourably with our line of sight, it will pass in front of the star performing a transit centred on phase 0.0.

Worked Example 2.1 demonstrates how the expected phase of transit is deduced from the radial velocity curve. The points in the orbit where the two objects are most closely aligned, as viewed from Earth, are known as the **conjunctions**. If the orbital inclination is close to $90°$, a transit will occur at **inferior conjunction** of the planet, meaning that the planet is closest to us. At the **superior conjunction** of the planet, for favourably oriented orbits, a secondary eclipse will occur, with the planet passing behind the star. For circular orbits the secondary eclipses and transits will always be separated by exactly $P/2$, and for circular orbits if one occurs, the other will too.

Exercise 2.1 (a) The first exoplanet transit to be discovered was that of HD 209458 b. This planet has an orbital period of 3.52 days, and its host star has mass $1.12\,M_\odot$ and radius $1.146\,R_\odot$. Using Equation 1.21, and Kepler's third law (Equation 1.1), calculate the probability that a system such as this will be observed to transit.

(b) What assumption(s) are implicitly made by using Equation 1.21? Comment on the likely validity of the assumption(s), and comment qualitatively on how your calculated probability is affected by the assumption(s). ■

Using the STARE telescope (Figure 2.1) the first complete exoplanet transit was knowingly observed on 9 September 1999; the light curve is shown in Figure 2.4.

This ground-breaking discovery by Charbonneau, Brown, Latham and Mayor ('Detection of planetary transits across a Sun-like star', 2000, *Astrophysical Journal*, **529**, 45) opened up the research area described in this book.

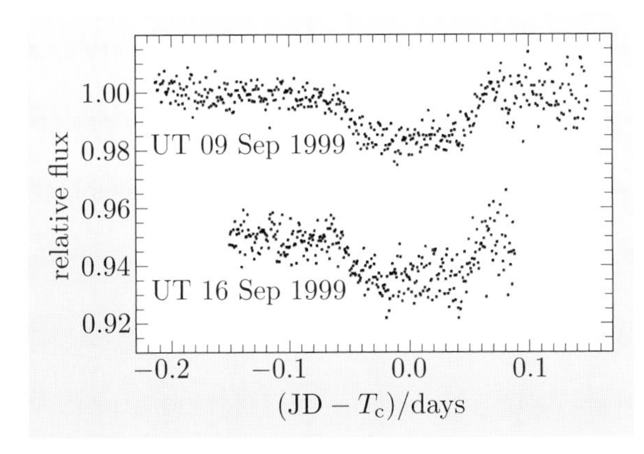

Figure 2.4 The first two complete exoplanet transit light curves knowingly observed. These data were taken with the STARE telescope. The transit is the 1.5% dip in brightness centred on the expected time of inferior conjunction, T_c. The data for 16 September 1999 have been offset.

The STARE observations were made after radial velocity variations indicated the presence of the planet HD 209458 b. The times at which transits occurred were as predicted by the radial velocity ephemeris. The data that were folded to produce Figure 2.3 make it clear that the orbital period of HD 209458 b is about 3.5 days. The two transits shown in Figure 2.4 were used to independently determine the period, subject to the assumption that it is roughly 3.5 days. The period can be measured from observations like these using the relationship

$$P = \frac{T_{\text{elapsed}}}{N_{\text{cycles}}}, \qquad (2.1)$$

where T_{elapsed} is the elapsed time between two observed occurrences of the same event, i.e. two mid-transit times, and N_{cycles} is the number of cycles between the two occurrences. The value obtained was consistent with, but less precise than, the result from the radial velocity data. The transit of 9 September 1999 was used to estimate the time, T_c, at which the centre of the transit occurs. In this case, the value of the **fiducial phase** from the transit photometry was more precise than that from the radial velocity data. The number of cycles observed determines how precisely a period can be measured: if the time elapsed between the beginning of cycle 1 and the end of cycle 100 is measured, the error in the period is the error in the elapsed time measurement divided by 100. If only one cycle is observed, the error in the period is equal to the error in the elapsed time measurement. When the first transits were observed, many cycles of radial velocity data had already been collected and analyzed.

- What is the value of N_{cycles} for the two transits shown in Figure 2.4?

❍ $N_{\text{cycles}} = 2$. The dates of the observations are 7 days apart, and we know $P \approx 3.5$ days.

The beginning and end of a transit are the **ingress** and **egress** respectively.

- What factor is most important in determining how precisely a fiducial phase can be determined?

❍ The sharpness of individual features in a cycle being observed. The beginning and end of a transit are sharp features in the photometric light curve, which allow the time of mid-transit, T_c, to be determined precisely.

A theoretical model of the shape of the transit light curve, fitted to the data of Figure 2.4, can reveal the radius, R_P, of HD 209458 b and the orbital inclination, i. We will discuss the details of this technique in Chapter 3. The currently accepted values (February 2009) for these quantities are $R_P = 1.359 \pm 0.015\,R_J$ and $i = 86.71^\circ \pm 0.05^\circ$; the best value for the radial velocity semi-amplitude is $A_{RV} = 84.67 \pm 0.70\,\mathrm{m\,s^{-1}}$.

Exercise 2.2 (a) The radial velocity semi-amplitude is

$$A_{RV} = \frac{2\pi a M_P \sin i}{(M_P + M_*)P\sqrt{1-e^2}}. \qquad \text{(Eqn 1.13)}$$

Assuming a circular orbit, and that $M_P \ll M_*$, rearrange this equation to give an expression for $M_P \sin i$ in terms of M_*, physical constants, and observable quantities. *Hint*: You will need to use Kepler's third law (Equation 1.1) to obtain a suitable expression for the semi-major axis, a.

(b) Using the data for the HD 209458 b system given above and in Exercise 2.1, and adopting $M_* = 1.12\,M_\odot$, use your expression to evaluate $M_P \sin i$ for HD 209458 b. You may ignore the empirical uncertainties and give your answer without an uncertainty estimate.

(c) Evaluate $\sin i$ using the orbital inclination given above, and consequently calculate the value of the mass of the planet HD 209458 b, giving your answer in kg, M_J and $M_\odot$.

(d) Comment, in the light of your value for M_P, on the applicability of the approximation $M_P \ll M_*$.

(e) In the case where the approximation $M_P \ll M_*$ is not justifiable, how would you work out a value for M_P? You need only describe what you would do; calculations are not required. ■

Figure 2.4 shows observations of a previously known exoplanet host: radial velocity variations had already revealed the planet's presence. The equipment is, however, capable of discovering previously unknown exoplanets by the transit technique. The $\sim$1% dip in brightness caused by a giant planet transit is easy to detect with small telescopes equipped with **CCD photometers**, and a single CCD image can simultaneously measure the brightness of all the stars within the frame. The data in Figure 2.4 came from a square CCD with 2000×2000 pixels, each of which covers a 10.8 arcsec square of sky. The number of arcsec per pixel is referred to as the **plate-scale**. The entire STARE image covers a field of view, W, of just over $6^\circ \times 6^\circ$. In contrast, some of the highest resolution optical images come from the Hubble Space Telescope, where plate-scales have been as small as $\sim$0.03 arcsec pixel^{-1}. Typically, ground-based telescopes' imaging instruments provide plate-scales $\sim$0.5 arcsec pixel^{-1}, and corresponding fields of view. Optical design constraints mean that small telescopes typically have larger fields of view.

- What is the advantage of using an optical system like STARE's, whose pixels each cover relatively large tiles of sky?

○ A larger region of sky falls within the field of view of the instrument, so a single image provides brightness measurements for a larger number of stars.

- Demonstrate that the STARE camera's 10.8 arcsec square pixels do indeed cover $6^\circ \times 6^\circ$ in total.

- Each pixel is 10.8 arcsec on a side, and there are 2000×2000 of them, so each edge of the CCD is of length 2000×10.8 arcsec, or 2.16×10^4 arcsec. There are 60 arcsec in 1 arcmin, and 60 arcmin in 1°, so this corresponds to $(2.16 \times 10^4)/3600^\circ$, or 6°.

Vital statistics of transit search programmes

As soon as HD 209458 b's transit was discovered, a number of teams began transit search programmes looking for previously undiscovered exoplanets. Many of these programmes use equipment similar to STARE. A summary of some notable programmes is given in Table 2.1; the PASS survey is a hypothetical survey, but the others are real.

Table 2.1 Selected transit search programmes, abstracted and updated from data compiled and prepared by Keith Horne in May 2005.

Name	D (cm)	$W^{0.5}$ (degrees)	N_{pix} (10^6)	N_{CCD}	Pixel (arcsec)	Sky mag.	Star mag.	d_{max} (pc)	N_{star} (10^3)
PASS	2.5	127.25	4.0	15	57.75	6.8	9.4	83	18
WASP0	6.4	8.84	4.0	1	15.54	9.6	11.8	246	2
RAPTOR	7.0	55.32	4.0	8	34.38	7.9	11.1	179	33
TrES	10.0	10.51	4.0	3	10.67	10.5	12.7	362	10
XO	11.0	10.06	1.0	2	25.00	8.6	11.9	258	3
HATnet	11.1	18.96	4.0	6	13.94	9.9	12.5	338	28
SuperWASP	11.1	30.44	4.0	16	13.7	9.9	12.5	338	74
RAPTOR-F	14.0	5.93	4.0	2	7.37	11.3	13.4	498	8
OGLE-III	130.0	0.59	8.0	8	0.26	17.1	18.7	3125	20

In Table 2.1, $W^{0.5}$ is the square root of the total field of view (for the STARE telescope this would be 6°); not all fields are square, so $W^{0.5}$ gives the most succinct meaningful indication of the area covered. N_{pix} is the number of pixels in a single CCD detector, while N_{CCD} gives the number of CCDs employed in the survey. The plate-scale of a single pixel is given in the column headed 'Pixel'. The sky magnitude indicates the sky brightness per pixel expressed as a magnitude (see below). The star magnitude gives an estimate of the faintest star around which a transit of a Jupiter-radius planet in a 4-day orbit will be clearly detected.

A crucial statistic is the expected rate of sky photons per pixel; we will see later how this affects the noise level that limits the detection of a transit. This number varies depending, obviously, on the sky brightness, which varies with time, location on the Earth, direction in the sky, light pollution, and proximity and phase of the Moon. Table 2.1 gives the sky magnitude per pixel assuming a fiducial value of the sky brightness: 15.6 mag arcsec^{-2}. This is an appropriate estimate for the optical sky at a good Earth-bound observatory. To convert from the sky magnitude per square arcsec to the sky photon rate per pixel, there are four steps:

1. Convert from the magnitude value per square arcsec to the equivalent flux density per square arcsec. The constants required to do this are given in the Appendix.
2. Divide the flux by the energy per typical photon, and multiply by the width of the **bandpass** (i.e. $\Delta\lambda$ for the range of wavelengths that are detected) to obtain the photon rate per square arcsec per unit collecting area of the telescope.
3. Multiply by the area of the pixel, i.e. the square of the value given in the 'Pixel' column of Table 2.1.
4. Multiply by the collecting area, A, of the telescope: this will give the number of sky photons per unit time falling on a single pixel.

The conversion from sky brightness per square arcsec to sky brightness per pixel has already been carried out in Table 2.1.

Exercise 2.3 (a) Estimate the expected rate of sky photons per pixel for the PASS transit survey, using the survey characteristics given in Table 2.1. You may assume that the CCD is sensitive to photons in the bandpass $550\,\text{nm} < \lambda < 850\,\text{nm}$. You will find the zero-points for the magnitude to flux conversions in Table A4 in the Appendix.

(b) Consequently, estimate the expected number, n_{sky}, of sky photons per pixel in a $10\,\text{s}$ exposure. ■

As we saw in Chapter 1, about 6% of nearby stars harbour planets; about a third of these are hot Jupiters, i.e. with orbital periods less than about 4 days. For these, the probability of a transit occurring can be over 10%. This means that roughly one in every thousand stars should host a transiting giant planet with a period of a few days. Consequently, wide-field photometry (simultaneously measuring the brightness of thousands, tens of thousands, or even hundreds of thousands of stars) is an effective way to discover new exoplanets. Furthermore, we can measure precise values for the transiting planets' masses and radii. Transit searches are an efficient way to discover exoplanets, and the planets revealed are potentially the most informative; the radii of other exoplanets will probably never be known. For these reasons, many wide-field transit search programmes were instigated in the early twenty-first century (see Table 2.1 for a few examples). One of the most successful of these has been SuperWASP. As the author is most familiar with SuperWASP, we will draw specific examples from it in the remainder of this chapter; the issues discussed are, however, generally applicable.

2.2 How many transiting planets will a survey find?

After the discovery of HD 209458 b's transit, astronomers seized the opportunity to discover perhaps thousands of transiting exoplanets. In this section we consider the expected number of transiting planet discoveries. Table 2.1 lists various planetary transit surveys, giving key characteristics of each. We can use these to estimate how many planets should be detectable by each survey. We consider factors required for this estimate in the following subsections. Our approach follows the paper 'Status and prospects of planetary transit searches: hot Jupiters galore' (2003, *ASP Conference Series*, **294**, 361–70) by Keith Horne.

2.2.1 $N(S)$ for standard candles of uniform space density

To begin our explanation of the expected haul of transiting planets, we first derive a general result that is used in many areas of astrophysics. Consider a population of objects, all of which have the same luminosity, L, which are uniformly distributed, with number density n_0, throughout a region of space. The observed **flux**, F, from any of these sources is given by

$$F = \frac{L}{4\pi r^2}, \tag{2.2}$$

where r is the distance to the source. Any astronomical equipment will have a limiting flux, S, below which sources are undetectable, so in general an observation will detect sources in the field of view with $F \geq S$. For our uniform population of '**standard candles**' with luminosity L, this implies a **limiting distance**, d_{max}, out to which the sources are detectable:

$$d_{\text{max}} = \left(\frac{L}{4\pi S}\right)^{1/2}. \tag{2.3}$$

Exercise 2.4 Derive Equation 2.3. ■

The volume of a spherical shell with radius r, and thickness $\mathrm{d}r$ is $\mathrm{d}V = 4\pi r^2\,\mathrm{d}r$, and the number of sources within this shell is the number density, n_0, multiplied by the volume of the shell, $\mathrm{d}V$. Thus the number of detectable sources is

We are using r for distance here instead of d to avoid having to integrate $d^2\,\mathrm{d}d$!

In our example we have a constant number density, so we could dispense with the integral, and simply use the volume of the sphere multiplied by n_0. Often in astronomy the number density is not constant, so we preserved the integral to illustrate the general case.

$$\begin{aligned} N(S) &= \int_0^{d_{\text{max}}} n_0\,\mathrm{d}V \\ &= \int_0^{d_{\text{max}}} 4\pi n_0 r^2\,\mathrm{d}r \\ &= \left[4\pi n_0 \frac{r^3}{3}\right]_0^{d_{\text{max}}} \\ &= \frac{4\pi n_0}{3}\,d_{\text{max}}^3, \end{aligned} \tag{2.4}$$

and substituting in for d_{max} from Equation 2.3 gives

$$N(S) = \frac{4\pi n_0}{3}\left(\frac{L}{4\pi S}\right)^{3/2}. \tag{2.5}$$

This tells us that for a uniform density population of standard candles, the number of sources brighter than a limiting flux S is proportional to $S^{-3/2}$. This relationship allows us to calculate how the number of detectable sources increases as the limiting flux for detection is decreased.

2.2.2 The duration of an exoplanet transit

We also require an expression for the duration of a transit in terms of the system parameters. This can be calculated from the orbital period and the fraction of the planet's orbit for which the planet appears in front of the stellar disc. Figure 2.5 illustrates this for a circular orbit. A distant observer positioned in the plane of the orbit will observe a transit while the planet passes between positions V and W. The lines from the edges of the star indicate the parallel paths of light rays travelling to this distant observer. They intersect the planet's orbit at points V and W. Assuming that the radius of the planet's orbit, a, is much greater than the radius of the star, R_*, the arc from V to W is almost the same length as the chord from V to W. Because the vertical lines are parallel, the length of the chord from V to W is simply twice the radius of the star, $2R_*$. For a circular orbit, the transit duration is therefore

$$T_{\text{dur}} = P \times \frac{\text{length of arc from V to W}}{2\pi a} \approx \frac{P \times 2R_*}{2\pi a},$$

so

$$T_{\text{dur}} \approx \frac{PR_*}{\pi a}. \tag{2.6}$$

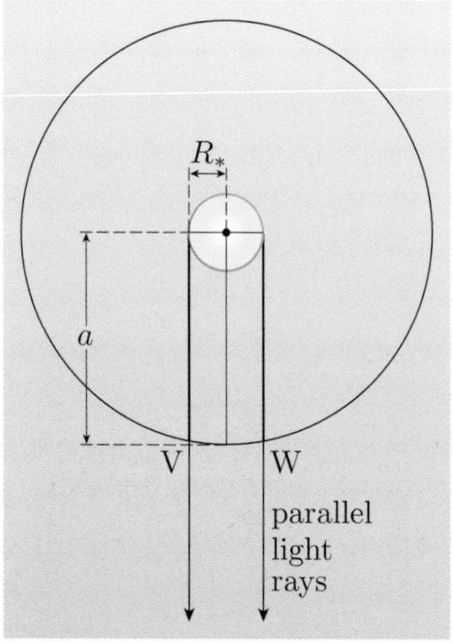

Figure 2.5 A schematic diagram showing a circular orbit of radius a around a star of radius R_*. The observer is in the direction of the bottom of the page, and the transit occurs while the planet is between positions V and W. The two outer vertical lines indicate the path taken by light rays on each side of the star to a distant observer. The rays are parallel. Points V and W are the intersections of these lines with the planet's orbit.

Exercise 2.5 (a) Derive the exact version of Equation 2.6.

(b) Thus derive an expression for the fractional deviation caused by the approximation in Equation 2.6 for the transit duration.

(c) Use Kepler's third law (Equation 1.1) to express the orbital radius, a, in terms of the stellar mass, M_*, using the assumption $M_* \gg M_{\text{P}}$.

(d) Calculate the percentage deviation in the approximated values for the transit duration for a star with mass $M_* = 1\,\text{M}_\odot$, radius $R_* = 1\,\text{R}_\odot$, and a planet with orbital period $P = 1$ day.

(e) State whether the approximation will be more or less exact as the orbital period of the planet increases, quantitatively explaining your reasoning.

(f) Consequently, discuss whether it is justifiable to use the approximation inherent in Equation 2.6 in estimates of yields for transit searches in the period range $P > 1$ day. ■

- Is there any situation where a transit with duration significantly different from that given by Equation 2.6 will be observed?

❍ Yes. In deriving Equation 2.6 we placed the observer in the plane of the orbit. If the observer is slightly out of the plane of the orbit, the planet will not pass across the centre of the star's disc. The transit can pass through any part of the star's disc, and can therefore have any duration from zero up to the value given by Equation 2.6.

To describe transits that don't follow a path passing through the centre of the stellar disc, it is useful to define the **impact parameter**, b, which is the shortest distance from the centre of the disc to the locus of the planet, as shown in Figure 2.6. The transit duration given by Equation 2.6 corresponds to an impact parameter, b, of zero. We will use this for our estimate of how many planets might be expected to be caught in the transit surveys. In the next chapter we will return to a more detailed discussion of the impact parameter.

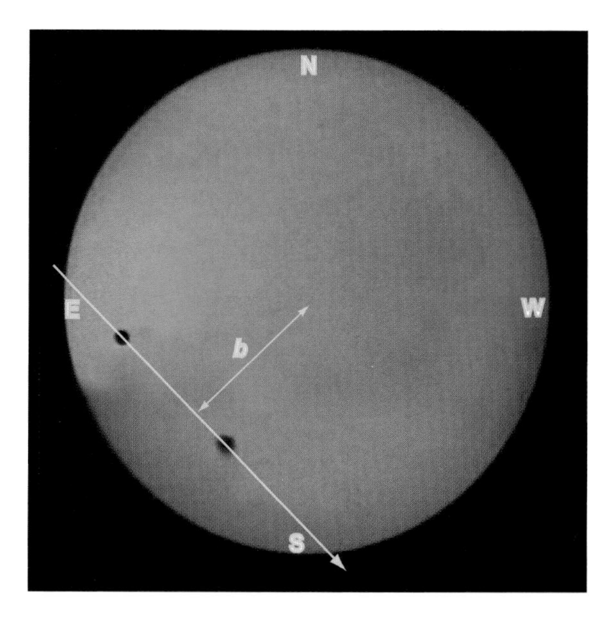

Figure 2.6 A composite of two images of the 2004 transit of Venus. The yellow arrow indicates the direction of the planet's path across the Sun's disc. The impact parameter, b, is the shortest distance from the centre of the disc to the locus of the planet, therefore $0 \leq b \leq R_P + R_*$.

In some texts, b is defined to be a dimensionless quantity, with $0 \leq b \leq 1 + (R_P/R_*)$.

Exercise 2.6 Calculate the transit duration for WASP-4 b, a planet with $P = 1.3382$ days, orbital semi-major axis 0.023 AU, and a host star of radius $1.15\,R_\odot$. You may assume that the impact parameter, b, is zero. ■

2.2.3 The signal-to-noise ratio

To discover a transiting planet, we need to detect the decrease in the brightness of the star as the transit occurs. Whether or not this is possible for a given instrument and a particular star depends primarily on the signal-to-noise ratio of the data. Equation 1.18 tells us that the depth of a transit is

$$\Delta F = \frac{R_P^2}{R_*^2} F, \tag{2.7}$$

where ΔF is the depth in flux units, and F is the flux from the star outside transit, as indicated in Figure 2.7. ΔF is the size of the signal that we wish to detect, and the amplitude of the noise can be estimated from the characteristics of the detector and Poisson or Gaussian statistics applied to the photon count. (A full discussion of statistics is beyond the scope of this book. You may already be familiar with the topic; if not, please accept the results that we draw from statistics.) When the star under consideration is at the faint limit, the sky background noise dominates over other noise sources. The number of sky photons can be expressed as

Sky noise does not necessarily always dominate, but it generally does for the surveys in Table 2.1.

$$n_{\text{sky}} = AQ\,\Delta\lambda\,\sigma_{\text{FWHM}}^2\,l_{\text{sky}}\,\Delta t. \tag{2.8}$$

Here Δt is the duration of the observation; A is the collecting area of the telescope; Q is the **quantum efficiency**, which gives the fraction of photons entering the telescope that are ultimately detected; $\Delta\lambda$ is the bandpass, i.e. the range of wavelengths that are detected; σ_{FWHM} indicates the angular size of the

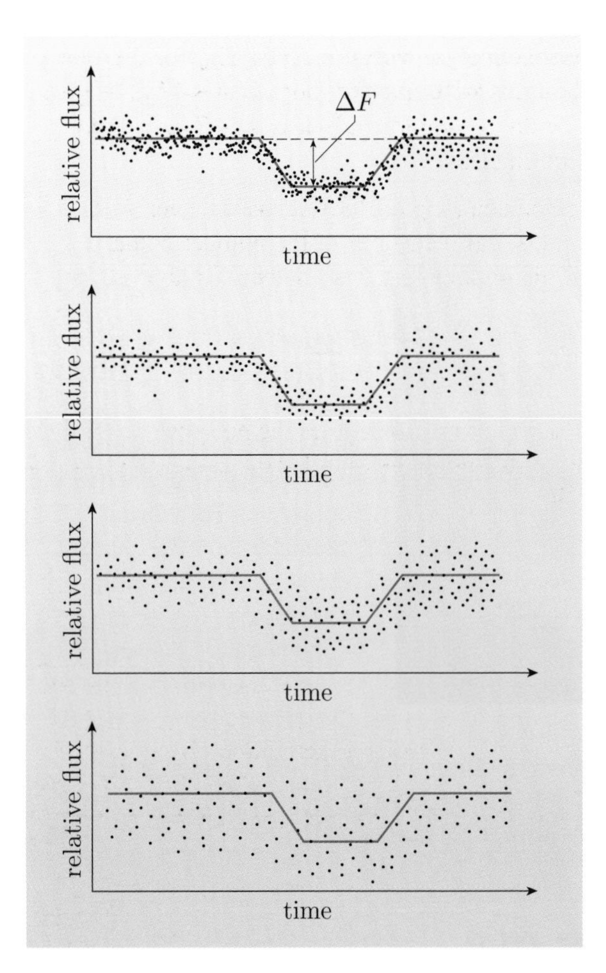

Figure 2.7 Examples of transit light curves with a range of values of signal-to-noise ratio. For very noisy data, the transit cannot be discerned; at moderate noise levels, the transit depth, ΔF, is imprecisely determined.

image of a star in units of pixels; and l_{sky} gives the number of sky photons per pixel per unit area per unit wavelength interval. We calculated l_{sky} for the PASS survey in Exercise 2.3. l_{sky} is a property of the sky at the location of the telescope; A, Q and $\Delta\lambda$ are properties of the instrumentation; and σ^2_{FWHM} depends on both the instrument and the site.

The **point spread function** describes the way in which the detected photons from a point source are spread out. In general, a one-dimensional slice through it is roughly Gaussian, and it is described by the width of the best-fitting Gaussian function, σ_{FWHM}. The detected photons from a point source are spread out as a result of refraction caused by irregularities in the Earth's atmosphere, which results in the **seeing**. At good sites the seeing is better than 1 arcsec. For conventional telescopes, generally the plate-scale of imaging instrumentation is designed so that it fully resolves the point spread function of the seeing disc, hence the plate-scale is generally $\sim$0.5 arcsec. This does not have to be the case, and almost all the entries in Table 2.1 have plate-scales larger than this: one is larger by a factor of 100. This instrumentation under-resolves the seeing disc, but as we will see, for transit surveys this is less important than covering a large area of sky.

The small change in brightness due to the transit is present for the duration of the transit. Because only a small fraction of the time is spent in transit, much more time will inevitably be spent measuring the brightness out of transit, so the measurement of the in-transit flux level will dominate the uncertainty in the determination of ΔF. Consequently, we will use $\Delta t \sim T_{\text{dur}}$ for the signal-to-noise calculation. Observations will never be more than $\sim 4T_{\text{dur}}$ as given by Equation 2.6, and observations with Δt significantly less than T_{dur} will not contribute much to the transit detection effort.

When a quantity is measured from the counting of discrete events, such as the arrival of individual photons, the statistical behaviour of the number count is determined by Poisson statistics. This implies that if we detect N photons, the statistical fluctuation will be $\sqrt{N}$:

$$\text{statistical uncertainty on number count, } N = \sqrt{N}. \tag{2.9}$$

This uncertainty in the number of sky photons determines the noise level. Pulling together the algebra, we derive the following expression for the noise (in photons) in the observation of a single transit:

$$\begin{aligned}\text{Noise} &\approx \sqrt{n_{\text{sky}}} \\ &\approx \sqrt{AQ\,\Delta\lambda\,\sigma_{\text{FWHM}}^2\,l_{\text{sky}}\,\Delta t} \\ &\approx \sqrt{AQ\,\Delta\lambda\,\sigma_{\text{FWHM}}^2\,l_{\text{sky}}\,\frac{PR_*}{\pi a}}.\end{aligned} \tag{2.10}$$

We can similarly evaluate the signal, ΔF, in terms of the number of photons detected, i.e. we can calculate Δn_{star}, the depth of the transit when the light curve is presented in photon rather than flux units. F is the flux from the star, and the observed flux is

$$F = \int_{\text{bandpass}} F_\lambda \, \mathrm{d}\lambda, \tag{2.11}$$

where F_λ is the flux per unit wavelength interval. The total flux from the star is given by Equation 2.2. This describes how starlight is uniformly spread over a sphere, so the flux diminishes with the inverse square of the distance. In reality, there is a further loss of photons caused by absorption and scattering by interstellar material. The simplest way to model this **interstellar extinction** of flux is to modify Equation 2.2 to

$$F_{\text{total}} = \frac{L_*}{4\pi d^2} \exp\left(-Kd\right), \tag{2.12}$$

where K is the **interstellar extinction coefficient**. The distribution of interstellar dust and gas is patchy, so this is only an approximation to the actual extinction for any given star. By using an appropriate average value for the extinction coefficient, we should be able to arrive at a reasonable approximation for a given survey. Of course, F_{total} and F_λ are related by

$$F_{\text{total}} = \int_0^\infty F_\lambda \, \mathrm{d}\lambda, \tag{2.13}$$

and by comparing Equations 2.11 and 2.13 we can define an efficiency parameter, η, to take account of the fact that some of the star's flux falls outside the bandpass:

$$\eta = \frac{F}{F_{\text{total}}}, \tag{2.14}$$

and obviously $0 < \eta < 1$.

The number of photons per unit area per unit wavelength interval per unit time is given by

$$\begin{aligned}\text{photon flux} &= \int_{\text{bandpass}} \frac{F_\lambda}{E_{\text{ph}}(\lambda)}\,\text{d}\lambda = \int_{\text{bandpass}} \frac{\lambda F_\lambda}{hc}\,\text{d}\lambda \\ &\approx \eta\, F_{\text{total}}\,\frac{\overline{\lambda}}{hc}. \end{aligned} \tag{2.15}$$

Here we have substituted in for the photon energy, E_{ph}, and defined a mean wavelength, $\overline{\lambda}$. Consequently, the number of photons detected from the star, n_{star}, will be

$$\begin{aligned} n_{\text{star}} &\approx AQ\,\Delta\lambda\,\eta\,F_{\text{total}}\,\frac{\overline{\lambda}}{hc}\,\Delta t \\ &\approx AQ\,\Delta\lambda\,\eta\,F_{\text{total}}\,\frac{\overline{\lambda}}{hc}\,\frac{PR_*}{\pi a} \\ &\approx AQ\,\Delta\lambda\,\eta\,\frac{L_*}{4\pi d^2}\exp(-Kd)\,\frac{\overline{\lambda}}{hc}\,\frac{PR_*}{\pi a}. \end{aligned} \tag{2.16}$$

Combining Equation 2.7 with the knowledge that the fractional depth of the transit is independent of the units adopted gives

$$\frac{\Delta n_{\text{star}}}{n_{\text{star}}} = \frac{\Delta F}{F}. \tag{2.17}$$

The signal, Δn_{star}, expressed in photon units, is

$$\begin{aligned} \Delta n_{\text{star}} &= \frac{R_{\text{P}}^2}{R_*^2}\,n_{\text{star}} \\ &= \frac{R_{\text{P}}^2}{R_*^2}\,AQ\,\Delta\lambda\,\eta\,\frac{L_*}{4\pi d^2}\,\frac{\overline{\lambda}}{hc}\,\frac{PR_*}{\pi a}\exp(-Kd). \end{aligned} \tag{2.18}$$

Thus the signal-to-noise ratio for a single transit is given by Equation 2.18 divided by Equation 2.10:

$$\begin{aligned} \frac{\text{S}}{\text{N}} &\approx \frac{R_{\text{P}}^2}{R_*}\,\frac{AQ\,\Delta\lambda\,\frac{\eta\overline{\lambda}}{hc}\,\frac{PL_*}{4\pi^2 a d^2}\exp(-Kd)}{\sqrt{AQ\,\Delta\lambda\,\sigma_{\text{FWHM}}^2\,l_{\text{sky}}\,\frac{PR_*}{\pi a}}} \\ &\approx \frac{R_{\text{P}}^2}{4(\pi R_*)^{3/2}}\left(\frac{AQ\,\Delta\lambda\,P}{l_{\text{sky}}\,a}\right)^{1/2}\frac{\eta\overline{\lambda}L_*\exp(-Kd)}{hcd^2\,\sigma_{\text{FWHM}}}. \end{aligned} \tag{2.19}$$

A planet detection would never be claimed on the basis of a single transit: one of the first requirements for any transiting planet candidate is that the transits repeat regularly. Poisson statistics imply that the signal-to-noise ratio improves with the number of transits observed. If N_{t} transits are observed, the signal-to-noise ratio is improved over that for a single transit by a factor $\sqrt{N_{\text{t}}}$. More transits will be observed if the time base of the observations, t, is increased, and the number caught depends on the fraction of the time when the survey is actually collecting data. This latter factor is expressed as the **duty cycle**, ξ, a fraction that varies from a maximum of 1 for uninterrupted coverage down to 0 for no coverage. If the observations have a duty cycle ξ, the accumulated coverage over the elapsed time, t, is simply ξt. On average the number of transits observed is given by

$$N_{\text{t}} = \frac{\xi t}{P}. \tag{2.20}$$

Now we have everything that we need to evaluate the signal-to-noise ratio that a particular instrument will achieve for a particular transiting star. Combining the signal-to-noise expression for a single transit with the factor $\sqrt{N_t}$ and Equation 2.20, we obtain the signal-to-noise ratio for transit detection in observations of duration t and duty cycle ξ:

$$\frac{\mathrm{S}}{\mathrm{N}} \approx \frac{R_\mathrm{P}^2}{4(\pi R_*)^{3/2}} \left(\frac{AQ\,\Delta\lambda\,P}{l_\mathrm{sky}\,a}\right)^{1/2} \frac{\eta\bar{\lambda}L_*\exp(-Kd)}{hcd^2\,\sigma_\mathrm{FWHM}} \times \sqrt{N_t}$$

$$\approx \frac{R_\mathrm{P}^2}{4(\pi R_*)^{3/2}} \left(\frac{AQ\,\Delta\lambda\,P}{l_\mathrm{sky}\,a}\right)^{1/2} \frac{\eta\bar{\lambda}L_*\exp(-Kd)}{hcd^2\,\sigma_\mathrm{FWHM}} \times \sqrt{\frac{\xi t}{P}}$$

so

$$\frac{\mathrm{S}}{\mathrm{N}} \approx \frac{R_\mathrm{P}^2}{4(\pi R_*)^{3/2}} \left(\frac{AQ\,\Delta\lambda\,\xi t}{l_\mathrm{sky}\,a}\right)^{1/2} \frac{\eta\bar{\lambda}L_*\exp(-Kd)}{hcd^2\,\sigma_\mathrm{FWHM}}. \qquad (2.21)$$

To discover a transiting planet, the survey must produce a signal-to-noise ratio sufficient to pick out the transits with a reasonable degree of confidence. The exact value of the signal-to-noise ratio required will depend on how many false positives can be tolerated and the efficiency of the **algorithms** used to analyze the data. A false positive is a light curve that the algorithm identifies as containing planetary transits, when it in fact does not. Noisier data will produce larger numbers of false positive candidates. If data with a low value of signal-to-noise ratio can be used effectively, then the survey distance d_max is high; conversely, if a high value of signal-to-noise ratio is required, then the survey distance d_max is low. This is a specific example of issues related to the limiting flux, S, which we discussed in a general way in Subsection 2.2.1.

- What is the maximum duty cycle for a survey using a single telescope?
- Generally, any given star is observable for only 10 hours a night or so, in which case the maximum duty cycle is $\frac{10}{24}$ or approximately 0.4.

The maximum duty cycle depends on the declination of the object and the latitude of the telescope. A star that is **circumpolar** from the location of the telescope can be observed all night. If the telescope is at one of the Earth's poles, a duty cycle of 1 can be attained for the winter months.

2.2.4 Survey volume and number of stars searched

For stars of a given luminosity, a combination of our work in Subsections 2.2.1 and 2.2.3 allows us to work out the distance, d_max, to which a given survey can detect transits. The volume of space searched is therefore the fraction of a sphere of radius d_max that is covered by the field of view. As illustrated in Figure 2.8, this volume is

$$V = \frac{\theta^2}{4\pi} \times \frac{4\pi\,d_\mathrm{max}^3}{3} = \frac{\theta^2\,d_\mathrm{max}^3}{3}, \qquad (2.22)$$

where θ^2 is the field of view in steradians.

This volume is comprised of a sum of individual shells:

We have again used the variable r for distance.

$$V = \int_0^{d_\mathrm{max}} \mathrm{d}V = \int_0^{d_\mathrm{max}} \frac{\theta^2}{4\pi} \times 4\pi r^2\,\mathrm{d}r$$

$$= \theta^2 \int_0^{d_\mathrm{max}} r^2\,\mathrm{d}r. \qquad (2.23)$$

Equation 2.23 rather than Equation 2.22 must be used if we are to calculate any sum over the survey volume of something that depends on distance.

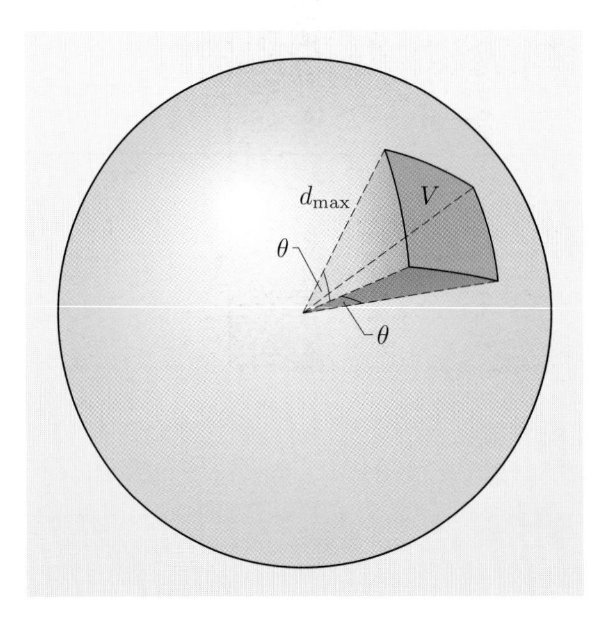

Figure 2.8 The volume, V, covered by astronomical observations with a survey distance d_{max} and a field of view θ^2 is the fraction $\theta^2/4\pi$ times the volume of the sphere.

In Subsection 2.2.1 we assumed that the density of sources was uniform throughout the region of space considered. Of course, this is far from the case for stars in our Galaxy, as Figure 2.9a shows. Position in the Galaxy is described by **Galactic longitude** and **Galactic latitude**, as illustrated in Figure 2.9b. The Sun is close to the **Galactic plane**, where the stellar density is n_0 and the Galactic latitude is $b_G = 0$. The density of stars declines as distance from the plane increases, and we can model this spatial variation in the stellar density as

$$\begin{aligned} n &= n_0 \exp\left(-\frac{h}{H}\right) \\ &= n_0 \exp\left(-\frac{d\,|\sin b_G|}{H}\right), \end{aligned} \tag{2.24}$$

where h is the distance above the plane, and H is the **scale-height** of the stellar density distribution.

Combining Equations 2.24 and 2.23, we can therefore obtain the total number of stars surveyed:

$$\begin{aligned} N_{star} &= \theta^2 \int_0^{d_{max}} n(r)\, r^2\, dr \\ &= \theta^2 \int_0^{d_{max}} n_0 \exp\left(-\frac{r\,|\sin b_G|}{H}\right) r^2\, dr. \end{aligned} \tag{2.25}$$

2.2.5 The number of transiting planets per star

Obviously, we don't yet know the number of planets per star in our Galaxy; this is one of the things that current research aims to discover. We discuss the exoplanet population in Chapter 4. Here we introduce some notation that we need below.

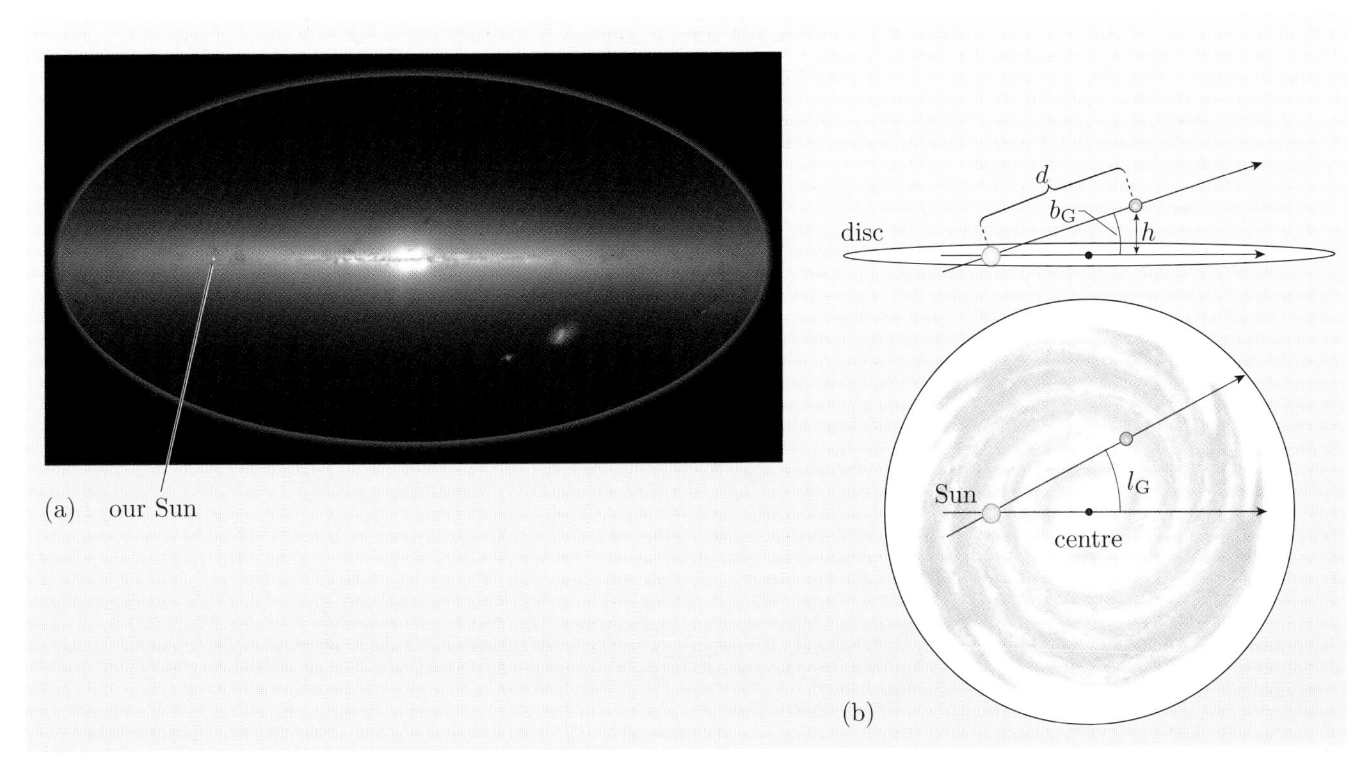

Figure 2.9 (a) A near-infrared image showing our own Galaxy. The bulge at the centre of the Galaxy appears prominent, and elsewhere the stellar density decreases with distance from the mid-plane of the Galaxy. The approximate *distance* of the Sun from the centre is indicated, though of course this is *NOT the position* of the Sun, since the image was taken from the Sun's position. (b) Position in the Galaxy can be indicated with Galactic longitude and latitude (l_G, b_G), as shown. The coordinates are centred on the Sun, and the Galactic mid-plane has $b_G = 0$. The height, h, above the mid-plane is $h = d \sin b_G$, as shown.

The number of planets per star will obviously vary from star to star, and the average number will probably depend on the properties of the star, i.e. the star's mass, M_*, etc. It is already established that Jupiter-mass planets are found preferentially around metal-rich stars. The distributions of planets' masses, M_P, and semi-major axes, a, have a profound influence on their detectability. The total number of planets per star, η_P, will be an integral over this two-dimensional space:

There are indications that the orbital periods of exoplanets may have an approximately uniform distribution in $\log a$.

$$\eta_P = \iint \alpha_P(a, M_P)\, da\, dM_P, \tag{2.26}$$

where $\alpha_P(a, M_P)$ is the unknown function describing the distribution of planet frequency over semi-major axis and planet mass, and the double integral is over all values of a and M_P. This function, α_P, and consequently η_P, is probably different for different types of star. We know that about 1% of F, G, K spectral type main sequence stars host hot Jupiter planets, and their functions α_P and η_P underlie and must be consistent with this fact.

Finally, we must bear in mind that η_P describes the total number of planets per star. The probability that any given planet will execute transits from our line of sight is given by Equation 1.21.

2.2.6 Scaling laws for the discovery of transiting exoplanets

Bringing together the previous subsections, we can now examine how survey design affects the expected number of planet discoveries. We begin with a general expression that includes all of the potential dependencies that we have considered above. The number of planets within the survey volume for stars of a particular type is given by

Again we use r for distance.

$$\begin{aligned} N_\mathrm{P} &= \theta^2 \int_0^{d_\mathrm{max}} \eta_\mathrm{P} n_0 \exp\left(-\frac{r\,|\sin b_\mathrm{G}|}{H}\right) r^2\,\mathrm{d}r \\ &= \theta^2 n_0 \eta_\mathrm{P} \int_0^{d_\mathrm{max}} \exp\left(-\frac{r\,|\sin b_\mathrm{G}|}{H}\right) r^2\,\mathrm{d}r, \end{aligned} \quad (2.27)$$

where n_0 is the number density of this type of star at the Galactic plane, and η_P has been assumed to be independent of distance.

- Why have we specified 'for stars of a particular type' in Equation 2.27?
- Because η_P is a function of the star's properties, and d_max depends crucially on the star's luminosity; the survey volume for luminous stars will be larger than that for dim stars.

To obtain the number of transiting planets for stars of a particular type, we must factor in the dependence of the transit probability on the orbital semi-major axis, so we must use the distribution of planet frequency, α_P, rather than the total number of planets per star, η_P:

$$\begin{aligned} \frac{\mathrm{d}N_\mathrm{P,trans}}{\mathrm{d}a\,\mathrm{d}M_\mathrm{P}} &= \theta^2 \int_0^{d_\mathrm{max}} \alpha_\mathrm{P}(a, M_\mathrm{P}) \frac{R_*}{a} n_0 \exp\left(-\frac{r\,|\sin b_\mathrm{G}|}{H}\right) r^2\,\mathrm{d}r \\ &= \theta^2 n_0 \frac{R_*}{a} \alpha_\mathrm{P}(a, M_\mathrm{P}) \int_0^{d_\mathrm{max}} \exp\left(-\frac{r\,|\sin b_\mathrm{G}|}{H}\right) r^2\,\mathrm{d}r, \end{aligned} \quad (2.28)$$

where we assumed again that the distribution of planet frequency as a function of planet mass and semi-major axis is independent of distance. There is no evidence for any such dependence (February 2009), so this is probably a good working assumption. If you are daunted by Equation 2.28 it is worth bearing in mind that the term $\mathrm{d}N_\mathrm{P,trans}/\mathrm{d}a\,\mathrm{d}M_\mathrm{P}$ is simply a number such that

$$N_\mathrm{P,trans} = \int_a \int_{M_\mathrm{P}} \frac{\mathrm{d}N_\mathrm{P,trans}}{\mathrm{d}a\,\mathrm{d}M_\mathrm{P}}\,\mathrm{d}a\,\mathrm{d}M_\mathrm{P}, \quad (2.29)$$

where $N_\mathrm{P,trans}$ is the total number of transiting planets expected to be discovered.

To simplify, we will ignore the density variation, though we know that it exists. If we do this, the exponential term is no longer needed and Equation 2.28 becomes

$$\frac{\mathrm{d}N_\mathrm{P,trans}}{\mathrm{d}a\,\mathrm{d}M_\mathrm{P}} = \frac{\theta^2\, d_\mathrm{max}^3\, n}{3} \frac{R_*}{a} \alpha_\mathrm{P}(a, M_\mathrm{P}). \quad (2.30)$$

- What does the symbol n represent in Equation 2.30?
- The average number density of stars of a particular type in the survey volume.

To evaluate the expected number of planets, we need an expression for each survey's limiting distance, d_max. The signal-to-noise ratio attained decreases with distance, and thus governs the value of $d_{\max}$ for stars of each spectral type. If we denote the limiting signal-to-noise ratio for which transits can be detected as $L_{\rm SN}$, Equation 2.21 allows us to relate $L_{\rm SN}$ and $d_{\max}$:

$$L_{\rm SN} \approx \frac{R_{\rm P}^2}{4(\pi R_*)^{3/2}} \left(\frac{AQ\,\Delta\lambda\,\xi t}{l_{\rm sky}\,a}\right)^{1/2} \frac{\eta\overline{\lambda}L_*\exp(-K d_{\max})}{hc\,d_{\max}^2\,\sigma_{\rm FWHM}}. \tag{2.31}$$

We want to extract an expression for $d_{\max}^3$ from Equation 2.31 to be substituted into Equation 2.30. Isolating $d_{\max}$ in Equation 2.31 would be tricky because it arises in the argument of the exponential function in the numerator, as well as in the denominator. We could get around this difficulty by ignoring the effect of the interstellar extinction, but that would remove an astronomically important factor from the discussion. Instead, we will derive an expression for $d_{\max}^3$ that depends on $\exp(-K d_{\max})$.

The first step is to multiply Equation 2.31 through by $d_{\max}^2/L_{\rm SN}$, obtaining

$$d_{\max}^2 \approx \frac{R_{\rm P}^2}{4(\pi R_*)^{3/2}} \left(\frac{AQ\,\Delta\lambda\,\xi t}{l_{\rm sky}\,a}\right)^{1/2} \frac{\eta\overline{\lambda}L_*\exp(-K d_{\max})}{hc\,L_{\rm SN}\,\sigma_{\rm FWHM}}. \tag{2.32}$$

Taking the square root and subsequently cubing this equation, we obtain an expression for $d_{\max}^3$:

$$d_{\max}^3 \approx \frac{R_{\rm P}^3}{8(\pi R_*)^{9/4}} \left(\frac{AQ\,\Delta\lambda\,\xi t}{l_{\rm sky}\,a}\right)^{3/4} \left(\frac{\eta\overline{\lambda}L_*}{hc\,\sigma_{\rm FWHM}\,L_{\rm SN}}\right)^{3/2} \exp\left(-\frac{3K d_{\max}}{2}\right). \tag{2.33}$$

Substituting for $d_{\max}^3$ in Equation 2.30, we obtain

$$\frac{{\rm d}N_{\rm P,trans}}{{\rm d}a\,{\rm d}M_{\rm P}} \approx \frac{R_{\rm P}^3}{8(\pi R_*)^{9/4}} \left(\frac{AQ\,\Delta\lambda\,\xi t}{l_{\rm sky}\,a}\right)^{3/4} \left(\frac{\eta\overline{\lambda}L_*}{hc\,\sigma_{\rm FWHM}\,L_{\rm SN}}\right)^{3/2} \times \exp\left(-\frac{3K d_{\max}}{2}\right) \times \frac{\theta^2 n}{3}\frac{R_*}{a}\,\alpha_{\rm P}(a, M_{\rm P}),$$

thus

$$\frac{{\rm d}N_{\rm P,trans}}{{\rm d}a\,{\rm d}M_{\rm P}} \approx \frac{\theta^2}{24\pi^{9/4}} \left(\frac{AQ\,\Delta\lambda\,\xi t}{l_{\rm sky}}\right)^{3/4} \left(\frac{\eta\overline{\lambda}}{hc\,\sigma_{\rm FWHM}\,L_{\rm SN}}\right)^{3/2} \times \frac{R_{\rm P}^3}{a^{7/4}}\,\alpha_{\rm P}(a, M_{\rm P}) \times \frac{nL_*^{3/2}\exp(-3K d_{\max}/2)}{R_*^{5/4}}, \tag{2.34}$$

where we have grouped the terms dependent on the survey, the planet and the star.

- Which terms in Equation 2.34 belong to each of the survey, the planet and the star?

- Everything on the first line is a property of the survey, or a constant; the terms between the two $\times$ signs are properties of the planet; the remaining terms depend on the star. (One could argue that the terms η and $\overline{\lambda}$ depend on the star as well as the survey.)

Exercise 2.7 Describe and explain with reference to Equation 2.5 how the dependencies in Equation 2.34 of the transiting planet haul on the following quantities arise: (a) L_*; (b) $L_{\rm SN}$. ■

Equation 2.34 is useful because it allows the trade-offs between the various equipment choices to be quantified. The number of planets detected depends on $\theta^2 A^{3/4}$, which means that the expected number of planets increases quickly with the sky area surveyed, and relatively slowly with the collecting area of the telescope. This expresses mathematically the statements that we made in Section 2.1 about the advantages of small telescopes in the search for transiting exoplanets.

2.2.7 Estimating the expected planet haul

The scaling laws derived in Subsection 2.2.6 make the dependencies on the various factors explicit. To evaluate the expected planet haul, it is easier to take a slightly different approach, working from the limiting magnitude for a particular survey. This magnitude corresponds to the limiting signal-to-noise ratio for detecting a typical hot Jupiter transit. These limiting magnitudes are given in Table 2.1 for solar-type stars. By using these magnitudes and the zero-points in the Appendix, we can work out the photon flux without needing to know d_{max}, η and K.

Worked Example 2.2

For the PASS survey, work out the number of transits, N_{t}, that would need to be observed to detect a planet transiting a host star at the survey's limiting magnitude. You may assume that the duty cycle of the observations is $\xi = 0.4$, the transit duration is $T_{\text{dur}} = 2.5\,\text{h}$, and the transit depth is 1% and can just be detected at a signal-to-noise ratio of 10 after combining the data from N_{t} transits. In addition to the information in Table 2.1, you may assume that $\overline{\lambda} = 700\,\text{nm}$, $\Delta\lambda = 300\,\text{nm}$, $Q = 0.5$, and that the light from the star falls entirely within 1 pixel (the PASS pixels are huge!).

Solution

We need to deduce how many transits are needed to achieve a signal-to-noise ratio of 10 at the limiting magnitude: 9.4. We will use Equation 2.17 to deduce the signal, Δn_{star}, in photon units. The results from Exercise 2.3 coupled with the general result from Poisson statistics (Equation 2.9) that if we detect N photons, the statistical fluctuation will be $\sqrt{N}$, give us the noise. Both the signal and the noise depend on the total observing time, which in our notation is $N_{\text{t}} \times \Delta t$.

Step 1: Calculate n_{star}.

The limiting magnitude is 9.4 and the mean wavelength is 700 nm, so we use the R band zero-point from the Appendix. This means that

$$\begin{aligned} F_{\lambda,\text{star}} &= 1.74 \times 10^{-11} \times 10^{-9.4/2.5}\,\text{W m}^{-2}\,\text{nm}^{-1} \\ &= 1.74 \times 10^{-11} \times 1.74 \times 10^{-4}\,\text{W m}^{-2}\,\text{nm}^{-1} \\ &= 3.02 \times 10^{-15}\,\text{W m}^{-2}\,\text{nm}^{-1}. \end{aligned}$$

We can use this to calculate the photon flux by dividing through by the mean energy per photon, $\overline{E}_{\text{ph}}$:

$$\begin{aligned}
\text{photon flux} &= \frac{F_{\lambda,\text{star}}}{\overline{E}_{\text{ph}}} \\
&= \frac{F_{\lambda,\text{star}}\,\overline{\lambda}}{hc} \\
&= \frac{3.02\times10^{-15}\,\text{W}\,\text{m}^{-2}\,\text{nm}^{-1}\times 700\,\text{nm}}{6.63\times10^{-34}\,\text{J}\,\text{s}\times 3.00\times10^{8}\,\text{m}\,\text{s}^{-1}} \\
&= 1.52\times10^{10}\times 700\times10^{-9}\ \text{photon s}^{-1}\,\text{m}^{-2}\,\text{nm}^{-1} \\
&= 1.06\times10^{4}\ \text{photon s}^{-1}\,\text{m}^{-2}\,\text{nm}^{-1}.
\end{aligned}$$

To work out the rate at which photons are detected from the magnitude 9.4 star by the PASS detector, we need to multiply the photon flux by the area, A, and the bandpass, $\Delta\lambda$, to obtain the rate at which detectable photons enter the telescope, and then by the quantum efficiency, Q, which gives the fraction of these photons that are detected:

$$\begin{aligned}
\frac{\text{d}n_{\text{star}}}{\text{d}t} &= \text{photon flux}\times A\times\Delta\lambda\times Q \\
&= 1.06\times10^{4}\times 4.91\times10^{-4}\times 300\times 0.5\ \text{photon s}^{-1} \\
&= 7.81\times10^{2}\ \text{photon s}^{-1}.
\end{aligned}$$

We are told to work with a transit duration of $\Delta t = 2.5\,\text{h}$, which is equivalent to 9000 s, and a duty cycle of $\xi = 0.4$. With these parameters the total number of star photons detected is

$$\begin{aligned}
n_{\text{star}} &= \frac{\text{d}n_{\text{star}}}{\text{d}t}\times N_{\text{t}}\times\Delta t\times\xi \\
&= 7.81\times10^{2}\,\text{s}^{-1}\times N_{\text{t}}\times 9000\,\text{s}\times 0.4 \\
&= 2.81\times10^{6}N_{\text{t}}.
\end{aligned}$$

Step 2: Calculate the signal in photons: Δn_{star}.

We are told that the transit depth is 1%, so

$$\Delta n_{\text{star}} = \frac{1}{100}n_{\text{star}} = 2.81\times10^{4}N_{\text{t}}.$$

Step 3: Calculate n_{sky}.

In Exercise 2.3 we obtained

$$\frac{\text{d}n_{\text{sky}}}{\text{d}t} = 1.72\times10^{4}\,\text{s}^{-1},$$

but we did not account for the quantum efficiency, Q. We need to multiply this rate by Q to correctly estimate the sky photon detection rate. To calculate the total number of sky photons contributing to the noise, we need to multiply the detection rate by the total observing time, which is again $N_{\text{t}}\times\Delta t\times\xi$:

$$\begin{aligned}
n_{\text{sky}} &= \frac{\text{d}n_{\text{sky}}}{\text{d}t}\times Q\times N_{\text{t}}\times\Delta t\times\xi \\
&= 1.72\times10^{4}\,\text{s}^{-1}\times 0.5\times N_{\text{t}}\times 9000\,\text{s}\times 0.4 \\
&= 3.096\times10^{7}N_{\text{t}}.
\end{aligned}$$

Step 4: Calculate the noise in photons: $\sqrt{n_{\rm sky}}$.

Since we are at the faint limit of the survey, the Poisson noise from the sky photon count dominates, and the noise is simply given by

$$\text{Noise} = \sqrt{n_{\rm sky}} = 5564\sqrt{N_{\rm t}}.$$

Step 5: Calculate the signal-to-noise ratio as a function of $N_{\rm t}$.

$$\frac{\rm S}{\rm N} = \frac{\text{Signal}}{\text{Noise}} = \frac{2.81 \times 10^4 N_{\rm t}}{5564\sqrt{N_{\rm t}}} = 5.05\sqrt{N_{\rm t}}.$$

Step 6: Substitute in the required signal-to-noise ratio to evaluate $N_{\rm t}$.

We are told that we can just detect the transits if the signal-to-noise ratio is 10 after combining $N_{\rm t}$ transits. Thus

$$\sqrt{N_{\rm t}} = \frac{1}{5.05}\frac{\rm S}{\rm N} = \frac{10}{5.05} = 1.98$$

so

$$N_{\rm t} = 3.92.$$

Therefore 4 transits would need to be observed.

2.3 Wide-field astronomy

The previous section shows that having an instrument that covers a large field of view is a huge advantage for efficient discovery of transiting exoplanets. Equation 2.34 shows that the predicted number of planet discoveries for a given number of weeks' observing the same region of sky is proportional to θ^2. Consequently, wide-field surveys have been particularly successful in the hunt for transits. Of 62 confirmed transiting planets known (October 2009), 38 have been discovered by the dedicated wide-field transit survey programmes SuperWASP, HAT, XO and TrES. For the remainder of this chapter we will describe the process of searching for transiting exoplanets, drawing examples from the SuperWASP transit search programme. This description is largely drawn from the papers 'The WASP project and the SuperWASP cameras' (2006, *Publications of the Astronomical Society of the Pacific*, **118**, 1407–18) by Don Pollacco et al. and 'Efficient identification of exoplanetary transit candidates from SuperWASP light curves' (2007, *Monthly Notices of the Royal Astronomical Society*, **380**, 1230–44) by Andrew Collier Cameron et al., but the problems and solutions apply generally.

2.3.1 Design considerations and hardware

Earlier we referred to the STARE equipment as a telescope, but essentially STARE is a large CCD detector mounted directly on high-quality wide-angle camera optics. This fact is more obvious in the SuperWASP hardware (Figure 2.10), where the commercially produced Canon camera lenses are immediately apparent. The WASP consortium operates two observatories, one in

(a)

(b)

Figure 2.10 (a) The eight Canon camera lenses that comprise WASP-North. The black machinery in the foreground is part of the Taurus mount. (b) A clearer view of the Taurus mount, which supports and moves the camera lenses. The lenses are fixed within the mount, each pointing in a slightly different direction so that each covers a distinct region of the sky.

each hemisphere, both equipped with eight cameras. Each observatory has a robotic mount that points at and tracks stars just as the mount of a conventional telescope does. The eight cameras are arranged within the mount so that they each point to different but slightly overlapping regions of sky. Each of the cameras images the sky onto a square 2048×2048 pixel CCD camera, with a plate-scale of 13.7 arcsec per pixel.

Worked Example 2.3

Use the information given above to do the following.

(a) Calculate the linear field of view (f.o.v.) of a single SuperWASP CCD camera in arcsec.

(b) Convert your value to arcmin and then degrees.

(c) Consequently, calculate the field of view of a single SuperWASP CCD camera in square degrees (deg^2).

(d) How many deg^2 can SuperWASP cover at any instant, given that there are a total of 16 SuperWASP CCD cameras deployed?

(e) Discuss quantitatively how many different pointings SuperWASP would need to make to cover the whole sky.

Solution

(a) There are 2048 pixels along a side of the CCD, and the plate-scale is $13.7\,\text{arcsec pixel}^{-1}$, so

$$\begin{aligned}\text{linear f.o.v.} &= 2048\,\text{pixel} \times 13.7\,\text{arcsec pixel}^{-1} = 2.806 \times 10^4\,\text{arcsec} \\ &= 2.81 \times 10^4\,\text{arcsec (to 3 s.f.)}.\end{aligned}$$

(b) There are 60 arcsec in 1 arcmin, so we have

$$\begin{aligned}\text{linear f.o.v.} &= 2.806 \times 10^4 \text{ arcsec}\\ &= \frac{2.806 \times 10^4}{60} \text{ arcmin} = 4.676 \times 10^2 \text{ arcmin}\\ &= 4.68 \times 10^2 \text{ arcmin (to 3 s.f.).}\end{aligned}$$

Similarly, there are 60 arcmin in a degree, so we have

$$\begin{aligned}\text{linear f.o.v.} &= 4.676 \times 10^2 \text{ arcmin}\\ &= \frac{4.676 \times 10^2}{60} \text{ deg}\\ &= 7.794 \text{ deg} = 7.79 \text{ deg (to 3 s.f.).}\end{aligned}$$

(c) The field of view in square degrees (a measure of solid angle) is simply the linear field of view squared, since the SuperWASP cameras themselves are square (cf. Figure 2.8). Consequently, a single SuperWASP camera covers

$$\text{f.o.v.} = 7.794^2 \text{ deg}^2 = 60.75 \text{ deg}^2 = 60.8 \text{ deg}^2 \text{ (to 3 s.f.).}$$

(d) There are 16 SuperWASP cameras, so assuming that they point at mutually exclusive (i.e. not overlapping) regions of sky, the total solid angle covered is

$$\text{f.o.v.} = 16 \times 60.75 \text{ deg}^2 = 972.0 \text{ deg}^2 = 972 \text{ deg}^2 \text{ (to 3 s.f.).}$$

(e) The two SuperWASP observatories are positioned in the north and south; each observatory can observe the celestial pole in its own hemisphere, and all declinations from the pole to beyond the celestial equator. Thus the entire sky is visible to SuperWASP over the course of a year. The total sky is a sphere, subtending a total solid angle of 4π steradians. The steradian is the usual unit used for solid angle, and 1 steradian is 1 radian squared. There are 2π radians in a full circle, just as there are 360° in a full circle. Using this, we can convert from steradians to degrees^2:

$$\begin{aligned}1 \text{ radian} &= \frac{360}{2\pi} \text{ deg},\\ 1 \text{ radian}^2 &= \left(\frac{360}{2\pi}\right)^2 \text{ deg}^2,\\ 1 \text{ steradian} &= 3283 \text{ deg}^2 \text{ (to 4 s.f.).}\end{aligned}$$

Consequently, the total area of the sky is

$$\begin{aligned}\text{total sky solid angle} &= 4\pi \text{ steradians} = 4\pi \times 3283 \text{ deg}^2\\ &= 4.126 \times 10^4 \text{ deg}^2.\end{aligned}$$

At any given moment, only half the sky is visible (half is below the horizon); but over the course of a night, almost the entire range of right ascensions can be observed (from stars just east of the Sun at sunset to stars just west of the Sun at sunrise). Consequently, we will assume that the whole celestial sphere is to be observed: half by SuperWASP North and half by SuperWASP South. The solid angle of the whole sky divided by the solid angle covered by a single SuperWASP pointing of 16 cameras gives the smallest possible number of pointings required to cover the whole sky:

$$\text{number of pointings} = \frac{4.126 \times 10^4 \,\text{deg}^2}{972 \,\text{deg}^2} = 42.4 \text{ (to 3 s.f.)}.$$

Of course, a fraction of a pointing makes little sense, so our answer needs to be rounded up. It is impossible to perfectly tile a sphere with squares, and the SuperWASP cameras have fixed offsets between the eight cameras giving overlap. Nonetheless, the number of pointings required to cover the whole sky will be only a few more than the number that we have calculated. SuperWASP needs a total of only ~50 pointings (25 in the north and 25 in the south) to cover the entire sky.

The standard SuperWASP exposure of 30 s yields a signal-to-noise ratio of 100 for stars of magnitude $V \sim 11.5$, and better precision than this for brighter stars. The time between the start of successive exposures at different pointings is about 1 minute. Thus SuperWASP can measure the brightness of *all* the stars in the sky brighter than $V \sim 11.5$ with about half an hour of observing time at each observatory. SuperWASP and the other wide-field surveys are performing photometry on a mind-boggling scale.

As we saw in Section 2.2, the number of transiting planet discoveries expected from a survey is proportional to the number of stars observed at better than the limiting signal-to-noise ratio. Each star needs to be observed frequently enough to be able to constrain the timing of the transit ingress and egress, but it would be inefficient to simply stare at a single field. Instead, the best strategy is to move between exposures to cover a larger area of sky. SuperWASP makes repeat observations of a given field every 6 minutes or so, following an automated pattern of tiles of sky between about 3.5 hours east and 3.5 hours west of the **meridian**. Thus, when a monitored field is ideally placed, a clear night will produce 7 hours of SuperWASP coverage with a **cadence** (the time between individual photometric measurements) of 6 minutes, i.e. 70 measurements. To achieve this, the SuperWASP mounts work extremely hard, slewing to a new area of sky more frequently than once a minute. SuperWASP restricts observations to fields within 3.5 hours of the meridian because as objects get further from the meridian, observations are made through an increasing **pathlength** of the Earth's atmosphere, which degrades the photometry: the seeing gets worse and the light is increasingly **reddened** and **extincted**.

Because of the restriction to observe only within a reasonable distance of the meridian, a typical star is followed for between 100 and 150 days. A season's data on a particular star typically consist of about 3000 measurements.

- Why are there not more like $100 \times 70 = 7000$ measurements?

- Fields are ideally placed for only a small fraction of a year: the orbital motion of the Earth causes the Sun to move in right ascension by 2 hours each month. On average, a given field is in the observing plan for only about 3.5 hours a night, and further data are lost due to bad weather and hardware failure.

Differential photometry

Generally, when studying variability in stellar light curves, astronomers perform **differential photometry**. The brightness of each star in the field is measured as a function of time. Most stars have effectively constant luminosity (i.e. their luminosity variations are generally much less than the measurement uncertainties), and consequently their apparent brightness at the top of the Earth's atmosphere is constant. Unfortunately for astronomy, the Earth's atmosphere transmits a variable fraction of the light passing through it, particularly if there are clouds. Light is lost to atmospheric absorption and scattering by atmospheric dust particles. The apparent brightness of any star measured by a ground-based telescope is affected by this. To a first approximation, the **atmospheric extinction** can be corrected by dividing the measured intensity of the target star by that of a known constant star that was simultaneously observed. This works if all the stars in question are close enough together that their light effectively takes the same path through the atmosphere, so the atmospheric transparency doesn't vary from star to star. A complicating factor is that the extinction is wavelength-dependent, so the fraction of light lost is dependent on the colour of the star.

Ideally, differential photometry on a target star is performed by dividing the light curve by a high signal-to-noise ratio light curve of a nearby constant star (or an ensemble of constant stars) that has an identical colour to the target. In practice, it is difficult to perfectly satisfy all these constraints, and it is usual to have some remaining extinction-related trends in the **differential light curve** that is obtained after dividing by the light curve of a constant star. In particular, it is usual for differential light curves to have upwards or downwards trends at high airmass.

2.3.2 Issues arising from wide-area coverage

A single SuperWASP CCD image is $60\,\mathrm{deg}^2$, which is much larger than a typical astronomical CCD image. Consequently, properties that are usually effectively identical for all stars in an image can vary appreciably across the field. For example, the **airmass**, the sky transparency, the sky brightness and the **heliocentric timing correction** all differ from one side of the field to the other. The airmass and the timing correction can be calculated exactly from each star's coordinates. However, for wide-field data like SuperWASP's, correcting for transparency variations is a more complicated, **iterative** procedure than that described in the box on differential photometry above.

The cameras used by SuperWASP and similar transit surveys were not originally designed for astronomy. The SuperWASP lenses are Canon photographic lenses optimized for use in theatre photography. They happen to be excellent for astronomical use: they have a large aperture, A; they have little **chromatic aberration**, so red light and blue light from a star are focused onto the same spot on the detector. These lenses were also immediately available commercially.

When a lens is used in photography, it is mounted on an SLR camera body, and focused to produce the desired image. In astronomy the targets are all effectively at the same infinite distance, so the desired focus never varies. Thus each lens is focused onto its CCD detector at the commissioning of the instrument, and in theory no further adjustments are required. In practice, however, the optical elements in the lenses are subtly temperature-dependent. As the temperature changes during the night, the focus and the **point spread function** (PSF) change. Generally, the detector is imperfectly aligned with the focal plane of the camera, so these changes are position-dependent: as the temperature changes, some parts of the field may become better focused, while others degrade. This defocusing is likely to cause stars to become blended with each other. This can be dealt with in the processing of the images, but it is complicated, and no matter how successful the processing, such defocusing will diminish the signal-to-noise ratio of the resultant photometry.

● What term in Equation 2.21 accounts for the size of the PSF?

❍ σ_{FWHM}.

2.4 From images to light curves

An example of a raw SuperWASP image is shown in Figure 2.11a. The raw data from the CCD cameras need to be corrected to remove the vignetting pattern caused by the camera optics, the characteristics of the detector, and the shutter correction (see Figure 2.11b). After this is done, the corrected images, known as the **reduced images**, should correctly indicate brightness as a function of sky position. Photometric measurements are made for as many of the objects in the

The reduced images are not smaller: the term arises from '**data reduction**', which is the name given to the process of correcting raw data for such instrumental effects.

Figure 2.11 (a) A raw SuperWASP image. The strong concentration of light towards the centre is caused by two effects. (i) The camera optics transmit light more efficiently at the centre of the field, and **vignette** the light around the outside. (ii) For short exposures, the time for the shutter to open and close is not negligible: the shutter is open for slightly longer at the centre of the image than at the edges. (b) These four images show the calibration corrections that are applied. The flatfield and the exposure map are divided into the raw image and remove the vignetting of the camera and the shutter correction, respectively. The bias and dark current are subtracted from the raw image and remove less obvious artefacts introduced by the CCD detector and the associated electronics. (c) After these corrections have been applied, the reduced image is obtained. In this example the vertical line in the lower half is an artefact of the very bright star that has saturated the detector.

field as possible. In principle, this is a very simple thing to do: one simply sums all the detected photons from each object. In practice, doing photometry on wide-field images is far from simple. Objects are blended due to the large plate-scale, making it difficult to assess how many of the detected counts are due to each individual star. The focus and the position on the CCD chip will probably change steadily throughout the night, so a robust technique is needed to identify each individual star and measure the flux to be attributed to it.

Entire papers and substantial fractions of the work for several PhD theses have been devoted to solving these problems. We will not go into the details here. The SuperWASP consortium developed a bespoke data reduction and analysis pipeline. One of the key strategies was to use previously known positions of stars, taken from a star catalogue with much more precise positions than SuperWASP itself can measure, and match these positions to the brighter stars in a SuperWASP field. Once the software has identified the exact position of the field in this way, it measures the brightness of all the stars in the catalogue that fall within the boundaries of the field. There can be over 200 000 stars in a single SuperWASP CCD frame.

The photometry is produced by simply counting up the number of photons detected within a circular aperture centred at the exact position of the star, and subtracting from this the number of sky photons that would be expected within this circle. Thus each photometric measurement from the frame is unambiguously associated with a particular star. The SuperWASP light curve for any given star is then simply all the photometric measurements for that object, plotted against time.

● How can the number of sky photons to be subtracted be estimated?

○ The sky brightness, i.e. number of sky photons per unit area, can be measured in blank patches of sky between stars. The sky brightness generally varies across the field, so a smooth two-dimensional function is fitted to the sky brightness measurements. This allows the expected sky brightness at any position to be estimated, and multiplying the value at any star's position by the area of the circular aperture gives the expected number of sky photons.

2.5 The SuperWASP archive

Processed SuperWASP photometry is stored in the SuperWASP archive. This archive is one of the key reasons why the SuperWASP project has been successful. Each camera produces $\sim 5 \times 10^8$ photometric data points per month; there are currently 16 cameras, and the project started collecting data in 2004.

Exercise 2.8 Estimate how many photometric data points the SuperWASP archive must organize and store. ■

The SuperWASP archive was designed to store the huge volume of data efficiently, and allow easy, quick, convenient access to any requested subset of data. SuperWASP researchers can retrieve data from the SuperWASP catalogue using a web browser; the time for the archive to extract the light curve of any object is ~ 1 s. An example light curve is shown in Figure 2.12. The results from analyses of the light curve data can be uploaded to the archive, making it an extremely powerful and flexible tool.

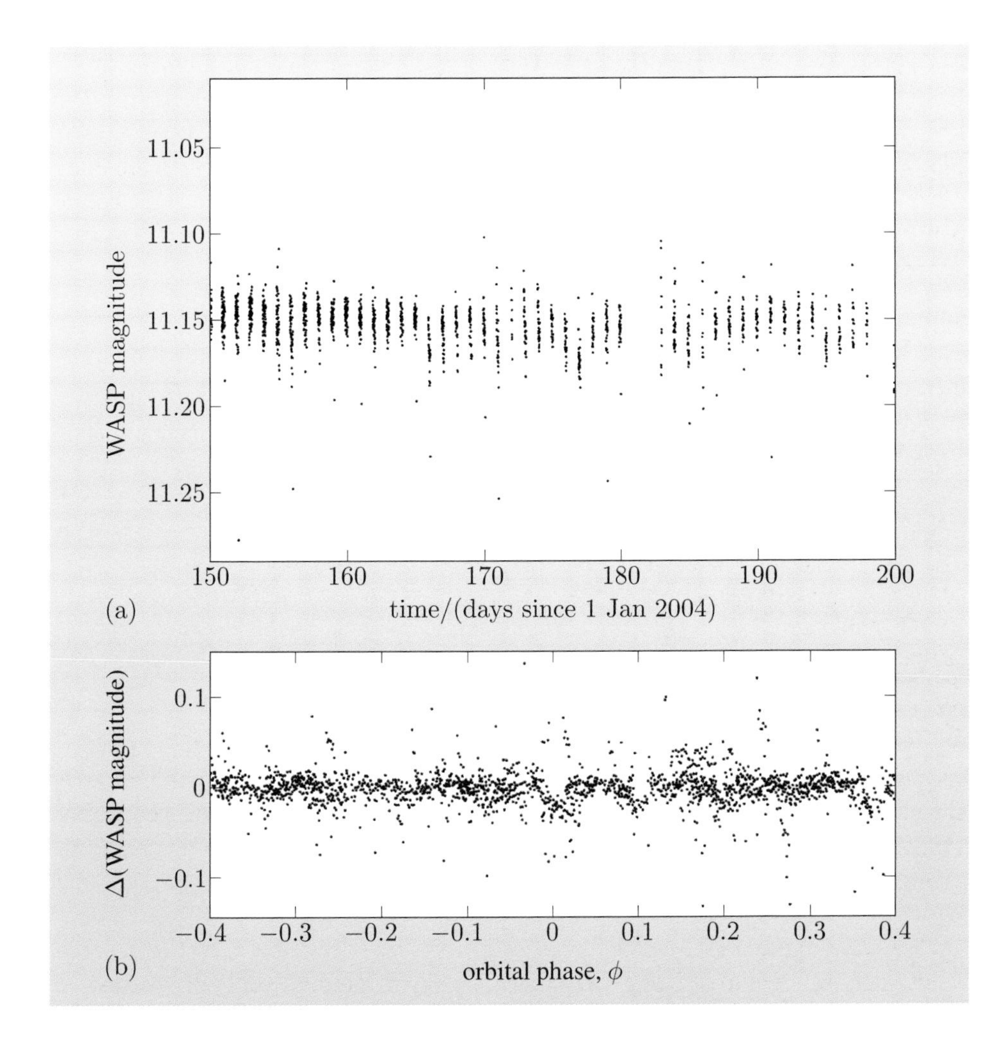

Figure 2.12 Typical WASP transit candidate data from 2004; this star has V = 11.15. (a) The raw SuperWASP light curve. (b) The light curve folded on the suspected planet's orbital period. The putative transit is the dip centred on phase 0.0.

2.6 Transit search methods

Figure 2.12b shows a transit candidate light curve folded on the suspected period. Candidates are identified by software that searches for possible transits. The SuperWASP catalogue contains tens of millions of stars, so transit candidate identification is performed automatically. The precise shape of a transit light curve contains a wealth of information, as we will learn in Chapter 3, but noise in Figure 2.12b obscures the shape of the transit. Consequently, the preliminary transit search uses the simplest possible approximation to the transit shape.

χ^2 fitting of models to data

In Section 1.3 we saw how a curve is fitted to radial velocity data so that parameters such as the amplitude of the reflex radial velocity of the star, A_{RV}, can be determined. The general process of fitting a function to empirical data, and then using the parameters of the functional fit to estimate the physical quantities of interest, is ubiquitous in science. In Section 1.3 we had several figures showing 'best-fitting' solutions, but we didn't define how these 'best-fits' were determined.

The Greek letter χ in the title of this box is pronounced kye, rhyming with sky and high. It is spelled 'chi'.

In empirical science, measurements should always be accompanied by estimates of the uncertainty. If a theory predicts that a quantity, Q, should have the value $Q = 1.9\,\text{eV}$ and an experiment measures the value, finding $Q = 2.0\,\text{eV}$, it is impossible to say whether the experiment has confirmed or falsified the theory without an estimate of the empirical uncertainty. If the empirical measurement was difficult to make, the uncertainties could be quite large: for example, the result could be $Q = 2.0 \pm 0.5\,\text{eV}$, in which case the measurement is perfectly consistent with the predictions of the theory. On the other hand, if the experiment produced the result $Q = 2.00 \pm 0.03\,\text{eV}$, then it suggests that the theory is incorrect. Since it is notoriously difficult to produce precise and robust uncertainty estimates, this result would probably not be enough to conclude that the theory is certainly wrong. If, however, the empirical result were $Q = 2.000\,00 \pm 0.000\,03\,\text{eV}$, then it should be safe to say that the theory has been falsified.

Generally, scientists assume that experimental results, for instance measurements of a value x, have randomly distributed empirical errors, and consequently have a bell-shaped distribution, described by the Gaussian function

$$f(x) = \frac{1}{\sqrt{2\pi}\sigma} \exp\left(-\frac{(x - x_0)^2}{2\sigma^2}\right), \tag{2.35}$$

where x_0 is the mean value obtained for x after a large number of independent measurements, and σ is the uncertainty in a single measurement. Figure 2.13 shows the shape of this probability distribution.

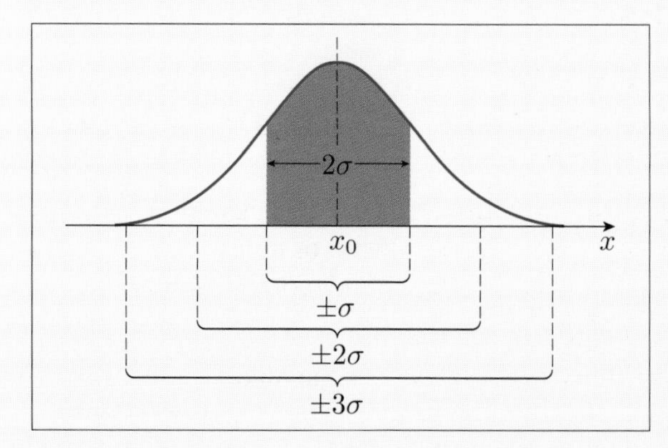

Figure 2.13 The Gaussian or normal distribution: a large number of measurements of a quantity, x, will produce a bell-shaped distribution of values. The width of the distribution is characterized by the standard deviation, σ, which indicates the uncertainty in the value of an individual measurement.

In astronomy, often quantities vary continuously as a function of time, for example the radial velocity measurements shown in Section 1.3. In this case it is impossible to make repeated measurements of the identical quantity. Instead, the empirical data consist of a time series: a set $\{x_i, \sigma_i, t_i\}$ of individual measurements (x_i), estimated uncertainty on each measurement (σ_i) and the time of the measurement (t_i). To this time series the astronomer will fit a function or model. Ideally, the choice of model is motivated by an understanding of the physics underlying the variation of the quantity being measured. For instance, Equation 1.16 is based on our understanding of orbital motion, and relates the fundamental parameters of a planetary system to the (measurable) reflex radial velocity of the host star. For any set of assumed parameters, we can use the model to calculate predicted values, μ_i, to compare with the measurements $x_i \pm \sigma_i$.

The appropriate way to gauge the goodness of the fit of a model with a particular set of parameters is to calculate a statistical quantity called χ^2 given by

$$\chi^2 = \sum_i \left(\frac{x_i - \mu_i}{\sigma_i}\right)^2. \qquad (2.36)$$

It is also possible to treat spectra in this way, in which case the empirical data are $\{x_i, \sigma_i, \lambda_i\}$.

Here the sum is over all the measurements in the set of data. If a particular measurement has a large uncertainty estimate, then the denominator is large, and the contribution to χ^2 is relatively small, even if the deviation between the measured value, x_i, and the corresponding model prediction, μ_i, is large. Because the quantity $x_i - \mu_i$ is squared, positive and negative deviations make equally weighted positive contributions to χ^2. If a model agrees well with the data, the deviations should be small, and consequently the best-fitting model is the one with the minimum value of χ^2. **Model fitting** in astrophysics is generally accomplished by a process of χ^2 **minimization**, as outlined below.

The process of using empirical astrophysical time series data to determine astrophysical quantities is therefore generally as follows.

- Measure empirical data $\{x_i, \sigma_i, t_i\}$.
- Adopt a physically motivated model that can predict values, μ_i, for the measured quantity, x_i, for given physical parameters.
- For a particular choice of input parameter values, calculate χ^2 and record its value; repeat this step for a new choice of input parameter combinations; repeat until all plausible parameter combinations have been covered.
- Adopt the set of input parameters that produces the minimum χ^2.

The box above describes the general process of model-fitting. In the initial SuperWASP transit search, the model has two brightness levels, the 'out of transit' level and the slightly lower 'in transit' level, and makes an instantaneous transition from one to the other. The deviations of this model from the true shape of a transit light curve are far smaller than the typical scatter in the empirical light curve. The model assumes that the transits occur periodically, and that $T_{\text{dur}} \ll P$. For each light curve, the software calculates χ^2 for models covering all possible combinations of periods, fiducial phases, depths and durations. The minimum χ^2 value found for each period, $\chi^2_{\text{min}}(P)$, is compared with the value of χ^2_{constant} for a model consisting of a flat, constant light curve. This reveals if the data are better fit by a model in which there are transits. If the observed light curve has periodic dips, then a model with appropriately placed transits will fit better and have a smaller value of χ^2. For each period, the software calculates

$$\Delta\chi^2(P) = \chi^2_{\text{min}}(P) - \chi^2_{\text{constant}}, \tag{2.37}$$

and uses this to generate a periodogram, e.g. Figure 2.14. For promising candidates, there should be sharp dips in the value of $\Delta\chi^2(P)$, indicating possible orbital periods for a transiting planet. In Figure 2.14 the best period indicated by the software is 3.64 days.

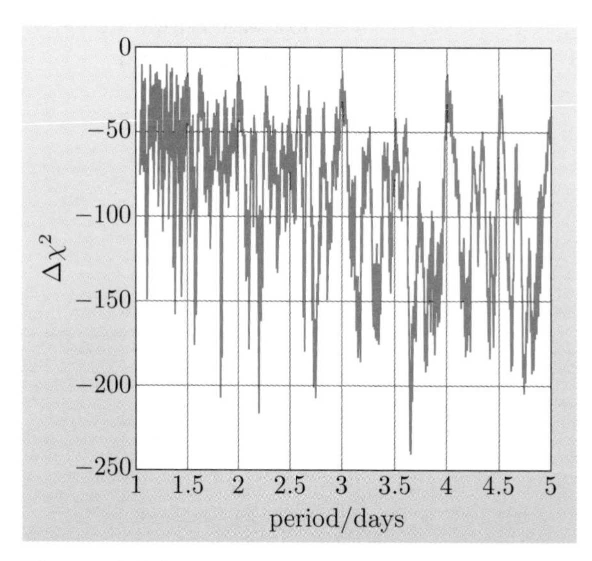

Figure 2.14 The WASP transit search periodogram for the candidate shown in Figure 2.12. The ordinate shows the difference, $\Delta\chi^2$, between χ^2 for the best-fitting box-shaped transit model and χ^2_{constant} for the model μ_i = constant. If there are periodic transit-like dips in brightness, the value of $\Delta\chi^2$ will dip sharply at the corresponding period. In this example, the best-fitting model has a period of 3.64 days.

- Since a planet has only one value of the orbital period, why is there more than one sharp dip in the periodogram?

○ $\Delta\chi^2(P)$ will be reduced wherever there are multiple dips in the light curve that occur at the same phase when the data are phase-folded on period P. For a transiting planet with period P, the periodogram might be expected to indicate P, nP and P/n, where n is any small positive integer. Some of the transits will line up when the data are phase-folded on any of these periods. Statistical fluctuations can also cause some periods to have large negative values of $\Delta\chi^2(P)$ purely by chance.

Exercise 2.9 Calculate the plausible range of the ratio T_{dur}/P. ■

Once sufficient data are collected on a particular field, transit-hunting software identifies candidates based on the fits described above. Unfortunately, many of these candidates are spurious. Some are simply chance alignments of random noise, but many are due to poor photometry at the beginning and end of nights when the airmass is high: these generally have periods that are close to integer days, i.e. $P = m$ days, where m is an integer. Some are **astrophysical mimics**: the light curve really does show transit-like dips, but the cause is not a transiting exoplanet.

2.7 Astrophysical mimics

The most common imposters identified as possible transiting planets are:

- blended eclipsing binary systems;
- grazing eclipsing binary systems with equal-mass components;
- transits by planet-sized stars.

These are illustrated schematically in Figure 2.15. With low signal-to-noise ratio photometry (cf. Figure 2.12) alone, it is impossible to distinguish these possibilities from genuine transiting planets.

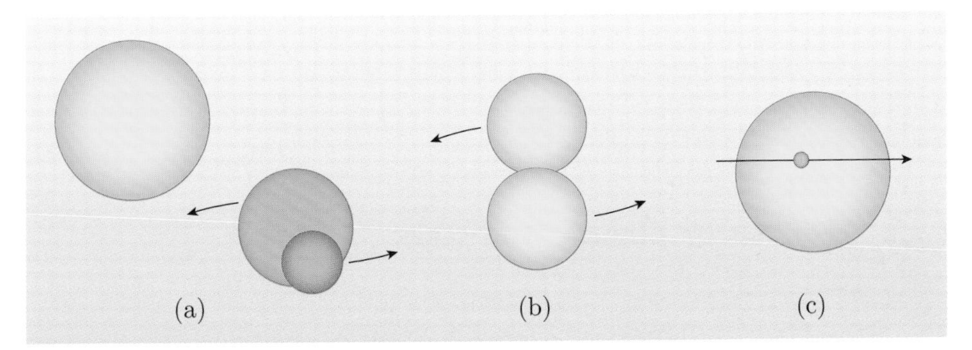

Figure 2.15 The three most likely astrophysical mimics. (a) A constant single star is blended with a fainter eclipsing binary along the same line of sight. The deep eclipses are diluted, producing a $\sim$1% dip. (b) A binary system comprising two similar-mass main sequence stars is viewed at an angle that produces grazing eclipses. (c) A brown dwarf or white dwarf star transits a main sequence star.

2.7.1 Blended eclipsing binary systems

This scenario is schematically illustrated in Figure 2.15a. The large plate-scale of wide-field surveys makes it likely that there will be more than one detectably bright object within a single pixel. Usually these **blends** are not physically associated with each other: they lie along the same line of sight at very different distances. An isolated star blended with a fainter deeply eclipsing binary will dilute the eclipse depth of the latter. This can produce a shallow dip similar to an exoplanet transit.

2.7.2 Grazing eclipsing binary systems

Usually, eclipsing binary stars produce deep eclipses that are easily distinguished from planetary transits. Because planets are always smaller than about $2\,R_J$, a dip of more than a few per cent in the light curve of a main sequence star cannot be produced by the transit of a planet. The depth of the transit is given by the ratio of radii of the host star and the transiting object:

$$\frac{\Delta F}{F} = \frac{R_P^2}{R_*^2}. \qquad \text{(Eqn 1.18)}$$

This equation assumes, however, that the entire disc of the transiting object lies inside the disc of the host star. It is perfectly possible that the orbital inclination is such that the nearer of the two objects just grazes the limb of the farther object, as illustrated in Figure 2.15b. If this happens, the depth of the dip is determined by the area that is occulted, and this is much less than the area of the disc of the nearer object.

In an eclipsing binary star system with circular orbits, if a grazing eclipse occurs at one conjunction it must also occur at the other conjunction, so a full orbital light curve contains two eclipses. In general, the depths of these two eclipses will differ: the eclipse of the brighter star will produce the deeper eclipse. The presence of alternating shallower and deeper eclipses is the hallmark of a grazing eclipse binary star system, and it is often possible to eliminate these from further consideration using the transit survey photometric data alone. If, however, the grazing eclipse binary system is comprised of two main sequence stars of similar mass, hence similar luminosity and radius, then the two eclipses will be indistinguishable, and these objects have transit survey photometric light curves just like those of genuine transiting planets.

2.7.3 Transits by planet-sized stars

Gas giant planets, white dwarf stars and brown dwarf stars are all about the same size. This is not a coincidence: all three classes of objects are supported by **degenerate electron pressure**, and the **equation of state** of degenerate matter causes the radii to be very similar, even though giant planets are less massive than brown dwarf or white dwarf stars. Since the transit light curve depends only on the size of the transiting object, it alone cannot discriminate between these three possibilities.

2.8 Candidate winnowing and planet confirmation

2.8.1 Follow-up observations

At their inception, the transit search survey programmes all planned follow-up observations of the candidate transiting exoplanets to eliminate the astrophysical mimics described in Section 2.7. The easiest to eliminate are the blended eclipsing binaries (Subsection 2.7.1): by observing a single transit with instrumentation giving a higher spatial resolution (i.e. using a smaller plate-scale), the blend can be resolved, and if there is a faint object with deep eclipses, this is easy to pick out.

The grazing eclipsing binary star systems can also be eliminated by observing a single transit at higher signal-to-noise ratio: in this case the shape of the dip in the light curve will discriminate between genuine transits and grazing eclipses. The former have flat-bottomed U-shaped transits, while grazing eclipses are V-shaped. To a good approximation, the amount of light lost at any time is proportional to the area of the stellar disc occulted: once a planet crosses the stellar limb, the occulted area is constant, leading to the flat-bottomed curve; in the case of a grazing stellar eclipse, the area occulted changes constantly, leading to the V-shaped curve.

The approximation here is to neglect stellar limb darkening. We will discuss the effects of this in Chapter 3.

Transits by planet-sized stars are very difficult to eliminate without obtaining a measurement of the mass of the transiting object from the star's reflex radial velocity curve. For this reason, a transiting planet is merely a candidate until its mass has been measured by the radial velocity technique. As we noted in Section 2.1, the equipment and telescope time required to perform these radial velocity measurements are extremely expensive and scarce compared to that required to obtain wide-field photometric survey data. For this reason, transit surveys try very hard to make the best possible use of the survey photometry to winnow out worthy candidates from imposters.

2.8.2 Tests performed on the survey photometry

The tests described below are performed by SuperWASP software, and the results are stored in the SuperWASP archive. Using this information, the SuperWASP consortium identifies only high-quality candidates for spectroscopic follow-up observations: roughly 20% of the candidates are planets, while the majority of the remainder are astrophysically interesting planet-sized secondary stars.

Resolved aperture blends

We have already discussed how blending can lead to spurious candidates. If the angular separation of the blend is comparable to the pixel size of the survey instrument, the wide-field survey photometry cannot eliminate candidates arising from blended eclipsing binaries. Spurious candidates arising from wider blends can be eliminated. SuperWASP uses circular apertures of radius 3.5 pixels to produce light curves. This radius is generally the best compromise between collecting all the light from the target and excluding light from other objects and the sky background. Some light from neighbouring objects will inevitably fall into the aperture: an example of how this can affect the light curve is shown in Figure 2.16. The upper panel shows the SuperWASP light curve of a candidate identified by the transit search software described in Section 2.6. In fact, this candidate is a non-variable star with no known planets. The dip in Figure 2.16a is caused by a deeply eclipsing star about 5 pixels away from the candidate; the light curve of the contaminating object is shown in Figure 2.16b. Light from the contaminating neighbour leaks into the candidate's aperture: during the neighbour's near-total eclipse, the contaminating light is diminished, and hence a dip appears in the candidate's light curve. These spurious candidates are eliminated by also performing photometry using a larger aperture. Contaminating objects make a larger contribution in the larger aperture. The difference between the two light curves will reveal the contamination, as Figure 2.16c shows.

Secondary eclipses

Many grazing eclipse binaries have two components of differing sizes and brightnesses. These can be eliminated by folding the data on twice the identified period and looking for differences in depth between odd and even 'transits'.

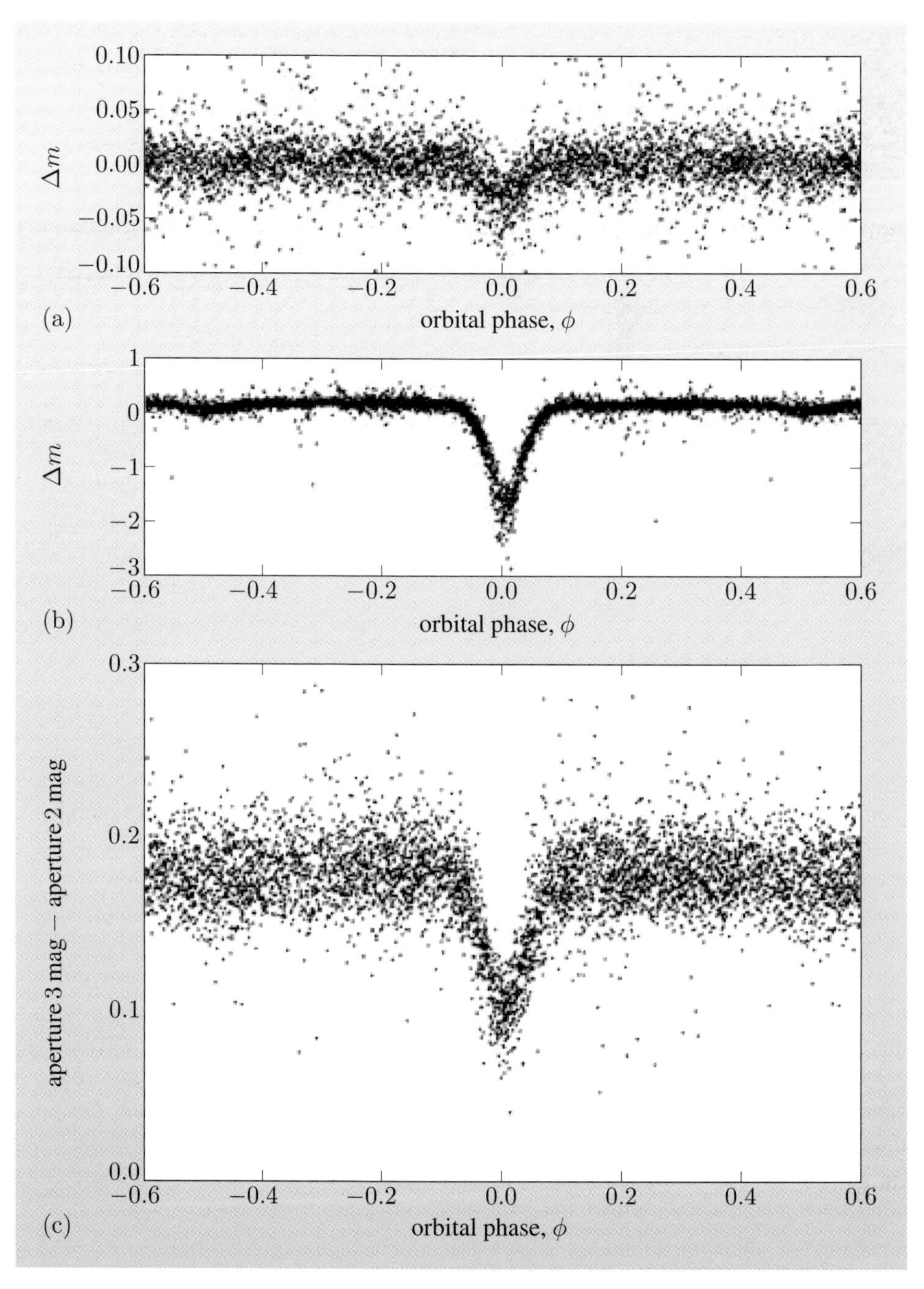

Figure 2.16 (a) The light curve of a spurious candidate identified by the SuperWASP transit search software. (b) The light curve of the deeply eclipsing binary 69 arcsec away from the candidate shown in (a). (c) The difference between the candidate's light curve derived from a 4.5-pixel aperture and the candidate's light curve derived from the standard 3.5-pixel aperture. The contaminating binary contributes more light in the larger aperture, and consequently the eclipse produces a strong signal. The plot in panel (c) is used to eliminate spurious candidates arising from contamination from nearby stars that are resolved by the survey images.

Ellipsoidal variables

A planet in a few-day orbit around a main sequence star does not exert sufficient gravitational effect on the star to distort its shape. Conversely, if two stellar mass objects are in such a short period orbit around their mutual centre of mass, they

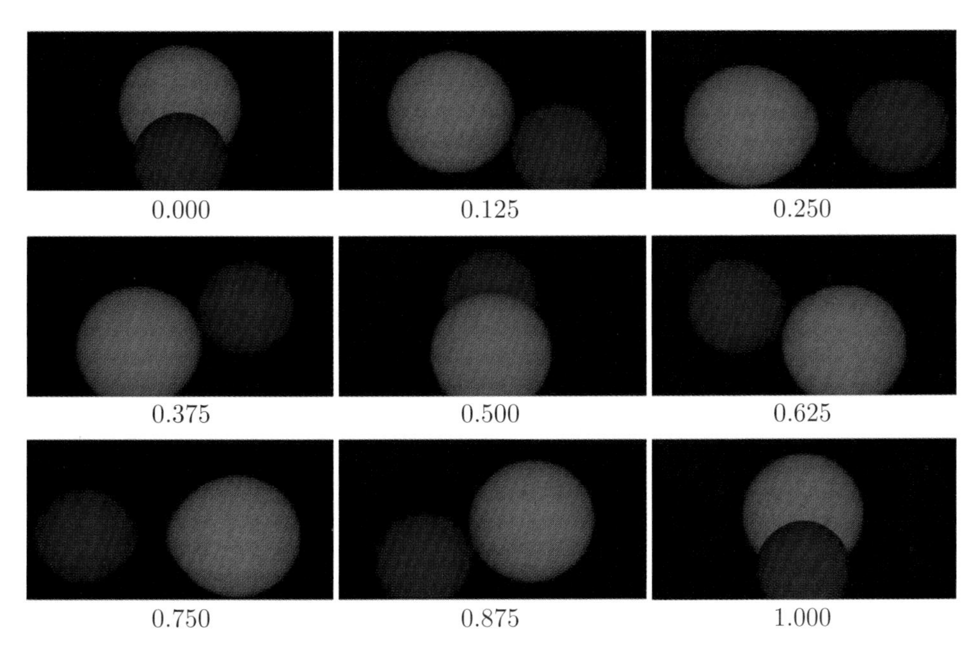

Figure 2.17 In a close binary comprised of two fluid stars, the Roche equipotential surfaces determine the shapes of the stars. The stars are distorted from spherical by a combination of centripetal acceleration and gravity. Their ellipsoidal shapes cause their apparent brightness to vary as they proceed around their orbit. Here the binary is shown at an eclipsing orientation, but the effect will occur for all inclinations except $i \approx 0$.

each exert an appreciable gravitational effect on the other. In this case, as shown in Figure 2.17, the shapes of the stars are determined by the **Roche equipotential surfaces**: fluid stars are distorted from spherical by the combination of the other star's gravity and the orbital centripetal acceleration. This effect elongates stars so that they appear bigger when viewed at quadrature (phases 0.25 and 0.75 for circular orbits) than when they are viewed at conjunction. This larger cross-sectional area at quadrature in turn means that the star appears slightly brighter at quadrature than at conjunction; a star that is distorted from spherical by the Roche equipotential geometry will consequently have a smooth double-peaked light curve, with peaks at phases $\phi = 0.25$ and $\phi = 0.75$. This type of variability is called **ellipsoidal variation** because the star's ellipsoidal shape causes it. The SuperWASP candidates are checked for any signs of such variations, and any showing it are likely white dwarf or brown dwarf eclipsing binaries or grazing eclipse binaries, and are eliminated from the planet candidate list.

Light curve modelling

We will discuss **light curve modelling** in detail in Chapter 3, where we will consider the many properties that can be deduced with surprising accuracy from a high-quality transit light curve. SuperWASP and other transit survey programmes do not provide these high-quality light curves, but analysis of the best-fit transiting planet model provides an important step in the candidate winnowing process. A planet will provide a transit of depth detectable by SuperWASP only if the host star is on or near the main sequence: giant stars are too big for a planetary transit to produce a detectable dip. Thus any genuine transits among the candidates will be found in the light curves of main sequence stars. Since all the SuperWASP objects are previously catalogued stars, their photometric colours are known.

Specifically, the SuperWASP analysis uses the J-H near infrared colour to indicate a main sequence mass and hence radius for the host star. In the SuperWASP candidate winnowing process, this preferred value for the stellar radius is adopted. The light curve is modelled using this, and the best-fit value for the stellar mass is examined. If the values are inconsistent with the **main sequence mass–radius relationship**, the candidate's priority for follow-up observations is downgraded.

- How is the stellar mass constrained by data showing transits?
- ❍ Kepler's third law allows us to relate a, P and $M_{\rm total}$. The time between successive transits tells us P, and $T_{\rm dur}$ gives us information on a.

2.8.3 Spectroscopic confirmation of candidates

Radial velocity measurements are made for candidates that survive the tests described above. Some candidates are eliminated with a single spectrum: if there are two sets of stellar spectral lines, the candidate is a **double-lined spectroscopic binary star**, and the 'transit' is probably a grazing eclipse. If the candidate has broad spectral lines, then it is rapidly rotating. This rapid rotation is likely to be the result of spin-up by a stellar mass companion, indicating the likelihood of a brown dwarf or white dwarf rather than a planetary transit. Pragmatically, it is very difficult to attain precise velocity measurements for stars with rotationally broadened lines. More candidates are eliminated after two spectra are taken: if the velocity difference is too large, this immediately rules out a companion of planetary mass.

Exercise 2.10 A planet transit candidate is observed spectroscopically to confirm whether or not it might be a genuine transiting planet. Two spectra are obtained at the beginning and end of a night, 6 hours apart, and reveal the host star to have the same spectral type as the Sun. The radial velocities of the two spectra differ by $10\,{\rm km\,s^{-1}}$. Could this reflex motion be due to the presence of a planet? ■

2.9 The pregnant pause: overcoming systematic errors

The process of identifying new transiting exoplanets is conceptually straightforward. The discovery of hot Jupiter exoplanets by radial velocity work, coupled with the high probability of transits occurring for such short period planets, led to the expectation that roughly one main sequence star in every 1000 should host a transiting hot Jupiter. Predictions were made that by 2009, thousands of transiting exoplanets would be known. Instead, only 62 have been identified (October 2009), some of which were first found by their radial velocity variations. It was extremely disappointing and slightly worrying that the expected plethora of transit discoveries did not occur promptly once the wide-field transit survey projects began analyzing their data.

The first planet to be discovered by the transit method was OGLE-TR-56 b, announced in 2002, three years after the discovery of transits by HD 209458 b.

Figure 2.18 shows the history of planet discoveries by the various methods. It shows that discoveries by the transit method took off in 2007, with only a handful of discoveries preceding 2007. The reason for this delay was that it took time to learn how to perform wide-field photometry effectively.

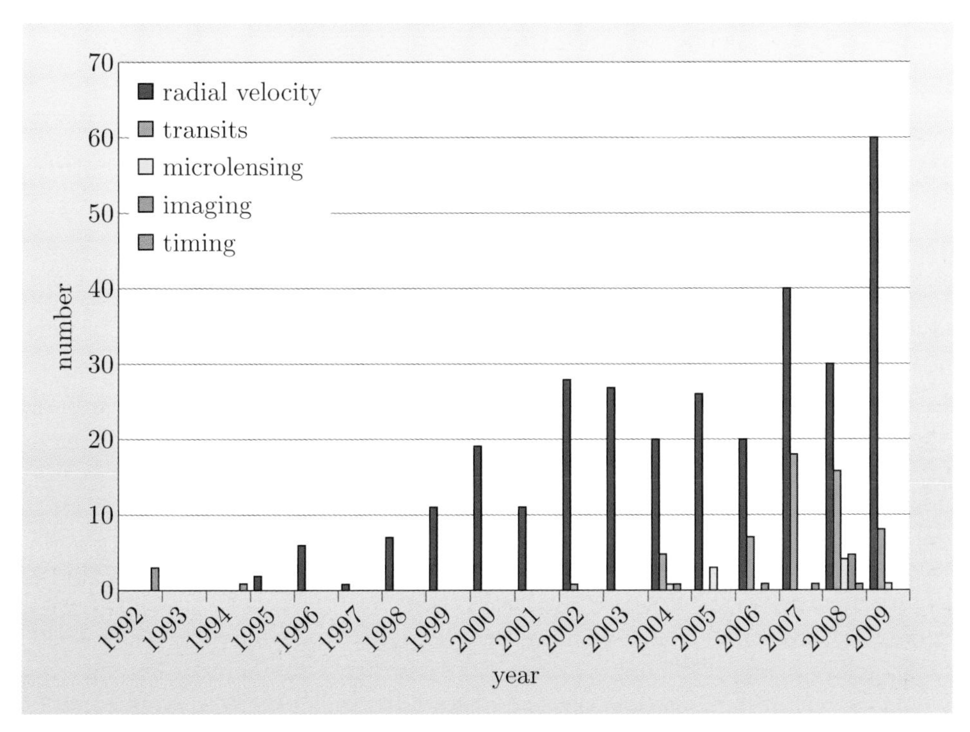

Figure 2.18 Planet discoveries over time, divided by method. Note that the data for 2009 are incomplete as the sample was taken in November 2009.

The SuperWASP light curves are corrected for atmospheric extinction by dividing each light curve by a simultaneous light curve formed from an ensemble of simultaneously observed constant stars in the field. Further colour-dependent corrections are then applied. In principle this should yield light curves corrected for atmospheric extinction, so a constant star would have a constant light curve. In practice the SuperWASP light curves suffer from **correlated noise**: sections of light curve are systematically higher or lower than they should be. At least two effects probably contribute to this:

- incorrectly calibrated transparency variations;
- temperature-dependent focus changes causing the PSF to vary with time and position in the field.

Figure 2.19 illustrates the second of these effects. Since apertures of the same size are used across the entire field, it is clear that **systematic errors** will be introduced into the light curve. In Figure 2.19a, the stars in the lower left will have almost all of their flux captured by an aperture of radius 3.5 pixels; in contrast, the stars in the upper right have larger PSFs, and the same 3.5-pixel aperture will consequently exclude some fraction of the stars' flux. Thus spurious position-dependent trends will be caused in the light curves of the stars in the field.

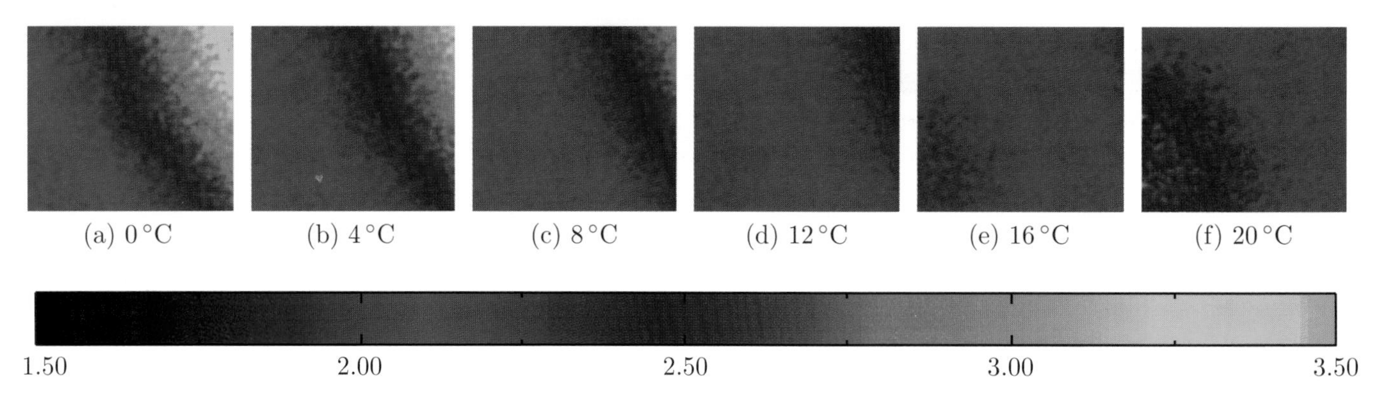

Figure 2.19 The SuperWASP images have temperature-dependent focus variations. Each panel illustrates the distribution of the FWHM of the PSF of stars in a particular SuperWASP camera in a single night. In panel (a), the temperature was 0°C and the stars in the lower left were well-focused, with FWHM of around 2 pixels. This pattern gradually changes with temperature, and in panel (f) the best-focused stars are towards the top right.

All our discussions of the data to this point have been based on Gaussian statistics (see the box in Section 2.6) and the properties of the instrumentation and the Earth's atmosphere. Each correction has been applied to remove a measured and well-understood artefact that would otherwise degrade the photometry. The effect illustrated in Figure 2.19 could perhaps be removed in a similar way, but this would be very difficult and could prove ineffective. Such a solution would vary the aperture size with both position and time, that is, using a different aperture size at different locations on the same image, and for each star at different times. This would cause new problems.

Instead, a less conceptually justifiable approach is used. An algorithm called '`SYSREM`' looks for common time-dependent and position-dependent trends in light curves. Having identified these trends, it removes them from the data. No attempt is made to understand the cause of the trends — the correction is simply applied. SuperWASP identifies and removes the four strongest such trends in the data. Figure 2.12 shows an example SuperWASP light curve before detrending using `SYSREM`; Figure 2.20 shows the same data before and after the `SYSREM` algorithm has been applied to an ensemble of light curves, including this one. It is clear that the algorithm has changed the value of each of the data points. The out-of-transit light curve is much flatter and has much lower point-to-point scatter in the `SYSREM`-corrected curve: this has been achieved by removing the common trends from this object and its neighbours. The transit remains. The transit-hunting described in Section 2.6 is most profitably performed on the `SYSREM`-corrected data. Systematic errors, such as those causing the dip around phase 0.1 in the pre-detrending light curve in Figure 2.20, cause many false-positive transit identifications. The plethora of candidates initially identified by wide-field transit search programmes delayed the confirmation of genuine transiting exoplanets.

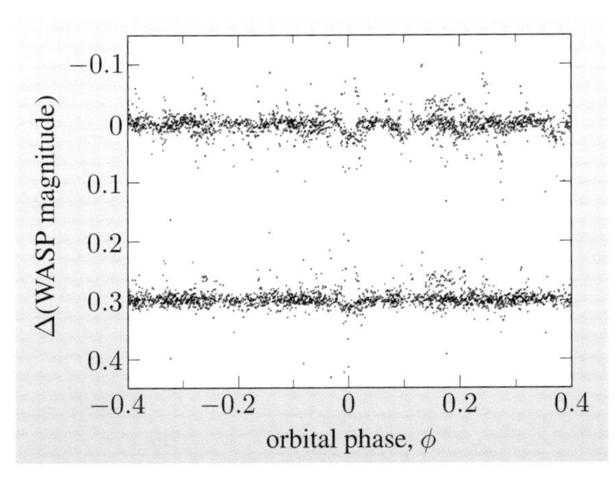

Figure 2.20 (a) A SuperWASP light curve folded on the period of the suspected transiting planet. This is the same data as shown in Figure 2.12. (b) These data post-SYSREM correction, folded in the same way.

Wide-field transit surveys are routinely identifying new transiting planets. Between them, the SuperWASP, XO, TrES and HAT projects have detected almost two-thirds of the known transiting exoplanets (October 2009), and their discoveries are all planets orbiting relatively nearby stars that consequently are bright enough for detailed follow-up studies. That such tiny, inexpensive telescopes are making such an important contribution to twenty-first century science is something to celebrate!

Summary of Chapter 2

1. The probability that a hot Jupiter, i.e. an exoplanet with an orbital period of 5 days or less, with a randomly oriented orbit, will transit its host star is $\sim 10\%$.
2. A precise orbital period is determined by observing many cycles and using

$$P = \frac{T_{\text{elapsed}}}{N_{\text{cycles}}} \qquad \text{(Eqn 2.1)}$$

   to obtain a value from the elapsed time between two widely separated occurrences of the same fiducial phase, for example two times of mid-transit.
3. Kepler's third law (Equation 1.1) gives the orbital semi-major axis in terms of the masses and the orbital period:

$$a = (G(M_* + M_P))^{1/3} \left(\frac{P}{2\pi}\right)^{2/3}. \qquad \text{(Eqn S2.2)}$$

   Since $M_P \ll M_*$, the orbital period and an estimate of the stellar mass, M_*, allow the semi-major axis to be deduced.

If you didn't encounter this in the chapter, it is because you skipped over the exercises, which are an *integral part* of the learning.

4. Radial velocity measurements determine $M_P \sin i$ in terms of the observables P and A_{RV}, and the host star mass, M_* (assuming $M_* \gg M_P$):

$$M_P \sin i = A_{RV} \left(\frac{M_*^2 P}{2\pi G}\right)^{1/3}. \qquad \text{(Eqn S2.6)}$$

For transiting planets, a precise value of i can be obtained by modelling the transit shape. For transiting planets with good radial velocity measurements, the largest contribution to the uncertainty in the value of the planet's mass, $M_{\rm P}$, is the uncertainty in M_*.

5. Iterative techniques can be used to refine the estimate of a value using an approximation as the starting point.

6. For instrumentation with a limiting flux S, the maximum distance at which a source of luminosity L can be detected is

$$d_{\rm max} = \left(\frac{L}{4\pi S}\right)^{1/2}. \qquad \text{(Eqn 2.3)}$$

Thus the volume of sky over which objects of a given luminosity can be detected depends critically on the limiting flux of the instrumentation. Consequently, for a hypothetical population of sources all with identical luminosity, L, the number of detectable sources is

$$N(S) = \frac{4\pi n_0}{3}\left(\frac{L}{4\pi S}\right)^{3/2}. \qquad \text{(Eqn 2.5)}$$

7. The duration of an exoplanet transit for $i = 90°$, i.e. an impact parameter $b = 0.0$, is

$$T_{\rm dur}(b = 0.0) \approx \frac{PR_*}{\pi a}. \qquad \text{(Eqn 2.6)}$$

This approximation is good to about 1% for a planet with a 1-day orbit, and the approximation improves as the orbital period increases.

8. For N detected photons, the Poisson noise is $\sqrt{N}$:

$$\text{statistical uncertainty on number count, } N = \sqrt{N}. \qquad \text{(Eqn 2.9)}$$

9. Interstellar extinction causes stars to appear dimmer than the inverse square law would predict. While the extinction is patchy, the effect can be approximated with the equation

$$F = \frac{L_*}{4\pi d^2}\exp(-Kd). \qquad \text{(Eqn 2.12)}$$

10. The signal-to-noise ratio for a single transit is

$$\frac{\rm S}{\rm N} \approx \frac{R_{\rm P}^2}{4(\pi R_*)^{3/2}}\left(\frac{AQ\,\Delta\lambda\,P}{l_{\rm sky}\,a}\right)^{1/2}\frac{\eta\overline{\lambda}L_*\exp(-Kd)}{hcd^2\,\sigma_{\rm FWHM}}. \qquad \text{(Eqn 2.19)}$$

A transit survey with duty cycle ξ, observing a transit host for elapsed time t, will on average catch $N_{\rm t}$ transits, where

$$N_{\rm t} = \frac{\xi t}{P}. \qquad \text{(Eqn 2.20)}$$

Observing $N_{\rm t}$ transits improves the signal-to-noise ratio over that for a single transit by a factor $\sqrt{N_{\rm t}}$:

$$\frac{\rm S}{\rm N} \approx \frac{R_{\rm P}^2}{4(\pi R_*)^{3/2}}\left(\frac{AQ\,\Delta\lambda\,\xi t}{l_{\rm sky}\,a}\right)^{1/2}\frac{\eta\overline{\lambda}L_*\exp(-Kd)}{hcd^2\,\sigma_{\rm FWHM}}. \qquad \text{(Eqn 2.21)}$$

11. Position in the Galaxy is indicated by Galactic longitude and Galactic latitude (l_G, b_G). These coordinates are centred on the Sun, which lies near the mid-plane of the Galaxy. The stellar density is a maximum at the Galactic mid-plane and falls off with scale-height, H.

12. The number of planets per star, η_P, is currently unknown, but a lower limit can be inferred from the facts that $\sim$1% of solar-type stars host hot Jupiter planets, and $\sim$6% of nearby stars host planets. The population of planets existing in our part of the Galaxy can be characterized with the (unknown) function $\alpha_P(a, M_P)$.

13. The number of planets around stars of a particular type expected from a transit search survey is

$$\frac{\mathrm{d}N_{\mathrm{P,trans}}}{\mathrm{d}a\,\mathrm{d}M_P} \approx \frac{\theta^2}{24\pi^{9/4}} \left(\frac{AQ\,\Delta\lambda\,\xi t}{l_{\mathrm{sky}}}\right)^{3/4} \left(\frac{\eta\overline{\lambda}}{hc\,\sigma_{\mathrm{FWHM}}\,L_{\mathrm{SN}}}\right)^{3/2}$$
$$\times \frac{R_P^3}{a^{7/4}}\,\alpha_P(a, M_P) \times \frac{nL_*^{3/2}\exp(-3Kd_{\max}/2)}{R_*^{5/4}}. \qquad \text{(Eqn 2.34)}$$

Everything on the first line is a property of the survey, or a constant; the terms between the two $\times$ signs are properties of the planet; the remaining terms depend on the star.

14. Astrophysical quantities are generally deduced by χ^2-fitting of a physically-motivated model to the data.

15. Wide-field surveys performed with small telescopes, i.e. diameters of $\sim$10 cm, are the most cost-effective way to find transiting planets. Such dedicated transit survey facilities are responsible for the majority of confirmed known transiting exoplanets.

16. The data volume and systematic errors are the two biggest challenges in large-scale transit hunting. State of the art database techniques are required. Stars rise and set on a timescale similar to that of a hot Jupiter transit; the Earth's atmospheric extinction is a limiting source of systematic error. To correct for systematic errors in wide-field photometry, an algorithm called SYSREM identifies and removes trends that are common to many light curves.

17. Three common types of astrophysical mimics can be mistaken for transiting planets: (i) a deeply eclipsing binary star blended with a brighter star; (ii) a grazing eclipse may produce a 1% dip, but the dip will be V-shaped rather than U-shaped with a flat bottom; (iii) a transiting planet-sized star. Types (i) and (ii), and transits by white dwarf stars, can be eliminated with photometry from telescopes of moderate size ($<$ 1 m diameter aperture). Transits by brown dwarf stars can be eliminated only by measuring the reflex radial velocity of the host star. Consequently, until reflex radial velocity measurements have been made, a transit candidate remains merely a candidate.

Chapter 3 What the transit light curve tells us

Introduction

We saw in Chapter 1 that the depth of a transit is proportional to the square of the ratio of the planet radius to the host star radius. In Chapter 2 we asserted that the transit light curve allows us to obtain a precise measurement of the orbital inclination, i, of the planet, and consequently allows an exact value of the planet's mass to be deduced from the radial velocity curve of the host star. Now we will explore the analysis of a transit light curve and learn exactly how the orbital inclination and other precise parameters for the planet and its host star are obtained.

In the first section of this chapter we will begin with Kepler's third law and use it to deduce how the transit duration depends on the orbital inclination, i, and the more fundamental parameters of the star and planet. In Section 3.2 we analyze the shape of the transit light curve, deriving quantitative models for the transit shape in the case of a uniform stellar disc and a limb darkened stellar disc. In Section 3.3 we step back from the analytical detail and summarize the logic behind the deduction of values for the radii, stellar and planetary masses, and orbital inclination. Finally, in Section 3.4 we discuss the light curve model fitting process that yields parameters for transiting planet systems.

This chapter underpins the key role of transiting exoplanets in the wider field of planetary astrophysics.

3.1 Kepler's third law and exoplanet orbits

Kepler's third law underlies much of astrophysics. It is a beautifully simple relationship that was deduced from observations of the Solar System planets, and can be derived from Newton's law of universal gravitation and Newton's second law of motion. Kepler's third law provides the basis for the quantitative analysis of planetary orbits, and is the starting point for analyzing exoplanetary transits.

3.1.1 The semi-major axis

One of the easiest things to measure once a transiting exoplanet is discovered is the orbital period, P (cf. Equation 2.1). The orbital period is related to the semi-major axis of the orbit, a, and the total mass of the two bodies via Kepler's third law:

$$\frac{a^3}{P^2} = \frac{G(M_* + M_\mathrm{P})}{4\pi^2}. \qquad \text{(Eqn 1.1)}$$

Generally $M_* \gg M_\mathrm{P}$, and we can obtain a fairly good estimate of the mass of the star from its spectral type. Consequently, the semi-major axis, a, is known approximately once the orbital period and the spectral type of the host star have been measured:

$$a \approx \left(GM_* \left(\frac{P}{2\pi} \right)^2 \right)^{1/3}. \qquad \text{(Eqn S2.2)}$$

3.1.2 The orbital speed

With values for a and P, the orbital speed can be deduced. This is simplest to derive for the case of a circular orbit, but is equally true for the general case of an elliptical orbit. Kepler's second law states that a planet orbiting in an elliptical orbit sweeps out equal areas in equal times, i.e. it moves fastest at periapsis (cf. Figure 1.14) when it is closest to the star, and slower when it is further from the star.

For a circular orbit, the orbital speed is a constant and is given by

$$v = \frac{2\pi a}{P}. \qquad (3.1)$$

If we are calculating the speed of the planet about the barycentre of the star–planet system, then the semi-major axis here is a_P, as defined in Section 1.2.

- Which fundamental conservation law of physics underpins Kepler's second law?

- The principle of conservation of angular momentum: since no net external torque is acting on the star–planet system, the total angular momentum remains constant.

- As written, with a in the numerator of the right-hand side, what speed does Equation 3.1 give?

- a is the semi-major axis of the astrocentric orbit, so Equation 3.1 gives the speed of the planet relative to the star.

- In applying Equation 3.1 to radial velocity measurements of the type discussed in Section 1.3, what quantity should be used for the semi-major axis?

- The radial velocity that can be measured is that of the stellar reflex orbit. The semi-major axis of this orbit relative to the barycentre should be used, i.e. a_*. This quantity is much smaller than the semi-major axis of the astrocentric orbit of the planet, a, or the semi-major axis of the barycentric orbit of the planet, a_P.

- What is the relationship between the three semi-major axes mentioned in the previous answer?

- $a = a_P + a_*$.

Exercise 3.1 (a) What is the orbital speed of the Earth around the Sun? (You may assume that the orbit is circular.)

(b) What would be the orbital speed of a planet in a circular orbit at a distance of 1 AU from a star of mass $0.5\,M_\odot$? ■

3.1.3 The orbital inclination, the impact parameter and the transit duration

The limb is the term used to refer to the edges of the disc of any astronomical body.

Phases of a transit

A transit has four **contacts**; these are illustrated in Figure 3.1. **First contact** is when the **limb** of the planet's disc first coincides with the limb of the star's disc as viewed by the observer. **Second contact** is when the entire disc of the planet is just within the stellar disc as viewed by the observer.

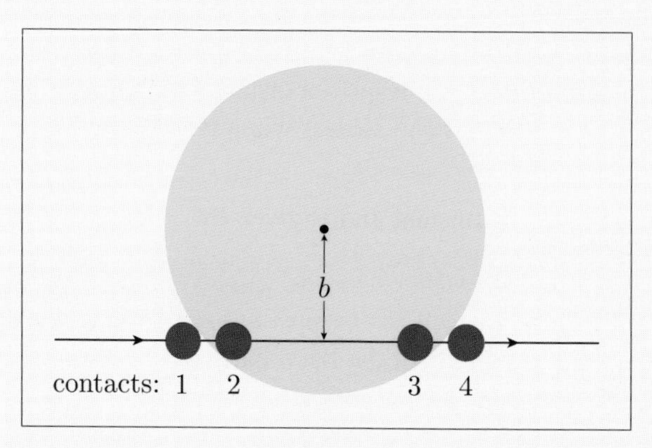

Figure 3.1 The position of the planet's disc relative to the star's disc at the four contact points. If the orbital motion of the planet is in the direction shown, the four planet discs correspond to first, second, third and fourth contacts, respectively. These contact points correspond to measurable features in the light curve.

Third contact occurs when the limb of the planet's disc coincides with the limb of the star's disc, i.e. the last instant when the entire disc of the planet is just within the stellar disc. **Fourth contact** occurs when the trailing limb of the planet's disc crosses the limb of the star's disc, i.e. the last instant of the transit.

In Subsection 2.2.2 we derived the duration of the transit that would be observed from an orbital inclination of $i = 90°$, or equivalently an impact parameter of $b = 0.0$:

$$T_{\text{dur}}(b = 0.0) = \frac{P}{\pi} \sin^{-1}\left(\frac{R_*}{a}\right). \qquad \text{(Eqn S2.8)}$$

We also noted that an approximate version of this equation (Equation 2.6) was sufficient for the purposes of the analysis in Chapter 2. Now we are going to examine how to use transit light curves to derive precise parameters, so we will use the exact version (Equation S2.8), and derive the dependence on the orbital inclination. Figure 1.19 allows us to derive the relationship between the impact parameter and the orbital inclination. In Figure 1.19 the observer is viewing along a horizontal line of sight, while the planet's orbital plane is inclined and makes an angle i with the vertical as shown. A version of Figure 1.19 emphasizing the trigonometry is reproduced here as Figure 3.2.

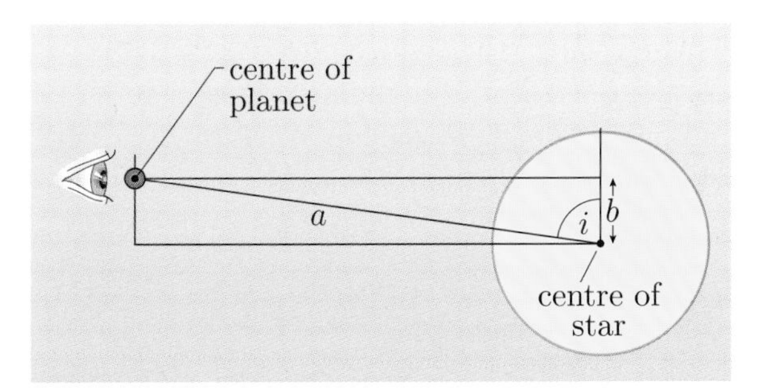

Figure 3.2 The geometry of Figure 1.19, emphasizing the trigonometry determining the impact parameter, $b = a\cos i$.

The impact parameter, b, is the vertical distance at mid-transit of the centre of the planet from the centre of the star as viewed by the observer. This distance is

$$b = a\cos i. \qquad (3.2)$$

Note in some other texts the impact parameter is defined as the dimensionless quantity

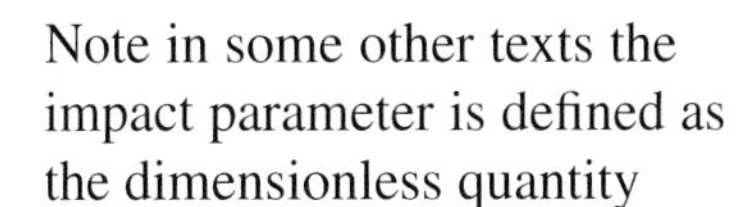

$$b = \frac{a\cos i}{R_*}.$$

The transit duration depends strongly on the impact parameter. Figure 3.3 shows the relevant geometry in a snapshot taken at fourth contact. The larger circle represents the disc of the star, while the smaller circle represents the disc of the planet.

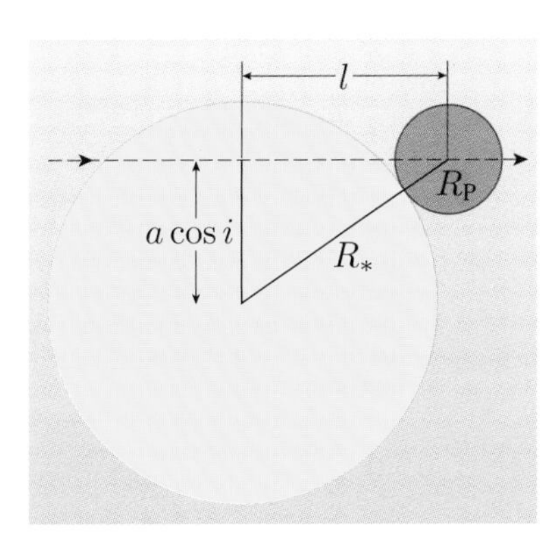

Figure 3.3 Pythagoras's theorem allows us to express the length l in terms of the impact parameter, b, and the radii of the star and planet.

Figure 3.3 shows a right-angled triangle that has a hypotenuse of length $R_* + R_P$ and a vertical side that is equal to the impact parameter, b, and therefore has length $a\cos i$. The third, horizontal, side of the triangle joins the positions of the centre of the planet's disc at mid-transit and fourth contact. By Pythagoras's theorem, the length of this side is

$$l = \sqrt{(R_* + R_P)^2 - a^2\cos^2 i}. \qquad (3.3)$$

The path of the planet is not precisely along this straight line, but is the flattest part of the extremely eccentric ellipse formed by the projection of the highly inclined orbit on the plane of the sky.

- For $a^2\cos^2 i > (R_* + R_P)^2$, the quantity under the square root is negative, and the length l would be a purely imaginary number. What physical situation does this correspond to?

- If $a^2\cos^2 i$ exceeds $(R_* + R_P)^2$, then the impact parameter is greater than the sum of the stellar and planetary radii, i.e. the planet's disc moves across the diagram above the star's disc without ever overlapping it, and no transit will occur.

Figure 3.4 illustrates how the transit duration can be deduced from the length l. During the transit, the planet moves along its orbit from point A to point B, subtending an angle α at the star. From the triangle formed by A, B and the centre of the star, $\sin(\alpha/2) = l/a$. With the angle α in radians, for a circular orbit the time elapsed is simply

$$T_{\text{dur}} = P\frac{\alpha}{2\pi} = \frac{P}{\pi}\sin^{-1}(l/a),$$

thus

$$T_{\text{dur}} = \frac{P}{\pi}\sin^{-1}\left(\frac{\sqrt{(R_* + R_{\text{P}})^2 - a^2\cos^2 i}}{a}\right). \tag{3.4}$$

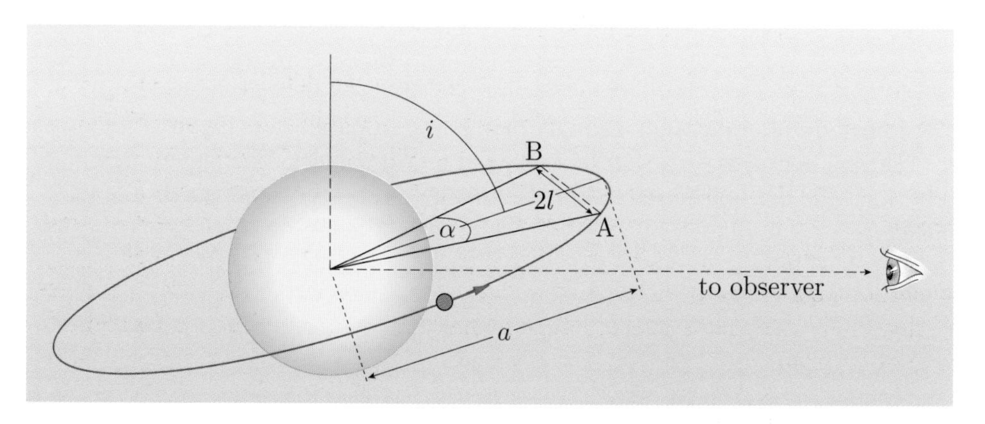

Figure 3.4 During transit, the planet moves from A to B around its orbit. If the orbit is circular, the distance around the entire orbit is $2\pi a$, and the arc length between A and B is $a\alpha$, with the angle α in radians. The distance along a straight line between A and B is $2l$. From the triangle formed by A, B and the centre of the star, $\sin(\alpha/2) = l/a$.

This is the exact general expression for the duration of a transit for an exoplanet in a circular orbit. For a non-circular orbit, the time taken to move from point A to point B in Figure 3.4 is not simply proportional to the angle α.

- What could we use instead of the above to work out the transit duration for an elliptical orbit?

- Kepler's second law tells us that equal areas are swept out in equal times. If the elliptical orbit is known, we can use this principle to work out the time between any two points on the orbit.

- Why is it appropriate to use a rather than a_{P} in Equation 3.4?

- Because it is the astrocentric motion that determines when the contact points occur. If we used a_{P}, the motion of the star would cause an underestimate of the transit duration.

Exercise 3.2 (a) Show that if $a \gg R_* \gg R_{\text{P}}$, then Equation 3.4 simplifies to

$$T_{\text{dur}} \approx \frac{P}{\pi}\left[\left(\frac{R_*}{a}\right)^2 - \cos^2 i\right]^{1/2}.$$

(b) Use this to work out the orbital inclination of a planet discovered around a Sun-like star (i.e. $M_* = 1\,\text{M}_\odot$, $R_* = 1\,\text{R}_\odot$) with $P = 6$ days and $T_{\text{dur}} = 2$ hours. Radial velocity measurements indicate that the orbit of the planet is circular.

(c) What would you deduce about a transit candidate with $T_{\text{dur}} = 4$ hours and other characteristics identical to those given above? ■

We showed in Exercise 3.2 that if we assume a circular orbit and $a \gg R_* \gg R_P$, then Equation 3.4 simplifies to

$$T_{dur} \approx \frac{P}{\pi}\left[\left(\frac{R_*}{a}\right)^2 - \cos^2 i\right]^{1/2}. \qquad (3.5)$$

If we know the spectral type of the star, we can infer approximate values of M_* and R_* by assuming that they are typical for the spectral type. Then the transit duration, T_{dur}, and the orbital period, P, are sufficient to deduce (via Equation 3.5) the approximate orbital inclination, i.

- How is the value of a to be used in Equation 3.5 determined from the quantities mentioned in the paragraph above?
- Kepler's third law gives a in terms of M_* and P.

Furthermore, once the orbital inclination is estimated in this way, we can use the time elapsed between first and second contact, t_{1-2}, to obtain a second estimate of the radius of the planet. At transit ingress the planet's disc moves from first contact to second contact as shown in Figure 3.1. The time taken to do this depends on the orbital inclination, the sizes of the two discs, and the orbital speed. The transit duration and the orbital period give us the first and last of these quantities, so a previously known value for R_* and a measurement of t_{1-2} implies a value of R_P. Of course, the transit depth has already given us an estimate of the ratio of the radii:

$$\frac{\Delta F}{F} = \frac{R_P^2}{R_*^2}. \qquad \text{(Eqn 1.18)}$$

Thus a light curve of sufficient quality to provide a precise measurement of t_{1-2} as well as the transit depth, ΔF, allows two independent estimates of the planet's radius, given the radius of the star. If these two estimates do not agree, then a new value can be assumed for R_* to see if it leads to more consistent results.

3.2 The shape of the transit light curve

In our discussion so far we have related measurements of particular features in the light curve to the physical parameters of the star and its transiting planet. The features discussed so far are:

- the transit depth, ΔF;
- the transit duration, T_{dur};
- the duration of ingress, t_{1-2} (or equivalently egress, t_{3-4}).

Figure 3.5 shows the first light curve of HD 209458 b's transit taken using the Hubble Space Telescope. The data have a sampling cadence of 80 s, and the out-of-transit data points for an individual measurement of the stellar flux, F, have a signal-to-noise ratio of almost 10^4: the scatter among the measurements of F is $1.1 \times 10^{-4} F$. This exquisite light curve provides the opportunity to precisely measure each of the three quantities listed above.

- Is the light curve in Figure 3.5 completely described by the three quantities listed above?
- ❍ No. The high precision light curve in Figure 3.5 clearly has a curved transit floor. This curvature is not characterized by any of the three quantities above.
- What does the shape of the transit floor imply about the brightness distribution of the stellar disc?
- ❍ The transit depth varies smoothly, being deepest at the centre, implying that most light is lost when the planet is farthest from the limb of the stellar disc. This implies that the stellar disc gets smoothly brighter from the limb to the centre.

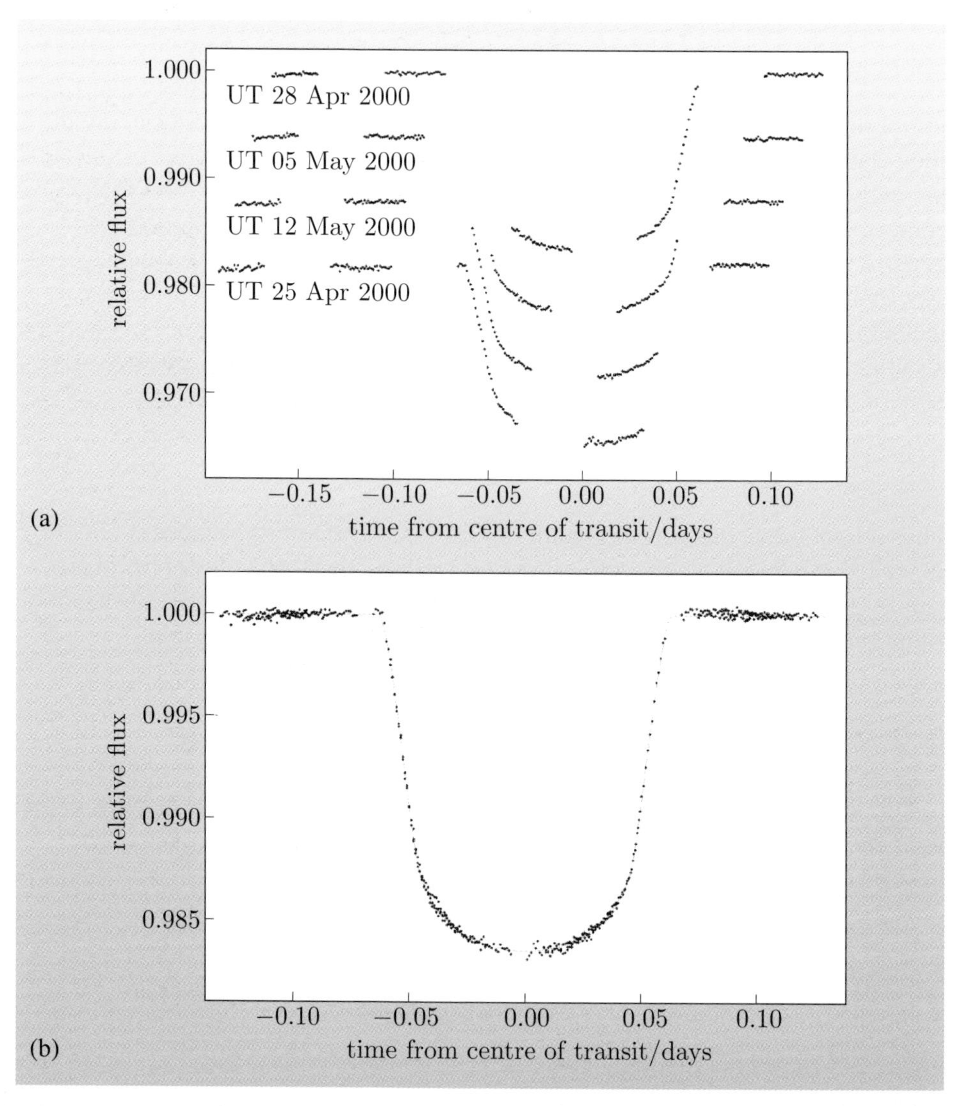

Figure 3.5 The first exquisitely high-precision transit light curve. (a) Hubble Space Telescope photometry of HD 209458 b during four different transits in April and May 2000. The four light curves have been offset from each other for clarity. The gaps in each individual light curve are caused by the Earth regularly occulting HD 209458 b as Hubble proceeds around its low Earth orbit. (b) The phased light curve obtained by combining the four sets of observations. The data have been normalized so that the out-of-transit level is exactly 1.000. The depth of the transit is 1.64%, and the noise level is orders of magnitude smaller than this.

Exercise 3.3 (a) Calculate the signal-to-noise ratio for the transit shown in Figure 3.5, where the signal is the transit depth, ΔF, and the noise is estimated by the scatter between out-of-transit points.

(b) Estimate the maximum scatter in the measurements of F that would allow detection of the curvature in the transit floor for the transit shown in Figure 3.5. ■

3.2.1 Limb darkening

Figure 3.5 makes it abundantly clear that the transit is not a flat-bottomed dip in the light curve. The curvature in the transit floor implies that a smoothly varying fraction of the stellar flux is lost during the transit, so

$$\frac{\Delta F}{F} = \frac{R_{\mathrm{P}}^2}{R_*^2} \qquad \text{(Eqn 1.18)}$$

cannot be strictly true unless one or both of the radii vary. It is extremely difficult to imagine any mechanism that would cause such a variation to occur at the time required to produce a symmetrical transit as viewed from a planet orbiting a random G type star; instead, we need to examine the assumptions implicit in Equation 1.18. As already noted, the shape of the transit floor suggests that the stellar disc is brightest at the centre and gets smoothly darker towards the limb. In reaching Equation 1.18 we implicitly assumed that the stellar disc has a uniform brightness, but in fact we know that this is not the case for the Sun or for stars in general, as discussed in the box on limb darkening below.

Limb darkening

Figure 3.6 shows a broad-band optical image of the Sun. There are some immediately obvious features in this image: the dark **sunspots**. The solar disc appears noticeably brighter at the centre, and appears progressively dimmer and redder towards the limb. This **limb darkening** is easily understood in terms of the radiative transfer physics of stellar atmospheres, and has important consequences for the shape of a transit light curve.

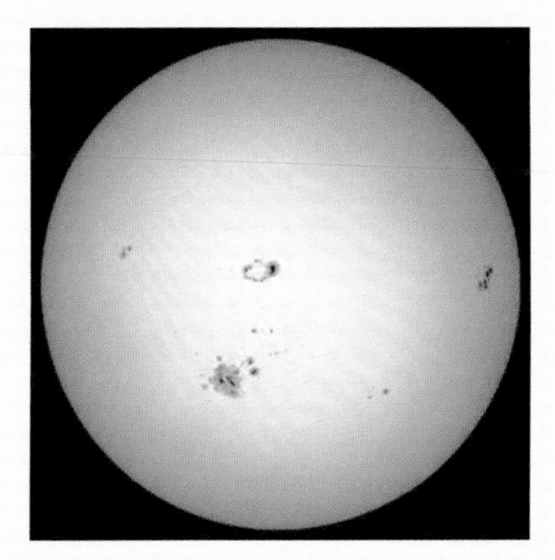

Figure 3.6 A NASA image of the Sun. The various dark spots are sunspots. The solar disc appears noticeably brighter at the centre, and appears progressively dimmer and redder towards the limb. This phenomenon is limb darkening.

Main sequence stars do not have solid surfaces: their outer layers are composed of plasma. The light that escapes from them comes from a variety of depths within the stellar atmosphere. The probability of escape of a photon emitted within a particular layer of the atmosphere is dependent on the **optical depth** of that layer. At a given frequency, ν, the optical depth, τ_ν, is the integral of the opacity, κ_ν, multiplied by the density, $\rho(s)$, along the path taken by a light ray; i.e. for light emitted at position X travelling towards an observer at infinity, as indicated in Figure 3.7, we have

$$\tau_\nu = \int_{\mathrm{X}}^{\infty} \rho(s)\, \kappa_\nu \,\mathrm{d}s, \tag{3.6}$$

where s indicates the position along the path.

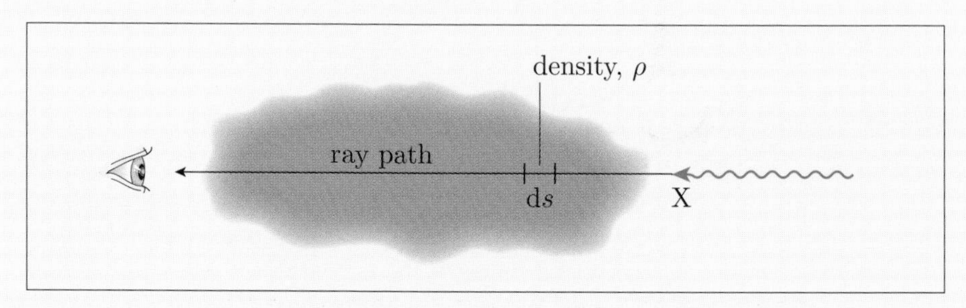

Figure 3.7 The optical depth is the integral along the ray path of the density multiplied by the opacity.

Because the opacity is dependent on the frequency (or equivalently the wavelength) of the radiation, the optical depth is also dependent on this. A particular physical depth in a stellar atmosphere will have a different optical depth depending on the frequency of radiation being considered. The probability of a photon travelling along the path without being absorbed or scattered is $\mathrm{e}^{-\tau_\nu}$, so the ratio between the emitted intensity, I_{emitted}, and the emergent intensity, I, is

$$\frac{I}{I_{\mathrm{emitted}}} = \mathrm{e}^{-\tau_\nu}. \tag{3.7}$$

At any position within the stellar atmosphere, photons are being continually emitted, absorbed and scattered. By definition, any photon detected by the observer at infinity travelled in the direction towards the observer. This means that a photon that emerges from the centre of the stellar disc is travelling radially outwards through the stellar atmosphere, as illustrated in Figure 3.8a.

On the other hand, a photon emerging from anywhere else on the stellar disc is travelling at an angle γ to the outwardly directed radius vector. The two photons indicated in Figure 3.8a have been emitted at the same depth in the stellar atmosphere, but to reach the observer, the photon emitted near the limb of the star must travel through a much greater path length, s, of the stellar atmosphere. For the photons emitted at depth h, the path length through the stellar atmosphere is

$$s \approx \frac{h}{\cos\gamma} = \frac{h}{\mu}, \tag{3.8}$$

where $\mu = \cos\gamma$ and the **plane-parallel** approximation has been made, i.e. we have assumed $h \ll R_*$ so the geometry is essentially that of Figure 3.8b.

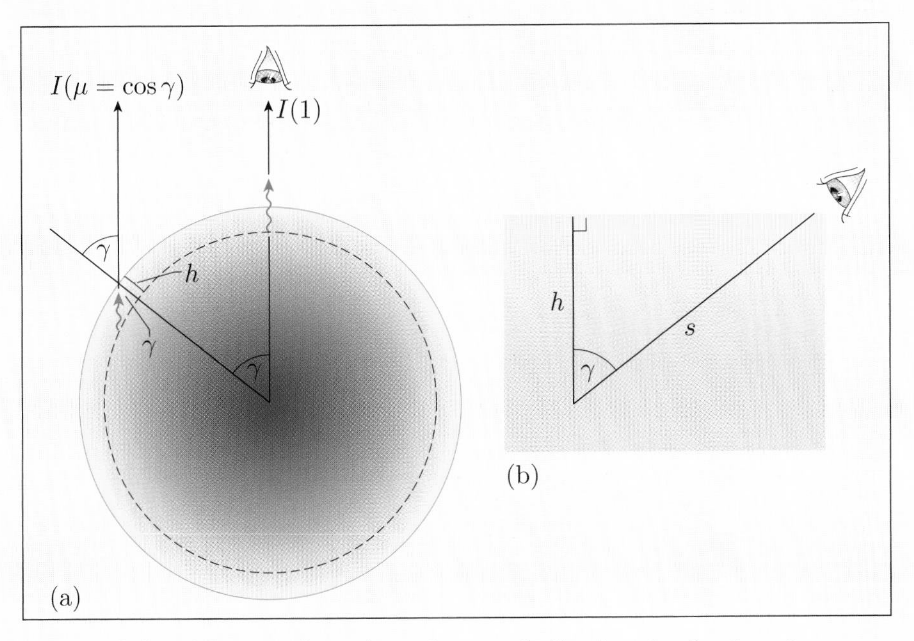

Figure 3.8 Cross-sections through a star indicating the depth and direction of travel of escaping photons. (a) Light travelling towards an observer beyond the top of the page at infinity is emitted in a radially outwards direction at the centre of the stellar disc. Conversely, light travelling towards this same observer from elsewhere on the disc is travelling at an angle γ to the radially outwards direction. The radially outwards direction minimizes the path length traversed from any depth, h. Light from a given depth is consequently more likely to escape for small angles γ. Thus the centre of the disc is brightest and the limb of the disc is faintest. (b) Light escapes only from the very outermost part of the star. The geometry is therefore effectively the plane-parallel atmosphere shown here, with $h \ll R_*$.

The optical depth for a given physical depth increases towards the limb of the star. A smaller fraction of the photons emitted at any particular depth escape to reach the observer from the limb of the star. Consequently, the stellar disc appears progressively dimmer towards the limb. The shorter path length travelled through the atmosphere by photons travelling radially outwards also means that more photons escape from deeper within the atmosphere at the centre of the disc. Generally, deeper layers are hotter, and radiation emitted there consequently approximates a bluer black body spectrum. This means that the disc appears progressively redder towards the limb.

The Sun is the only star for which we can currently measure the limb darkening directly and in detail. Unfortunately, the exact form of the limb darkening is a function of both opacity and emissivity at each depth within the atmosphere. These are dependent on the wavelength being considered, and the composition and thermodynamic properties of the atmosphere at each point. Thus the limb darkening will be dependent on the spectral type of the star, and on its detailed composition. Some empirical results for the solar limb darkening at two different wavelengths are shown in Figure 3.9.

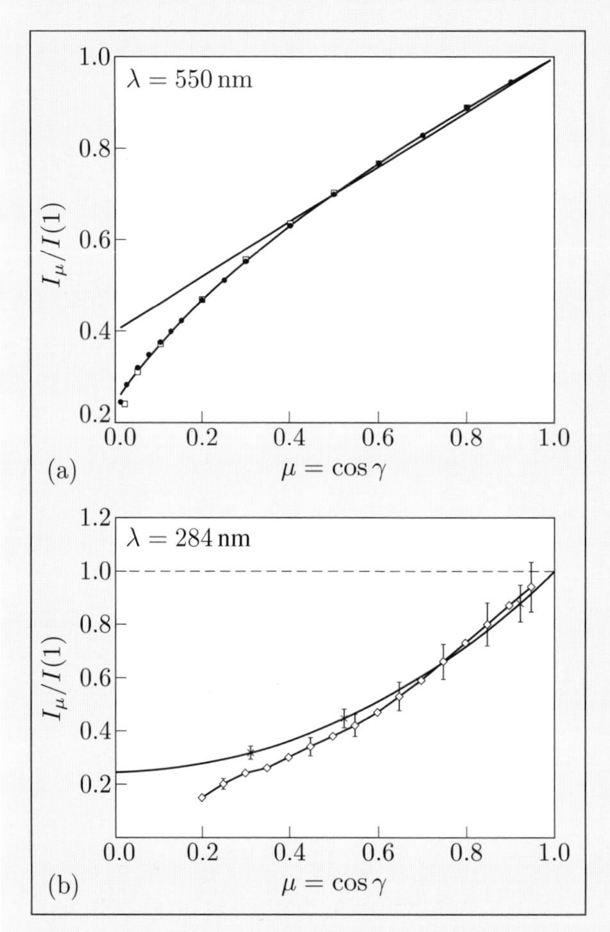

Figure 3.9 Limb darkening for the Sun. Each panel shows $I(\mu)$ as a function of μ for a particular wavelength. (a) Squares are measurements made at 550 nm; dots are the corresponding results from solar models. The solid lines give linear and logarithmic limb darkening relationships (see text). (b) At 284 nm, two alternative sets of measurements are shown; the upper one produced results for all angles shown. The crosses with error bars indicate the measurement uncertainties for selected angles. For $\mu < 0.55$, the two sets of measurements do not agree terribly well.

Even for the Sun, and for a single wavelength, it is difficult to obtain definitive results. In Figure 3.9a the dots show results from a computational model of radiative transfer through the solar atmosphere. The results agree well with the measurements. We cannot make direct observations of limb darkening in stars of spectral type and composition different from the Sun, so the results from computational models are generally used to guide our assumptions about their limb darkening. Researchers often refer to these assumptions as **limb darkening laws**, though they are simply curves adopted as adequately good approximations to the (unknown) stellar limb darkening profile. Limb darkening laws are not laws in the same sense as

Kepler's laws or the laws of thermodynamics. Figure 3.9a shows two relationships, i.e. two possible limb darkening laws, between $I(\mu)$ and μ, plotted as solid lines. The first of these is the simplest linear limb darkening relationship:

$$\frac{I(\mu)}{I(1)} = 1 - u(1-\mu), \tag{3.9}$$

where u is the **limb darkening coefficient** that governs the gradient of the intensity drop between the centre and limb of the disc. The second relationship illustrated in Figure 3.9a is a logarithmic relationship:

$$\frac{I(\mu)}{I(1)} = 1 - u_l(1-\mu) - \nu_l\,\mu \ln \mu, \tag{3.10}$$

which has two limb darkening coefficients, u_l and ν_l.

Two other relationships that may be adopted are the quadratic law

$$\frac{I(\mu)}{I(1)} = 1 - u_\text{q}(1-\mu) - \nu_\text{q}(1-\mu)^2 \tag{3.11}$$

and the cubic law

$$\frac{I(\mu)}{I(1)} = 1 - u_\text{c}(1-\mu) - \nu_\text{c}(1-\mu)^3, \tag{3.12}$$

with their coefficients u_q, ν_q and u_c, ν_c, respectively. The more complex relationships given in Equations 3.10, 3.11 and 3.12 have more flexibility, allowing them to fit empirical data more closely, but the cost of this is having two coefficients that must be determined somehow, or simply fixed arbitrarily.

All of these relationships have been used in modelling exoplanet transit light curves. For a particular star, any one of them with the appropriate limb darkening coefficient(s) might provide the best match to the actual (*a priori* unknown) stellar limb darkening.

The wavelength-dependence of the stellar limb darkening is immediately apparent in the multicolour transit light curves shown in Figure 3.10. The longest-wavelength data are shown in the uppermost transit curve, which has a very gently curving transit floor, while the shortest-wavelength transit has a transit floor that is so curved that it is impossible to discern the phases of the second and third contact points from visual inspection of it alone. Each of the observed transits in Figure 3.10 has a best-fitting model transit overplotted. These models are shown overplotted without offsets in Figure 3.11a, clearly showing the wavelength-dependence of transit shape. Figure 3.11b shows the counts spectrum from which these light curves were constructed, and indicates the wavelength ranges that comprise each light curve.

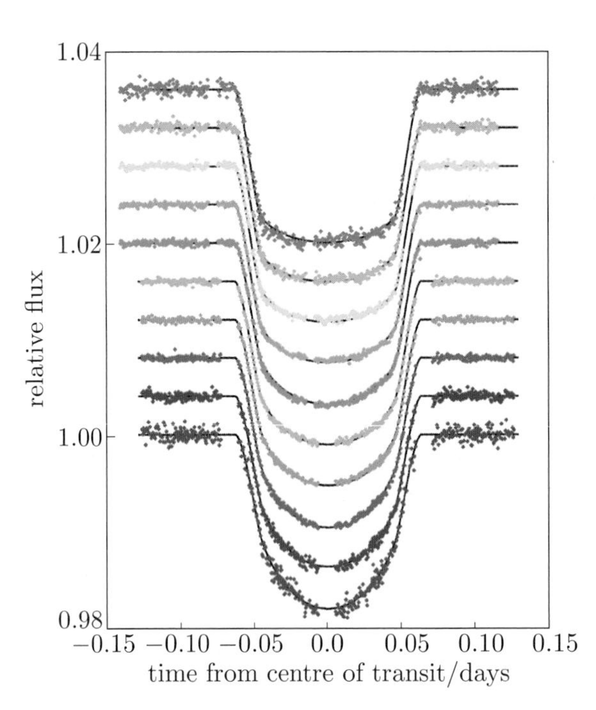

Figure 3.10 Multicolour transit light curves for HD 209458 b taken with the Hubble Space Telescope. All the curves have been normalized to an out-of-transit flux level of 1.00, and the curves have been offset for clarity. The longest wavelength data are at the top, and wavelength decreases for each successive curve. The wavelength ranges corresponding to each colour-coded curve are indicated in Figure 3.11. The solid lines are the model fits to the data.

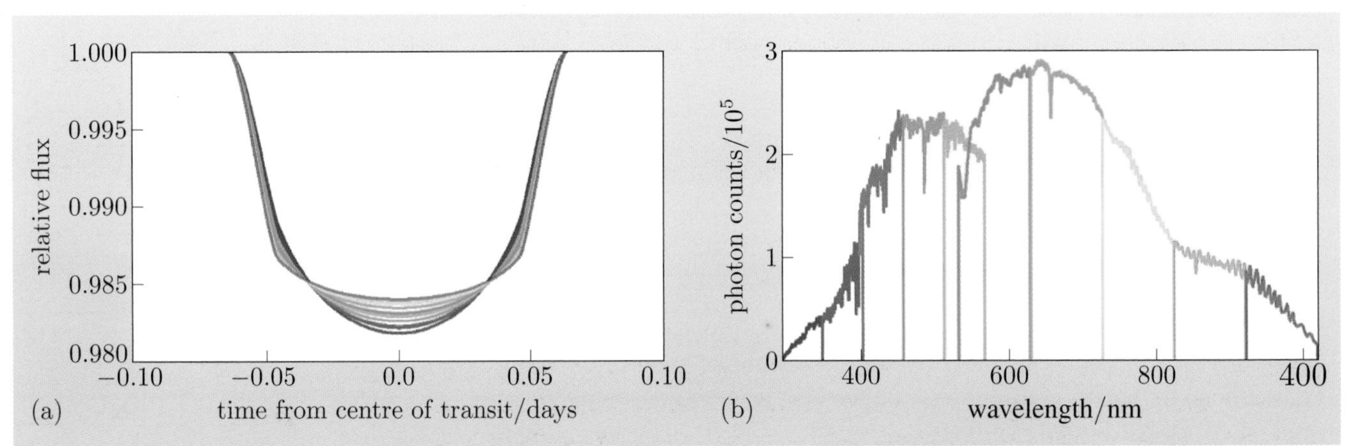

Figure 3.11 (a) The fits to the multicolour transit light curves for HD 209458 b plotted without the offsets. The transit is most flat-bottomed for the reddest wavelengths, where limb darkening is weaker. (b) The counts spectrum of the data used to construct these light curves.

Exercise 3.4 In the paper 'Homogeneous studies of transiting extrasolar planets — I. Light curve analyses' (2008, *Monthly Notices of the Royal Astronomical Society*, **386**, 1644–66) by John Southworth, he fits five different limb darkening laws to the transit light curves of 14 transiting exoplanets. For the exoplanet WASP-1 b, the limb darkening coefficients are determined as follows for the I band light curve:

linear law: $u = 0.215$;

logarithmic law: $u_l = 0.14,\ \nu_l = -0.12$;

quadratic law: $u_q = 0.29,\ \nu_q = -0.13$.

For light emerging close to the limb of the planet and therefore travelling at an angle $\gamma = 80°$ to the outwardly directed radius vector, what is the fractional intensity, $I(\mu)/I(1)$, in each case? Does it make a difference which limb darkening law is adopted? ■

3.2.2 The eclipsed area as a function of time

In Figure 3.10 we showed the results of model transit light curves computed and fit to the observed data. An essential step in computing a model transit light curve is to derive a general expression for the area of the star that is occulted at each instant of the transit event. We have seen in Subsection 3.2.1 that the intensity of the disc of the star is a function of μ, or alternatively we could express this intensity distribution as a function of r, the distance from the centre of the stellar disc. We will derive an expression for the occulted area, A_e, that is amenable to being modified to allow for this axially symmetric intensity distribution. The approach follows that of the paper 'Analytic light curves for planetary transit searches' (2002, *Astrophysical Journal Letters*, **580**, L171–5) by Kaisey Mandel and Eric Agol.

The first step in the derivation is to obtain a general expression for the projected separation of the centres of the stellar and planetary discs as a function of the orbital parameters and time. For simplicity we restrict ourselves to circular orbits, and for conciseness we will use the orbital angular speed, ω, where

$$\omega = \frac{2\pi}{P}. \tag{3.13}$$

The orbital phase angle (in radians) at a time t is therefore ωt, and this is related to the orbital phase, ϕ, by

$$\phi = \frac{\omega t}{2\pi}. \tag{3.14}$$

The fiducial phase $\phi = 0$ coincides with inferior conjunction of the planet, i.e. mid-transit ($\omega t = 0$; cf. Figure 3.12a). The stellar orbit is inclined to the line of sight, with orbital inclination i. At inferior conjunction, the projected separation between the centres of the star and planet is $b = a\cos i$, and at any other instant, the separation $s(t)$ is given by the right-angled triangle shown in Figure 3.12b. The locus followed by the planet across the disc of the star is not a straight line, but is part of the ellipse formed by the projection of the orbit on the plane of the sky. The planet moves in a circular orbit, so in astrocentric coordinates, the component of its displacement in the plane of the sky is $a\sin\omega t$, while the other component of its true displacement is $a\cos\omega t$. The full displacement in the plane of the sky is observed, while the other component of the displacement is foreshortened, so that the observed displacement is $a\cos i\cos\omega t$, as illustrated in Figure 3.12b. Using Pythagoras's theorem on the right-angled triangle in Figure 3.12b, we have

$$s^2(t) = (a\sin\omega t)^2 + (a\cos i\cos\omega t)^2. \tag{3.15}$$

Taking the square root and simplifying, we obtain

$$s(t) = a\left(\sin^2\omega t + \cos^2 i\cos^2\omega t\right)^{1/2}. \tag{3.16}$$

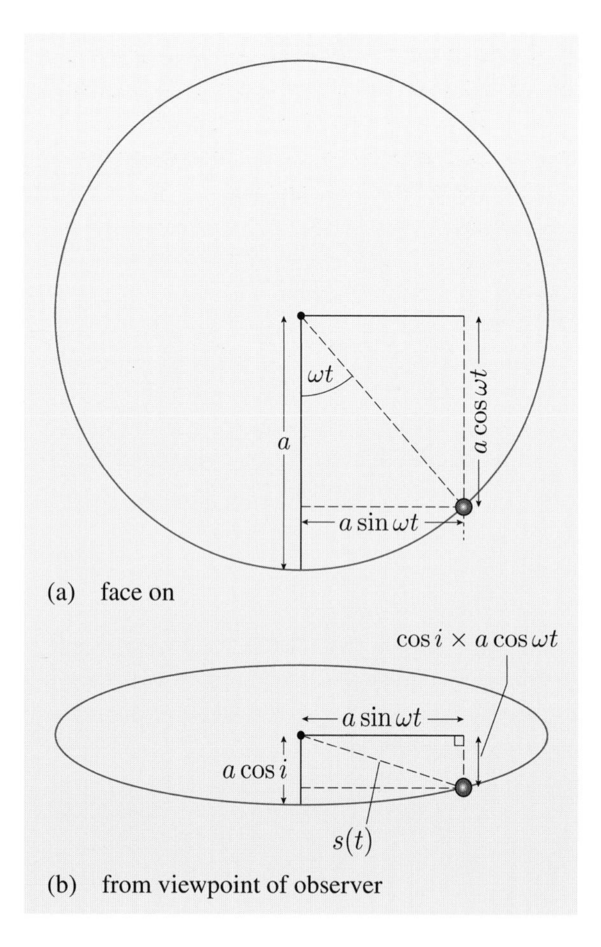

Figure 3.12 (a) The orbit as viewed from above. The displacement between the centres of the star and planet is a vector addition of a component $a \sin \omega t$ in the plane of the sky, and an orthogonal component $a \cos \omega t$. (b) The orbit as viewed along the line of sight of the observer. The projected separation of the centres of the star and planet is $s(t)$. The component of this in the plane of the sky is not foreshortened, but the other component is foreshortened by a factor of $\cos i$.

Exercise 3.5 A planet transits across a star like the Sun ($M_* = 1\,\mathrm{M}_\odot$, $R_* = 1\,\mathrm{R}_\odot$) with an impact parameter $b = 3\,\mathrm{R}_\odot/4$. The inclination angle is $i = 86.5°$. Calculate the position of the planet at three epochs: (i) 1 hour before mid-transit, (ii) at mid-transit, (iii) 1 hour after mid-transit. Then use this information to sketch the locus of the planet transit across the stellar disc. (*Hint*: Use Kepler's third law to calculate the orbital period of the planet, then calculate the orbital phase corresponding to the three times. Use Equation 3.16 to work out the horizontal and vertical components of the planet's position in terms of the stellar radius.) ■

Figure 3.13a shows the two overlapping discs and the variables that we will use to calculate the eclipsed area, A_e, in the difficult case where the planet's disc covers part of the stellar limb. The origin is placed at the centre of the star's disc. Two angles, α_1 and α_2, are indicated. The distance $s(t)$ is parameterized in terms of the stellar radius,

$$s(t) = \xi R_*, \tag{3.17}$$

and the ratio of the radii of the two discs is defined to be equal to p, so we also have

$$R_\mathrm{P} = pR_*. \tag{3.18}$$

There are several shaded areas indicated in Figure 3.13b. The light blue area is a **sector** of the stellar disc, while the dark blue area is a sector of the planet's disc.

The green area is a large triangle formed by an extension of the stellar sector. The area indicated in purple is half of the area that we actually wish to calculate, while the area in red is an irregular shape formed by subtracting the stellar disc sector from the large triangle. The (desired) purple shape is the planetary disc sector with the irregular red shape subtracted. Using this geometric reasoning, we can write a pseudo-equation for the desired area $A_{\rm e}$:

$$\begin{aligned} A_{\rm e} &= 2 \times (\text{small dark blue sector} - \text{red shape}) \\ &= 2 \times (\text{small dark blue sector} \\ &\qquad - (\text{large green triangle} - \text{large light blue sector})) \\ &= 2 \times (\text{small dark blue sector} \\ &\qquad + \text{large light blue sector} - \text{large green triangle}). \end{aligned} \tag{3.19}$$

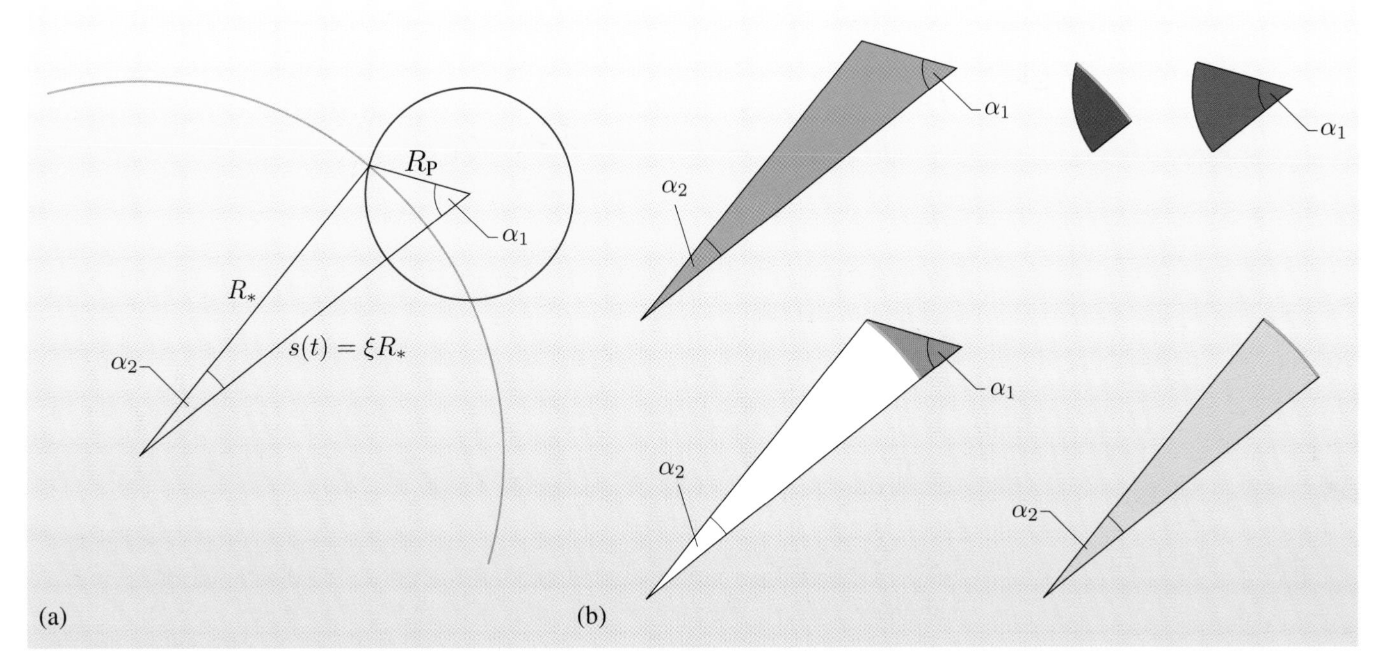

Figure 3.13 The partially overlapping discs of the star and the planet. The area of the stellar disc occulted by the planet during ingress and egress can be calculated using the geometry of the shapes shaded in colour here.

Each of the shapes on the right-hand side of Equation 3.19 has an area that is easy to express algebraically:

$$\text{area of small dark blue sector} = \pi R_{\rm P}^2 \times \frac{\alpha_1}{2\pi} = \frac{p^2 R_*^2 \alpha_1}{2}, \tag{3.20}$$

$$\text{area of large light blue sector} = \pi R_*^2 \times \frac{\alpha_2}{2\pi} = \frac{R_*^2 \alpha_2}{2}, \tag{3.21}$$

where α_1 and α_2 are in radians. The area of the large green triangle can be written down using the sine rule:

The sine rule allows us to apply $\text{area} = \dfrac{\text{base} \times \text{perp. height}}{2}$, where perp means perpendicular.

$$\begin{aligned} \text{area of large green triangle} &= \frac{R_* \times \xi R_* \sin\alpha_2}{2} \\ &= \frac{\xi R_*^2}{2} \sin\alpha_2. \end{aligned} \tag{3.22}$$

The cosine rule relates the cosine of an angle to the sides of an arbitrary triangle containing that angle.

The angles α_1 and α_2 can each be evaluated using the cosine rule, referring to the large green triangle in Figure 3.13:

$$\cos\alpha_1 = \frac{p^2 + \xi^2 - 1}{2\xi p}, \tag{3.23}$$

$$\cos\alpha_2 = \frac{1 + \xi^2 - p^2}{2\xi}. \tag{3.24}$$

Finally, by Pythagoras's theorem,

$$\sin\alpha_2 = \frac{\sqrt{4\xi^2 - (1 + \xi^2 - p^2)^2}}{2\xi}. \tag{3.25}$$

Using these results, we can express Equation 3.19 as

$$\begin{aligned} A_e &= 2 \times \left(\frac{p^2 R_*^2 \alpha_1}{2} + \frac{R_*^2 \alpha_2}{2} - \frac{\xi R_*^2 \sqrt{4\xi^2 - (1 + \xi^2 - p^2)^2}}{4\xi} \right) \\ &= R_*^2 \left(p^2 \alpha_1 + \alpha_2 - \frac{\sqrt{4\xi^2 - (1 + \xi^2 - p^2)^2}}{2} \right). \end{aligned} \tag{3.26}$$

Equation 3.26 gives an expression for the eclipsed area of the star for the case where the planet intersects the limb of the star.

- What value does A_e have when the disc of the planet falls entirely within the disc of the star?
- In this case $A_e = \pi R_P^2$, or in the notation introduced above, $A_e = \pi p^2 R_*^2$.
- What value does A_e have when the disc of the planet falls entirely outside the disc of the star?
- In this case none of the star is occulted and $A_e = 0$.

To write down a general expression for A_e, we need to express the cases of the planet's disc falling entirely within and outside the stellar disc in terms of our variables ξ and p. We can use Equations 3.16 and 3.17 to express ξ in terms of time (or equivalently orbital phase), and thus obtain $A_e(t)$. In Subsection 3.1.3 we argued that the time between first contact and second contact gives a measure of the size of a transiting planet, but we did not develop this idea mathematically. The work that we are doing here provides the relevant mathematics.

The planet falls entirely outside the stellar disc if the distance between the centres of the two discs exceeds the sum of their radii, i.e.

$$A_e = \begin{cases} 0 & \text{if } R_* + R_P < s, \\ 0 & \text{if } R_*(1 + p) < \xi R_*, \\ 0 & \text{if } 1 + p < \xi. \end{cases}$$

The planet falls entirely within the stellar disc if the distance between the centres of the two discs is less than the difference of their radii, i.e.

$$A_e = \begin{cases} \pi R_P^2 & \text{if } R_* - R_P \geq s, \\ \pi p^2 R_*^2 & \text{if } R_*(1 - p) \geq \xi R_*, \\ \pi p^2 R_*^2 & \text{if } 1 - p \geq \xi. \end{cases}$$

Gathering this together, we have

$$A_{\rm e} = \begin{cases} 0 & \text{if } 1+p<\xi, \\ R_*^2\left(p^2\alpha_1+\alpha_2-\frac{\sqrt{4\xi^2-(1+\xi^2-p^2)^2}}{2}\right) & \text{if } 1-p<\xi\leq 1+p, \\ \pi p^2 R_*^2 & \text{if } 1-p\geq\xi. \end{cases} \quad (3.27)$$

The eclipsed area, $A_{\rm e}$, is thus a known function of R_*, p and $\xi(t)$, i.e. $A_{\rm e}(R_*, p, \xi)$, or equivalently of R_*, $R_{\rm P}$ and $s(t)$, i.e. $A_{\rm e}(R_*, R_{\rm P}, s)$.

Worked Example 3.1

A planet of radius $0.1\,{\rm R}_\odot$ transits across the face of a star like the Sun ($M = 1\,{\rm M}_\odot$, $R = 1\,{\rm R}_\odot$). The semi-major axis of the orbit is $20\,{\rm R}_\odot$, and the inclination of the orbit is $i = 89.5°$. At a time 2.0 h before the mid-transit, the planet is crossing the limb of the star.

(a) Determine the period of the planetary orbit, and hence the phase corresponding to the time mentioned above.

(b) Calculate the ratio of the radii of the planet and star, p, and the ratio of the distance of the planet from the centre of the star's disc to the radius of the star, ξ, at the time mentioned above.

(c) Verify that the planet is indeed just crossing the limb of the star at this time.

(d) Calculate the eclipsed area of the star at this time.

Solution

(a) Using Kepler's third law (Equation 1.1), the period of the planetary orbit is

$$\begin{aligned} P_{\rm orb} &= \left(\frac{4\pi^2 a^3}{GM_*}\right)^{1/2} \\ &= \left(\frac{4\pi^2\times(20\times 6.96\times 10^8\,{\rm m})^3}{6.673\times 10^{-11}\,{\rm N\,m^2\,kg^{-2}}\times 1.99\times 10^{30}\,{\rm kg}}\right)^{1/2} \\ &= 8.95\times 10^5\,{\rm s}. \end{aligned}$$

The orbital period is therefore 249 hours (or 10.4 days).

2.0 hours before the mid-transit corresponds to a phase offset of

$$\omega t = \frac{2\pi t}{P_{\rm orb}} = \frac{2\pi\times 2.0\,{\rm h}}{249\,{\rm h}} = 0.051\text{ radians.}$$

To four significant figures, the phase offset is 0.050 52 radians, which one could argue for rounding to either 0.050 radians or 0.051 radians as we have done. $A_{\rm e}$ (calculated below) would be $0.0124\,{\rm R}_\odot^2$ rather than $0.0105\,{\rm R}_\odot^2$ if the phase offset were rounded to 0.050 radians.

(b) From Equation 3.18, the ratio of the radii of the planet and star is

$$p = \frac{R_{\rm P}}{R_*} = 0.1.$$

From Equation 3.17, the ratio of the distance of the planet from the centre of the star's disc to the radius of the star is $\xi = s(t)/R_*$. Now, the distance $s(t)$

is given by Equation 3.16 as

$$\begin{aligned} s(t) &= a\left(\sin^2\omega t + \cos^2 i \cos^2\omega t\right)^{1/2} \\ &= 20\,\mathrm{R}_\odot \times \left(\sin^2(0.051\,\mathrm{rad}) + \cos^2(89.5°)\cos^2(0.051\,\mathrm{rad})\right)^{1/2} \\ &= 1.025\,\mathrm{R}_\odot. \end{aligned}$$

So

$$\xi = \frac{1.025\,\mathrm{R}_\odot}{\mathrm{R}_\odot} = 1.025.$$

(c) The condition for the planet to be crossing the limb of the star is that $1 - p < \xi \leq 1 + p$. Since $1 - p = 0.9$ and $1 + p = 1.1$, the condition is satisfied, and the planet is indeed crossing the limb of the star.

(d) To calculate the fractional area of the star that is occulted at this time, we need the second case from Equation 3.27. To solve that, we first need to calculate the angles α_1 and α_2, which may be obtained from Equations 3.23 and 3.24:

$$\begin{aligned} \alpha_1 &= \cos^{-1}\left(\frac{p^2 + \xi^2 - 1}{2\xi p}\right) = \cos^{-1}\left(\frac{0.1^2 + 1.025^2 - 1}{2 \times 0.1 \times 1.025}\right) \\ &= 1.272 \text{ radians}, \end{aligned}$$

$$\begin{aligned} \alpha_2 &= \cos^{-1}\left(\frac{1 + \xi^2 - p^2}{2\xi}\right) = \cos^{-1}\left(\frac{1 + 1.025^2 - 0.1^2}{2 \times 1.025}\right) \\ &= 0.0957 \text{ radians}. \end{aligned}$$

So the eclipsed area is

$$\begin{aligned} A_\mathrm{e} &= R_*^2\left(p^2\alpha_1 + \alpha_2 - \frac{\sqrt{4\xi^2 - (1 + \xi^2 - p^2)^2}}{2}\right) \\ &= \mathrm{R}_\odot^2 \times \left((0.1^2 \times 1.272) + 0.0957 - \frac{\sqrt{(4 \times 1.025^2) - (1 + 1.025^2 - 0.1^2)^2}}{2}\right) \\ &= \mathrm{R}_\odot^2 \times (0.0127 + 0.0957 - 0.0979) \\ &= 0.0105\,\mathrm{R}_\odot^2. \end{aligned}$$

The eclipsed area is therefore a fraction of the stellar disc equivalent to $(0.0105\,\mathrm{R}_\odot^2)/(\pi\,\mathrm{R}_\odot^2) = 0.0033$ or 0.33%. For comparison, the fully eclipsed area would be 0.1^2 or 1% of the stellar disc.

3.2.3 The light lost from an axially symmetric stellar disc

We use the simpler variable name, r, to describe the normalized radial coordinate needed in some intricate algebra later.

The total flux, F, from the stellar disc is the intensity, I, integrated over the surface of the disc. We restrict our analysis to axially symmetric intensity distributions, so $I = I(r')$, where r' is measured from the centre of the stellar disc, as shown in Figure 3.14. Thus

$$F = \int_{\text{disc}} I(r')\,\mathrm{d}A = \int_{r'=0}^{r'=R_*} I(r')\,2\pi r'\,\mathrm{d}r'. \tag{3.28}$$

Some authors use the *astrophysical flux*, which is the disc-averaged specific intensity, and differs from our definition by a factor of π.

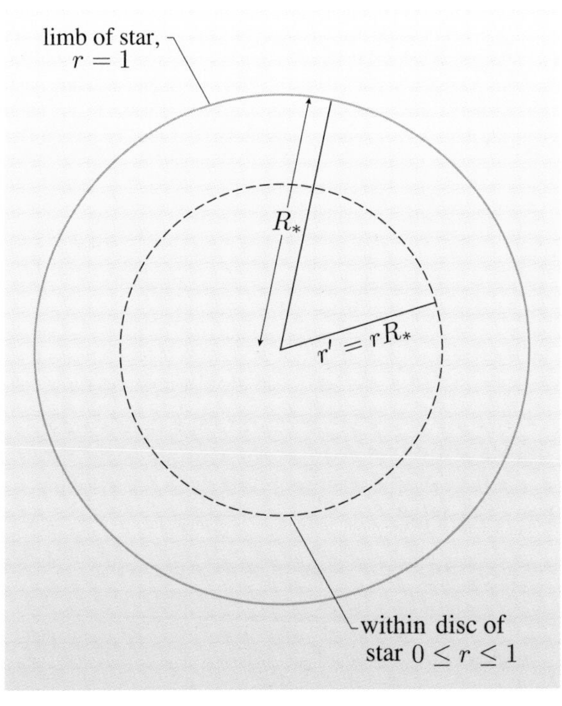

Figure 3.14 The variables r' and $r = r'/R_*$ used to describe locations on the axially symmetric stellar disc.

- Why have we imposed axial symmetry, rather than assuming a general case where I has an azimuthal-dependence as well as a radial-dependence?

- Our knowledge of the brightness distribution of the Sun and our analysis of stellar atmospheres suggest that the stellar disc will be limb darkened so that $I = I(r')$, but we have no reason to expect a substantial breaking of the axial symmetry.

For a stellar disc with uniform brightness, the intensity is a constant, $I = I_0$ across the entire disc, and Equation 3.28 simplifies:

$$\begin{aligned} F &= \int_{r'=0}^{r'=R_*} I_0\,2\pi r'\,\mathrm{d}r' \\ &= 2\pi I_0 \int_{r'=0}^{r'=R_*} r'\,\mathrm{d}r' \\ &= 2\pi I_0 \left[\frac{r'^2}{2}\right]_{r'=0}^{r'=R_*} \\ &= \pi I_0 R_*^2. \end{aligned} \tag{3.29}$$

We can now write down an expression for ΔF as a function of time. It is simply the missing flux that is occulted by the planet's disc, i.e.

$$\Delta F = \int_{\text{occulted area}} I(r')\,\mathrm{d}A, \tag{3.30}$$

where the integral is over the area of the stellar disc that is occulted by the planet. Performing this integral over the irregular purple shape shown in Figure 3.13 is not in general a straightforward step, but for the simplest case of a uniform stellar disc it is easy: Equation 3.30 becomes

$$\Delta F = \int_{\text{occulted area}} I_0\, \mathrm{d}A = I_0 \int_{\text{occulted area}} \mathrm{d}A$$
$$= I_0 A_\mathrm{e}. \tag{3.31}$$

With this rather trivial mathematical step, we have completed the work of deriving the analytic description of the shape of a transit light curve, albeit in the simplified case of a planet in a circular orbit transiting an idealized star whose disc has uniform brightness.

Exercise 3.6 (a) Derive an expression for the flux, $F(t)$, observed when a planet of radius R_P, with orbital inclination i, in a circular orbit of radius a and period P, transits an idealized star of radius R_* that has negligible limb darkening. You may use any of the relationships derived so far in this book. You do not need to exhaustively make the substitutions to produce a single expression relating $F(t)$ to i, a and P, but you need to define all the symbols used in your expression for $F(t)$ in terms of these parameters.

(b) Verify that for $p = 0.1$ and $\xi = 0.2$, your expression is consistent with

$$\frac{\Delta F}{F} = \frac{R_\mathrm{P}^2}{R_*^2}. \tag{Eqn 1.18}$$

(c) Assume that the orbital inclination has already been determined from the transit duration, T_dur, as we saw in Subsection 3.1.3, and that the period, P, is known. Use your expression to explain how R_P can be deduced in terms of R_* from measurements of the timings of first and second contacts, i.e. show that the ratio R_P/R_* can be determined by a method that is entirely independent of the transit depth.

(d) Parts (b) and (c) illustrate two independent methods for determining the ratio R_P/R_*. Discuss which of these two methods is most affected by the limb darkening that will occur in a real star. ■

The next bit is clever. Figure 3.14 shows a normalized axial coordinate, r, defined such that it is 0 at the centre of the stellar disc and 1 at the limb, i.e.

$$r = \frac{r'}{R_*}. \tag{3.32}$$

Figure 3.15 demonstrates how the eclipsed area is composed of a series of partial annuli, one of which is shown. It is clear from the figure that the eclipsed area within the circle of radius rR_* is exactly the area that would have been eclipsed were the star of radius rR_* rather than of radius R_*. (If this is not immediately apparent, imagine the figure without the circle of radius R_*, and compare it to Figure 3.13.) Therefore we can use the function that we have already derived for A_e with a change of variables to allow us to perform the integration over the axially symmetric stellar disc. If your abstract reasoning skills are not on form, you may wish to skip the following derivation and simply accept the result. The element of area shaded in red in Figure 3.15 is the additional eclipsed area $\mathrm{d}A(r)$

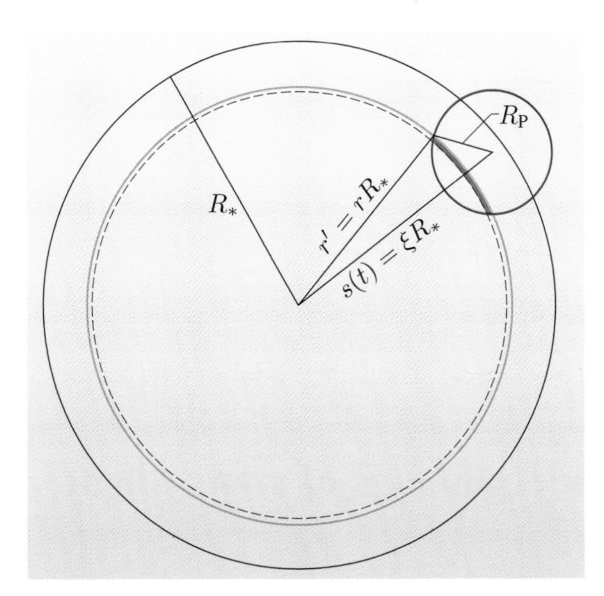

Figure 3.15 The eclipsed area A_e is made up of a series of partial annuli. Within each of these annuli $I(r)$ is constant, so ΔF for an axially symmetric brightness distribution can be evaluated by summing over these annuli.

when incrementing r by an amount $\mathrm{d}r$. But we have already observed that this element area is also the difference between A_e appropriate for a star of radius $R_* = r'$ and for a star of radius $R_* = r' + \mathrm{d}r'$. Consequently, the area shaded in red is

$$\mathrm{d}A(r) = \frac{\mathrm{d}A_e}{\mathrm{d}r'}\,\mathrm{d}r'. \tag{3.33}$$

We can also write an expression for the eclipsed area for a star of radius rR_*, i.e. a star for which the annulus shown in Figure 3.15 marks its outer boundary. To do this we will need to change variables, as we used variables p and ξ, which were normalized with respect to the radius of the star. Since $p = R_P/R_*$, changing the radius of the star to rR_* implies that p becomes p/r; similarly, ξ becomes ξ/r. Consequently, the eclipsed area of our hypothetical smaller star is given by

$$\begin{aligned}\int_0^{r'=rR_*} \frac{\mathrm{d}A_e}{\mathrm{d}r'}\,\mathrm{d}r' &= A_e\left(rR_*, \frac{p}{r}, \frac{\xi}{r}\right)\\ &= r^2 A_e\left(R_*, \frac{p}{r}, \frac{\xi}{r}\right).\end{aligned} \tag{3.34}$$

Here we have made use of the fact that the only explicit occurrence of R_* in Equation 3.27 is a factor of R_*^2 multiplying the entire expression in each non-zero case. The implicit occurrences of R_* as the unit length in our normalized coordinate system have been taken care of by our change of variables. Comparing Equations 3.33 and 3.34, it is clear that the integrand on the left-hand side of the latter is equal to the right-hand side of the former. Consequently, we can say that

$$\mathrm{d}A(r) = \frac{\mathrm{d}}{\mathrm{d}r}\left[r^2 A_e\left(R_*, \frac{p}{r}, \frac{\xi}{r}\right)\right]\mathrm{d}r. \tag{3.35}$$

This expression gives us exactly what we need to perform the integral required to obtain the flux deficit, ΔF, for a transit across a limb darkened star.

Going back to the completely general expression for ΔF, Equation 3.30, we can say for the limb darkened axially symmetric case that

$$\begin{aligned}\Delta F &= \int_{r=0}^{r=1} I(r)\,\mathrm{d}A(r)\\ &= \int_{r=0}^{r=1} I(r)\,\frac{\mathrm{d}}{\mathrm{d}r}\left[r^2 A_{\mathrm{e}}\left(R_*, \frac{p}{r}, \frac{\xi}{r}\right)\right]\mathrm{d}r. \end{aligned} \tag{3.36}$$

This expression therefore allows us to calculate the exact shape of a transit light curve by specifying the limb darkening law, the sizes of the star and planet, and the orbital parameters P, a and i.

3.3 The analytic underpinnings of transit light curve fitting

With ample justification it has been said that the core activity of physicists is fitting models to data. In constructing a mathematical model, the important factors governing a measured value must be identified and quantified, and equations specifying how these factors combine to produce the measurable quantity must be derived. In Section 3.2 we constructed a mathematical model to predict transit light curves. In Section 2.6 we briefly discussed the process of model-fitting; it is through this process that the parameters of transiting exoplanet systems are deduced from the observed data. In the next section we will describe how the observed transit light curve is used in conjunction with a mathematical model to determine the parameters of the observed system. Note that the model derived in Section 3.2 is not the *only* way to calculate a theoretical transit light curve for a limb darkened star. For example, a completely complementary approach is to build a model star–planet system in a computer simulation. In such a simulation, the star is covered with a grid and the geometry of the problem is used to calculate whether the light emitted by each grid point is visible to the observer at each orbital phase. This approach has strengths and weaknesses compared to employing the analytic model of Section 3.2; we focus on the latter purely because it *is* analytic, and therefore we can write down the salient equations and analyze it. In this section we show how the various features of the transit light curve allow the parameters of the star and planet to be deduced; in the next section we will illustrate how the model-fitting to accomplish this is performed.

3.3.1 The scale of the system

It is clear that the depth of the transit is dependent on the ratio of the two radii. As we explored in Exercise 3.6, there is a further constraint on the radius of the planet from the time between first contact and second contact, but this does not yield an absolute value for R_{P} either. In the exercise we used the timings to determine the ratio of the two radii. We could extend the work that we did in Exercise 3.6 to express the size of the orbit, a, in terms of either of the radii. Consequently, the transit alone does not determine the absolute size of anything in the system, it simply determines the ratios $R_{\mathrm{P}} : R_* : a$. As we noted in Section 3.1, once a transiting system has been identified, we have an excellent method for determining the orbital period, P, and knowing this, Kepler's third law gives us an expression (Equation S2.2) for a that depends on the total mass, or to a good approximation on the stellar mass, M_*. Astronomers have been studying stars for centuries, and consequently we have reliable ways of estimating their properties based on their spectral type. Adopting values for M_* and P therefore fixes the scale of the system, by providing a value (in physical units) for a.

3.3.2 The radii and consistency checks

With a fixed in physical units, analysis of the transit light curve alone can yield values for R_{P} and R_*. The values obtained can be checked in several ways.

- First, the value for R_* should be consistent with the stellar properties deduced from the spectral type. Indeed, we discussed in Subsection 2.8.2 how a consistency check of this type is used in the SuperWASP candidate-winnowing process.
- As we mentioned above in Subsection 3.3.1, there are two independent estimates of R_{P} from two different properties of the transit light curve; if the solution is reliable, these should be mutually consistent.
- The transit duration should be consistent with the values of R_* and a, and the shape of the transit should be consistent with the impact parameter required to match the duration.

3.3.3 The orbital inclination and the mass of the planet

The transit duration coupled with the values of the two radii, R_{P} and R_*, yields a value for the orbital inclination through Equation 3.4. Values for M_* and P, coupled with a measurement of the stellar reflex orbital radial velocity, A_{RV}, give us $M_{\mathrm{P}} \sin i$, as we have seen in Equation S2.6. Using the orbital inclination from the transit duration thus allows the actual value of M_{P} to be determined. Thus for transiting planets we can obtain the mass, the radius and the orbital inclination.

3.4 Parameter determination from transit light curve fitting

Figure 3.16 shows the results of model-fitting to the data of Figure 2.4 folded on orbital phase. The stellar parameters were fixed at $M_* = 1.1\,\mathrm{M}_\odot$ and $R_* = 1.1\,\mathrm{R}_\odot$, and a linear limb darkening law of the form of Equation 3.9 was adopted. The best-fitting values for the planet HD 209458 b found with these assumptions are formally $R_{\mathrm{P}} = 1.27 \pm 0.02\,\mathrm{R_J}$ and $i = 87.1^\circ \pm 0.2^\circ$. We will discuss the uncertainty estimates from model fitting in the box 'Uncertainty estimates and ...' later in this section.

3.4.1 Light curve features to be matched

In Section 3.3 we examined analytically how the various parameters can be deduced from properties of the light curve. In the process of light curve

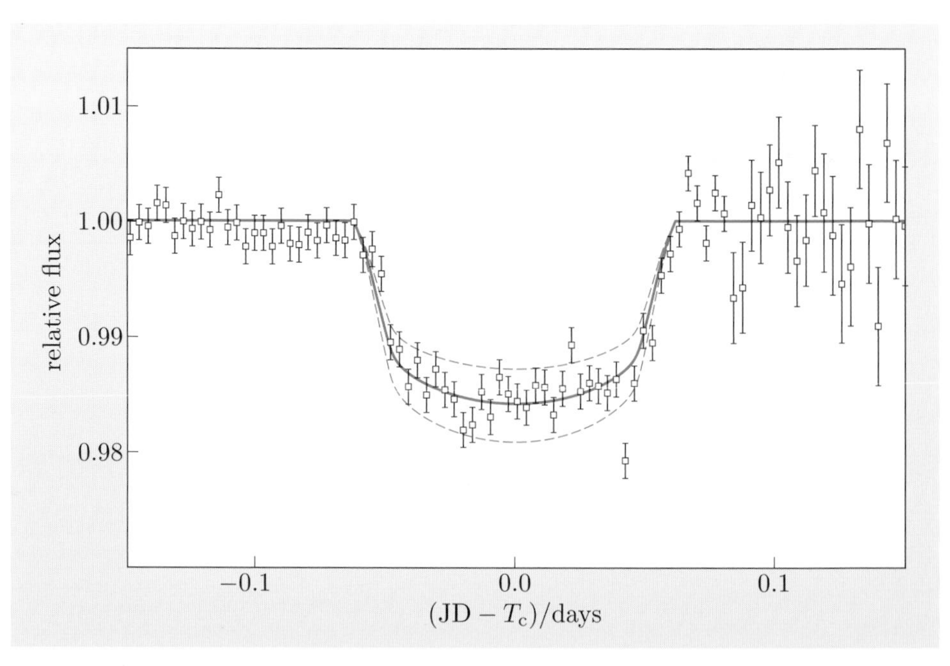

Figure 3.16 The first model fit to an observed exoplanet transit light curve. The observed data are for the transit of HD 209458 b shown in Figure 2.4, and the solid and dashed lines are model transit light curves calculated assuming $M_* = 1.1\,\mathrm{M}_\odot$ and $R_* = 1.1\,\mathrm{R}_\odot$. The solid line corresponds to $R_\mathrm{P} = 1.27\,\mathrm{R_J}$ and $i = 87.1°$, and provides a good fit to the data, while the dashed lines show the models for planet radii 10% smaller and 10% larger.

modelling, these properties all simultaneously influence the parameters of the best fit. Consequently, instead of obtaining two independent values for R_P, one from the depth and another from the slope of ingress and egress, the best-fit parameter will be the one that produces the minimum χ^2 by producing the overall light curve that best matches the observations. Figure 3.17 schematically illustrates the features that can be measured in a high signal-to-noise ratio transit light curve. These are as follows.

David Kipping's paper arXiv:1004.3819 discusses seven different definitions of transit duration.

- The depth, ΔF, measured from the out-of-transit level to the mid-transit level.
- The transit duration, which can be defined in various ways. The quantity T_dur that we derived in Equation 3.4 gives the time between first and fourth contacts. T_dur depends on R_P, R_* and i. An alternative definition, which is labelled T_tran in Figure 3.17, measures the time between mid-ingress and mid-egress, which is approximated by the time at which the centre of the planet's disc crosses the limb of the star, and is therefore independent of R_P.
- The duration of ingress, t_{1-2} (or equivalently egress, t_{3-4}). As we have already discussed, this depends on R_P. It also depends on the orbital inclination: for central transits, the motion of the planet is along the normal to the limb of the star, so the distance moved between first and second contact is a minimum. For larger impact parameters, the motion of the planet is at an angle to the normal to the stellar limb and therefore the distance travelled and the elapsed time between first and second contacts are longer.
- The curvature of the light curve between second and third contacts depends on the stellar limb darkening, and on the locus that the planet takes across the stellar disc, which in turn depends on R_* and i.

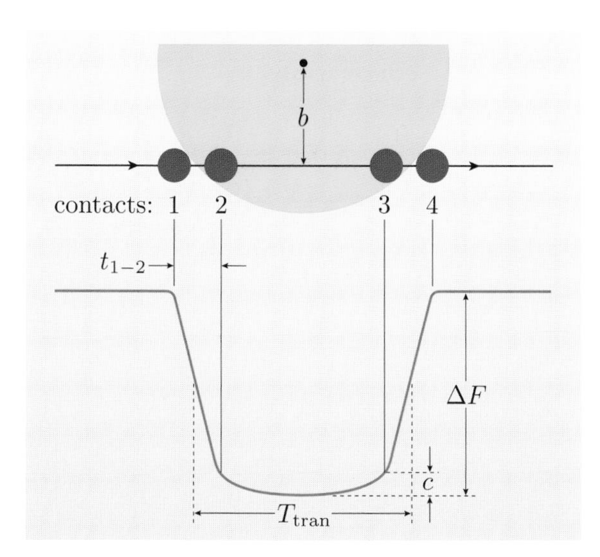

Figure 3.17 Properties of the transit light curve.

In the process of fitting a model light curve to observed data, the parameters of the model are adjusted until the model is able to fit the combination of all of these features.

Exercise 3.7 Sketch transit light curves for the following two contrasting transits:

(a) A large planet ($R_{\rm P}/R_* \sim 0.1$) with a large impact factor ($b \sim 0.9$) and very little limb darkening.

(b) A small planet ($R_{\rm P}/R_* \sim 0.01$) with a small impact factor ($b \sim 0$) and significant limb darkening. ■

3.4.2 The process of light curve fitting

The exact procedure used in fitting a model light curve can be varied as appropriate to the data in hand, to the prior knowledge about the system, and to the preferences, tools and expertise of the researcher. In general, a mathematical model (e.g. that described in Equation 3.36) is adopted, the stellar mass is assumed to fix the scale of the system, and a limb darkening law is adopted. In some cases, the limb darkening coefficient(s) may be fixed, but generally the limb darkening coefficients are treated as free parameters. For low signal-to-noise ratio data like those shown in Figure 3.16, the data cannot define the shape of the curvature of the transit floor to high precision, so the simplest limb darkening law, Equation 3.9, should be used. This linear limb darkening relationship has a single limb darkening coefficient, while the other relationships (Equations 3.10, 3.11 and 3.12) each have two coefficients, which must be either arbitrarily fixed or determined from the fit to the data. Following these initial assumptions, a process of χ^2 minimization is carried out to determine the parameters that produce the best fit to the observed transit light curve. In this way, best-fitting values of $R_{\rm P}$, R_* and i are determined.

Uncertainty estimates and parameter correlations in model-fitting

A complete discussion of this subject is well beyond the scope of this book, but we summarize some of the key points here. In Figure 3.16 the solid line shows the best-fit model, $\mu(t, \text{parameters})$, and we can see by eye that it does do a good job of matching the data points $(x_i \pm \sigma_i)$. Examination of the figure reveals that the model does not pass through all of the $\pm 1\sigma$ uncertainty estimates for the individual data points. This is as expected. Figure 2.13 shows us that we expect roughly a third of all measurements to deviate by more than 1σ, so a well-fitting model should pass through roughly two-thirds of the $\pm 1\sigma$ error bars.

Assuming that the uncertainty estimates are correct and the measurement uncertainties are Gaussian distributed, there is therefore an 'expected' goodness of fit. This is characterized by the 'reduced chi-squared' or χ_N^2. For a fit of a model with n_p free parameters to n_x data points with Gaussian uncertainties,

$$\chi_N^2 = \frac{\chi^2}{n_x - n_p}, \tag{3.37}$$

where χ^2 is given by

$$\chi^2 = \sum \left(\frac{x_i - \mu_i}{\sigma_i} \right)^2. \qquad \text{(Eqn 2.36)}$$

With the assumptions above, χ_N^2 should have a value of

$$\chi_N^2 = 1 \pm \sqrt{\frac{2}{n_x - n_p}} \tag{3.38}$$

for a model that fits the data.

The best-fit parameters correspond to the minimum value of χ^2, and we can use the variation of χ^2 as the parameters are adjusted to assess the uncertainty on the fitted values. If, for example, changes in the orbital inclination over an interval Δi make negligible difference to the value of χ^2, then the fitting procedure is unable to discriminate between values of i in this range. To quantify and generalize this statement, if varying one particular parameter over a defined interval while holding all the other parameters constant produces a change in χ_N^2 of $\ll 1$, then the fit cannot discriminate meaningfully between parameter values within the interval. Consequently, again assuming that the uncertainty estimates are correct and the measurement uncertainties are Gaussian distributed, the procedure for estimating the uncertainty in a single fitted parameter is as follows.

1. Find the values of the model parameters that produce the minimum value of χ^2; call this value χ^2_{min}.
2. Check that $\chi_N^2 \approx 1$ for these values of the parameters.
3. Hold all but the chosen parameter constant at their best-fitting values.
4. Vary the chosen parameter over a range around its best-fit value, and record the χ^2 corresponding to each input value.

5. Plot a graph of χ^2 versus the chosen parameter. The $\pm 1\sigma$ error estimate is the range within which

$$\chi^2 = \chi^2_{\text{min}} + 1. \tag{3.39}$$

Sometimes the parameters are correlated, i.e. the data cannot tightly constrain an individual parameter, but can constrain a combination of two parameters. This can be examined by making a χ^2 contour map. Figure 3.18 shows an example: the contours represent the 1σ, 2σ and 3σ confidence intervals in the (R_{P}, i)-plane. If the two parameters, R_{P} and i, were completely independent of each other in their influence on the χ^2 statistic, the intervals would be more-or-less circular. In Figure 3.18 we can see that the contours are skewed, with higher values of R_{P} corresponding to lower values of i; i.e. these two parameters are correlated.

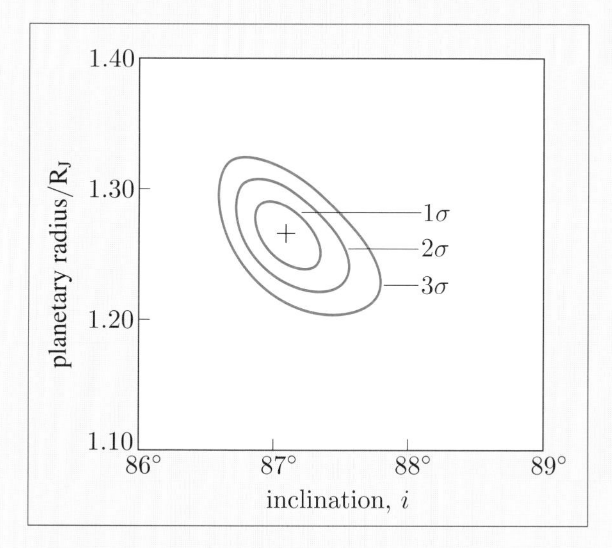

Figure 3.18 The 1σ, 2σ and 3σ confidence intervals in the (R_{P}, i)-plane from the fits to the data shown in Figure 3.16.

The results for R_{P}, R_* and i need to be critically examined. The most obvious check is to verify that R_* is consistent with the mass assumed for M_*. If it is not consistent, then fitting with a different limb darkening law, or changing the value of fixed limb darkening coefficients, or changing the assumed input stellar mass, or all of the above, should be explored. The box above describes how the formal errors in the deduced parameters may be estimated, assuming a Gaussian distribution in the measurement uncertainties. Realistic estimates of the uncertainties in the parameters of the fit should include the full range of possible outcomes with various justifiable input assumptions. Consequently, it is generally the case that the uncertainty in R_{P} is dominated by the uncertainty in the characteristics of the host star (i.e. M_*, R_* and limb darkening). Even when these factors are carefully considered, the true uncertainties on the parameter values are still likely to exceed these formal estimates, because the measurement uncertainties in the high signal-to-noise light curves are generally not Gaussian distributed; instead, they are dominated by systematic errors.

The source of these systematic errors undoubtedly includes effects that arise in the instrumentation: subtle changes in the state of the optics or detector can lead to changes in sensitivity. These can be at least partially characterized and removed. As more transiting systems are studied intensively and precise parameter determinations are made, it appears that some of the 'errors' may actually be caused by variability of the star itself.

Table 3.1 Selected results from transit light curve fitting: parameters for planets orbiting host stars that appear brightest from Earth.

Planet	$M_P \sin i$ (M_J)	R_P (R_J)	i (degrees)	Refs
HD 209458 b	$0.685^{+0.015}_{-0.014}$	$1.32^{+0.024}_{-0.025}$	86.677 ± 0.060	1, 2, 3
HD 189733 b	1.13 ± 0.03	1.138 ± 0.027	85.76 ± 0.29	4, 1, 5
HD 149026 b	$0.359^{+0.022}_{-0.021}$	$0.654^{+0.060}_{-0.045}$	$85.3^{+0.9}_{-0.8}$	1, 6
HD 17156 b	$3.212^{+0.069}_{-0.082}$	$1.023^{+0.079}_{-0.055}$	$86.2^{+2.1}_{-0.8}$	7
HAT-P-2 b	9.09 ± 0.24	$1.157^{+0.073}_{-0.092}$	$86.72^{+1.12}_{-0.87}$	8
HD 80606 b	3.94 ± 0.11	1.029 ± 0.017	89.285 ± 0.023	9, 10
WASP-18 b	10.43 ± 0.40	1.165 ± 0.077	86 ± 2.5	11, 12
WASP-7 b	$0.96^{+0.12}_{-0.18}$	$0.915^{+0.046}_{-0.040}$	$89.6^{+0.4}_{-0.9}$	13
HAT-P-11 b	0.081 ± 0.009	0.452 ± 0.020	88.5 ± 0.6	14, 15
WASP-14 b	$7.725^{+0.43}_{-0.67}$	$1.259^{+0.080}_{-0.058}$	$84.79^{+0.52}_{-0.67}$	16
XO-3 b	11.79 ± 0.59	1.217 ± 0.073	84.2 ± 0.54	17
HAT-P-8 b	$1.52^{+0.18}_{-0.16}$	$1.50^{+0.08}_{-0.06}$	$87.5^{+1.9}_{-0.9}$	18
HAT-P-1 b	0.524 ± 0.031	1.225 ± 0.059	86.28 ± 0.20	19
WASP-13 b	$0.46^{+0.06}_{-0.05}$	$1.21^{+0.14}_{-0.12}$	$86.9^{+1.6}_{-1.2}$	20

References for Table 3.1:

1: Torres et al., 2008, *ApJ*, **677**, 1324–42
2: Knutson et al., 2007, *ApJ*, **655**, 564–75
3: Southworth et al., 2007, *MNRAS*, **379**, L11–15
4: Boisse et al., 2009, *A&A*, **495**, 959–66
5: Winn et al., 2007, *AJ*, **133**, 1828–35
6: Nutzman et al., 2009, *ApJ*, **692**, 229–35
7: Winn et al., 2009, *ApJ*, **693**, 794–803
8: Pál et al., 2010, *MNRAS*, **401**, 2665–74
9: Pont et al., 2009, *A&A*, **502**, 695–703
10: Fossey et al., 2009, *MNRAS*, **396**, L16–20
11: Southworth et al., 2009, *ApJ*, **707**, 167–72
12: Hellier et al., 2009, *Nature*, **460**, 1098–100
13: Hellier et al., 2009, *ApJ*, **690**, L89–91
14: Dittmann et al., 2009, *ApJ*, **699**, L48–51
15: Bakos et al., 2010, *ApJ*, **710**, 1724–45
16: Joshi et al., 2008, *MNRAS*, **392**, 1532–8
17: Winn et al., 2008, *ApJ*, **683**, 1076–84

18: Latham et al., 2009, *ApJ*, **704**, 1107–19
19: Johnson et al., 2008, *ApJ*, **686**, 649–57
20: Skillen et al., 2009, *A&A*, **502**, 391–4

Table 3.1 summarizes selected published planet transit results. More exhaustive data can be found at various locations on the internet. Despite the rather pessimistic discussion above concerning the error estimates, it is truly amazing that we can measure the radii *of planets that we can't even see* to such precision. These results are one of the great triumphs of modern astrophysics.

Summary of Chapter 3

1. The four contact points of a transit occur when one of the limbs of the planet crosses one of the limbs of the host star.

2. The impact parameter, b, is the closest approach of the centre of the planet's disc to the centre of the star's disc, and is given by

$$b = a \cos i. \qquad \text{(Eqn 3.2)}$$

3. For a circular orbit, the duration of a transit viewed at orbital inclination i is

$$T_{\text{dur}} = \frac{P}{\pi} \sin^{-1}\left(\frac{\sqrt{(R_* + R_\text{P})^2 - a^2\cos^2 i}}{a}\right). \qquad \text{(Eqn 3.4)}$$

4. The optical depth at frequency ν is the path integral of the density, $\rho(s)$, multipled by the opacity, κ_ν:

$$\tau_\nu = \int_\text{X}^{\infty} \rho(s)\, \kappa_\nu \, \text{d}s. \qquad \text{(Eqn 3.6)}$$

Radiation of intensity I_emitted is attenuated when passing through material of optical depth τ_ν:

$$\frac{I}{I_\text{emitted}} = \text{e}^{-\tau_\nu}. \qquad \text{(Eqn 3.7)}$$

5. Stellar discs appear limb darkened, with maximum intensity at the centre. Radiation travelling radially outwards has escaped, on average, from deeper, hotter layers than radiation travelling obliquely, so the centre of the disc is generally the most blue. Consequently, limb darkening is stronger at shorter wavelengths. Limb darkening depends on wavelength, structure of the star and detailed composition, so varies from star to star. It can be directly measured only for the Sun, and is calculated using computational models for other stars. The common limb darkening 'laws' adopted for the intensity emitted at an angle $\cos^{-1}\mu$ to the stellar radius vector are the linear law

$$\frac{I(\mu)}{I(1)} = 1 - u(1-\mu), \qquad \text{(Eqn 3.9)}$$

the logarithmic law

$$\frac{I(\mu)}{I(1)} = 1 - u_l(1-\mu) - \nu_l \mu \ln \mu, \qquad \text{(Eqn 3.10)}$$

and the quadratic law

$$\frac{I(\mu)}{I(1)} = 1 - u_{\rm q}(1-\mu) - \nu_{\rm q}(1-\mu)^2. \qquad \text{(Eqn 3.11)}$$

6. Due to limb darkening, transit light curves are not generally flat-bottomed. Consequently,

$$\frac{\Delta F}{F} = \frac{R_{\rm P}^2}{R_*^2} \qquad \text{(Eqn 1.18)}$$

provides only an estimate of the flux deficit, ΔF, during transit. High signal-to-noise ratio light curves allow the measurement of the duration of ingress, t_{1-2}, and egress, t_{3-4}, and the curvature of the transit floor, as well as the duration, $T_{\rm dur}$, and the flux deficit at mid-transit, ΔF. Planet parameters are determined by model fits to the light curve, and the best fit should simultaneously reproduce all aspects of the observed transit.

7. The locus of the centre of the planet's disc relative to the centre of the star's disc is a small part of the foreshortened astrocentric orbit. The distance between the centres of the two discs is

$$s(t) = a\left(\sin^2 \omega t + \cos^2 i \cos^2 \omega t\right)^{1/2}, \qquad \text{(Eqn 3.16)}$$

where

$$\omega = \frac{2\pi}{P}$$

and $t = 0$ at inferior conjunction of the planet.

8. The eclipsed area, $A_{\rm e}$, of the stellar disc is given by

$$A_{\rm e} = \begin{cases} 0 & \text{if } 1+p < \xi, \\ R_*^2\left(p^2\alpha_1 + \alpha_2 - \dfrac{\sqrt{4\xi^2 - (1+\xi^2-p^2)^2}}{2}\right) & \text{if } 1-p < \xi \leq 1+p, \\ \pi p^2 R_*^2 & \text{if } 1-p \geq \xi, \end{cases} \qquad \text{(Eqn 3.27)}$$

where

$$p = \frac{R_{\rm P}}{R_*}, \quad \cos\alpha_1 = \frac{p^2+\xi^2-1}{2\xi p}, \quad \cos\alpha_2 = \frac{1+\xi^2-p^2}{2\xi},$$

$$\xi = \frac{a}{R_*}\left(\sin^2 \omega t + \cos^2 i \cos^2 \omega t\right)^{1/2} \quad \text{and} \quad \omega = \frac{2\pi}{P_{\rm orb}}.$$

9. The flux from a star out of transit is given by

$$F = \int_0^{R_*} I(r')\, 2\pi r'\, {\rm d}r'. \qquad \text{(Eqn 3.28)}$$

During transit, the flux drops by an amount

$$\Delta F = \int_{\text{occulted area}} I(r')\, {\rm d}A. \qquad \text{(Eqn 3.30)}$$

For a limb darkened star, this integral can be evaluated using the relationship

$$\Delta F = \int_{r=0}^{r=1} I(r) \frac{\rm d}{{\rm d}r}\left[r^2 A_{\rm e}\left(R_*, \frac{p}{r}, \frac{\xi}{r}\right)\right] {\rm d}r, \qquad \text{(Eqn 3.36)}$$

where r is the normalized axial coordinate, $r = r'/R_*$. This yields the exact shape of a transit light curve once the limb darkening law, the sizes of the star and planet, and the orbital parameters P, a and i are specified.

10. A transit light curve alone cannot determine the absolute size of anything, but it can determine the ratios $R_P : R_* : a$. Kepler's third law combined with knowledge of $M_{total} \approx M_*$ and P allows a to be calculated.
11. A transit light curve fit should be consistent with values of M_* and R_* inferred from the host star spectral type. The uncertainty in the precise host star characteristics can be the largest contributor to the uncertainties in the planet parameters M_P, R_P and i.
12. The best-fit parameters are determined by χ^2 minimization. The $\pm 1\sigma$ error estimate on any single parameter is the range within which

$$\chi^2 = \chi^2_{min} + 1, \qquad \text{(Eqn 3.39)}$$

with all the other parameters held constant at their best-fit values. The χ^2 contour map for two parameters reveals whether the fit can independently constrain them. If the contours are skewed, the parameters are correlated: the fit cannot constrain them independently.

Chapter 4 The exoplanet population

Introduction

Transiting exoplanets are important because for these particular planets we can precisely measure their basic physical and orbital characteristics. We should not lose sight of the larger goal, however. We are interested in exoplanets in general because they inform our ideas about how planetary systems form and evolve, and thus increase our understanding of the formation, evolution and context of our own Solar System. The existing body of exoplanet research allows us to begin to quantify the prevalence, orbital properties and physical characteristics of the exoplanet population. It is vital to remember that all the methods for detecting exoplanets are subject to strong **selection effects**.

This chapter begins with an examination of these selection effects in Section 4.1. New exoplanets are being discovered continuously, so the known population will have grown beyond that described here by the time you read this. We give a brief summary of the properties of the known exoplanet population in Section 4.2. We examine the physics and the various parameters which determine the size of a giant planet in Sections 4.3 and 4.4. In the light of this, we consider the known radii of exoplanets in Section 4.5. Finally, in Section 4.6 we discuss inferences about the underlying Galactic exoplanet population.

4.1 Selection effects

We mentioned the selection effects for four of the detection methods in Chapter 1. We summarize them here, and add the corresponding selection effects for detection by microlensing. It is easy to describe the selection effects in prose, but rather more difficult to precisely quantify them mathematically. It is, of course, this mathematical quantification that is required to make inferences about the underlying population of exoplanets from the statistical properties of the known sample.

4.1.1 Direct imaging

This method is viable only for bright planets in distant orbits around nearby faint stars. Because large planets are brighter than small planets in the same orbit, it favours large planets, i.e. gas giant planets. Because the thermal emission from gas giant planets is partially powered by *Kelvin–Helmholtz contraction*, as we will see in Section 4.4, it is most sensitive to young planets. For the foreseeable future this method will be limited to a small subset of nearby exoplanets. Despite this, it is important, because being able to actually see the planet removes many ambiguities about the planet's properties.

4.1.2 Astrometry

Astrometry is most sensitive to massive planets in wide orbits around nearby low-mass stars. There are many astrometry projects aiming to test whether low-mass stars within a few parsecs of the Sun host planets in wide orbits, but so far only a single planet (GJ 879 b) has been confirmed by astrometry and none has been discovered (May 2009).

4.1.3 Radial velocity

The **radial velocity** (RV) method has discovered the majority of known exoplanets, so its selection effects are the most important to understand. There are a number of factors that feed into them. We have already examined the quantitative dependence of the reflex RV amplitude on the masses and orbital parameters:

$$A_{\mathrm{RV}} = \frac{2\pi a M_{\mathrm{P}} \sin i}{(M_{\mathrm{P}} + M_*)P\sqrt{1-e^2}}, \qquad \text{(Eqn 1.13)}$$

which reveals that massive close-in planets with eccentric orbits around low-mass host stars produce the biggest RV amplitudes. There are two further factors:

- The sharpness, strength and stability of the stellar spectral lines governs the precision to which an RV measurement can be made. The HARPS spectrograph at ESO has state of the art wavelength precision, stable at around $0.1\,\mathrm{m\,s^{-1}}$ over the course of a day. Stars themselves are not stable at this precision. They oscillate due to the presence of **acoustic waves**; Figure 4.1 shows some examples. The amplitudes and periods of these waves are dependent on the detailed structure and composition of the star, and generally there will be many oscillations simultaneously present, causing an evolving wave form and amplitude. Consequently, the precision for which an RV measurement can be made depends on the spectral type, structure and composition of the star as well as its apparent brightness.
- To detect a long-period planet, observations must be made over a long baseline. Excluding the preliminary estimates of the orbital periods of the planets detected in 2008 by the imaging method, the longest-period planet known (May 2009) is 55 Cnc d, which has $P_{\mathrm{orb}} = 14.3$ years. 55 Cnc is a nearby metal-rich G8V star, bright enough to be seen with the naked eye. Only three other known exoplanets have $P > 10$ years. As the baseline of observations generally increases, this number will grow.

4.1.4 Transits

As we derived in Chapter 1, the probability of a randomly oriented planet transiting its host star is given by

$$\text{geometric transit probability} = \frac{R_* + R_{\mathrm{P}}}{a} \approx \frac{R_*}{a}, \qquad \text{(Eqn 1.21)}$$

so the transit method strongly selects for close-in planets. In Chapter 2 we derived an expression, Equation 2.34, for the expected number of planet discoveries for transit search programmes. This expression quantifies some selection effects for transit search surveys. These are:

- a strong selection effect in favour of large planets in the factor R_{P}^3;
- a strong selection effect in favour of close-in planets in the factor $a^{-7/4}$;
- the ubiquitous limiting flux factor, $L_*^{3/2}$, in favour of luminous host stars;
- a selection effect in favour of small host stars through the factor $R_*^{-5/4}$.

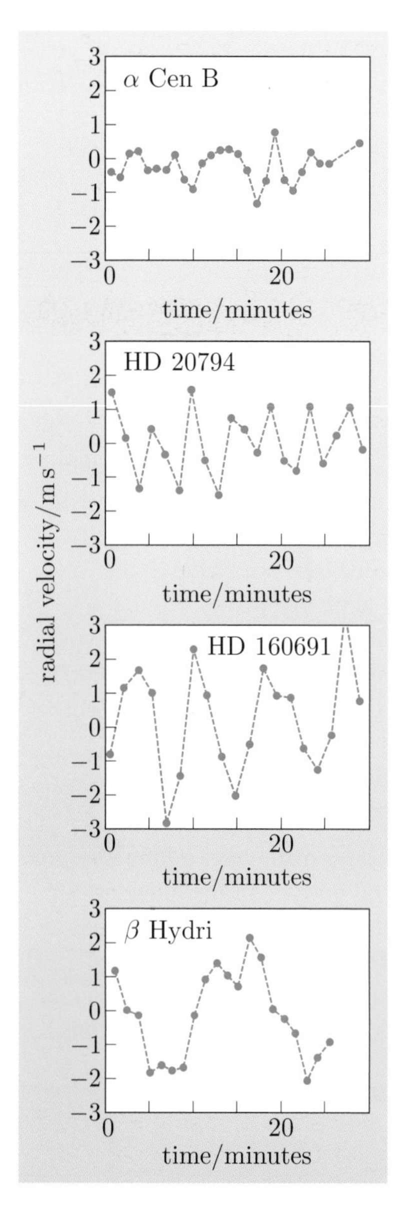

Figure 4.1 Radial velocity measurements for four stars of varying spectral type. All four show short-period oscillations that are intrinsic to the star. The periods and amplitudes are different for each of the stars shown.

Reducing Equation 2.34 to a proportionality, we have

$$\text{transit detection probability} \propto R_\mathrm{P}^3\, a^{-7/4} L_*^{3/2} R_*^{-5/4}. \tag{4.1}$$

Less quantifiable is the selection that arises as a result of crowding in wide-field transit search programmes. The WASP transit search avoids the Galactic plane because it is too densely populated to be effectively studied with wide-field cameras. Since the population of stars in the plane of the Galaxy differs from that at high Galactic latitude, this factor will bias the population of planets discovered towards that of high Galactic latitude. This bias is ameliorated by the limiting flux selection effect: most planet hosts discovered by WASP and other wide-field surveys will be brighter than V $\approx$ 12, and these are generally quite close to the Sun and the Galactic plane, as illustrated in Figure 1.21. Thus even though the WASP host stars don't include low Galactic latitude objects, they are mostly members of the **Galactic disc stellar population**.

In addition to wide-field surveys, there are a number of pencil-beam surveys, which attain the signal-to-noise ratio necessary to detect transits of stars significantly fainter than $V \approx 12$. These surveys study only a small area of sky, and each may consequently have selection effects arising from the population(s) of stars within that field. For example, CoRoT studies two fields in the Galactic plane, as illustrated in Figure 4.2.

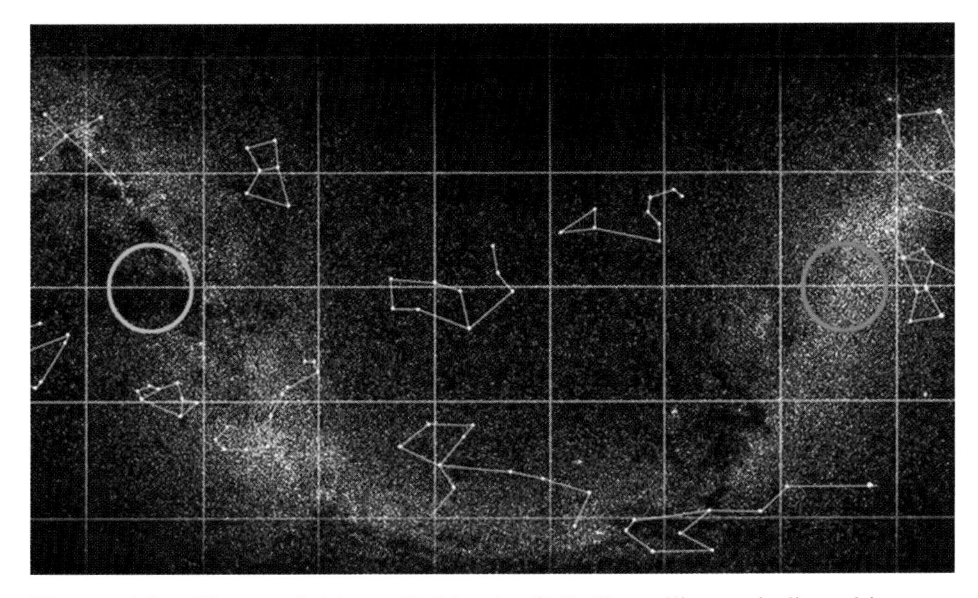

Figure 4.2 The two fields studied by the CoRoT satellite are indicated in circles. The Milky Way is the bright band of stars crossing the sky, and several major constellations are indicated.

4.1.5 Microlensing

The microlensing technique is most sensitive to planets with orbits of 1 AU or wider. Since the properties of the microlensing planets cannot generally be precisely measured or verified, these planets warrant relatively little discussion here.

4.2 The demographics of the known exoplanets

It is a fairly straightforward exercise to examine the published properties of the known exoplanets, particularly if the data can be downloaded from a homogeneous source on the internet. We have generated figures from data provided (in November 2009) by the Exoplanet Encyclopaedia internet site, and used these to generate many of the figures that appear in this section. We referred to the primary literature to check the discovery methods and augment the data. We imposed a limit $M_P \sin i < 13\,M_J$ in the sample that we discuss. It is important to realize (a) that these data are subject to all the selection effects discussed in Section 4.1 above, and (b) that these demographics will change as more planets are discovered. Only by understanding and fully correcting for the selection effects can we draw conclusions about the properties of the Galaxy's population of planets.

This limit is in accordance with the IAU definition of a planet and excludes objects thought to have ignited deuterium burning.

4.2.1 Known population as a function of discovery method

Figure 2.18 showed the history of planet discoveries by each method. In Figure 4.3 we compare the demographics of the host stars of these exoplanets. The first panel shows the apparent brightness distribution, which is probably the dominating factor in how easy it is to study a particular star; the second two panels show the mass and **metallicity** distributions of the planet host stars. These two characteristics play strong roles in selection effects and are of prime importance in models of planet formation.

Figure 4.3a shows that on average the planet host stars are brightest for the imaging method, decreasing in apparent brightness for the RV, transit and timing methods. The peak in the host star number at V $\approx$ 12 for the transiting method is because this is the limiting magnitude for WASP and similar wide-field survey techniques. This illustrates some of the selection effects inherent in these three detection methods: each picks out a different subset of the exoplanet host star population. The most noticeable feature in Figure 4.3b, namely the peak at $M_* \approx 1\,\mathrm{M_\odot}$, is a result of the selection of solar-type stars as the preferred targets of RV surveys. The transiting planet distribution does not show a peak at $M_* \approx 1\,\mathrm{M_\odot}$, but each bin has a rather low population.

● What is the expected random fluctuation in a number count, n, in a histogram bin?

❍ $\sqrt{n}$, from Poisson statistics.

● So which bins have the biggest fractional uncertainties, σ_n/n?

❍ Low number count bins, since $\sigma_n/n = 1/\sqrt{n}$.

● The transit method has monitored more stars than the RV method, so why are there more known RV planet hosts at V $\approx$ 8 than there are known transiting planet hosts?

❍ There are not all that many main sequence stars in the sky with V $\approx$ 8. Only about one in a thousand is expected to host a transiting giant planet, so only very few transiting planet hosts with V $\approx$ 8 are expected to exist.

● Each of the transiting planets in the sample has been confirmed by RV work, and the probability of a randomly oriented orbit transiting is $\lesssim 10\%$, so why are there so few known RV planets at $10 < \mathrm{V} < 14$?

❍ It is possible to make RV detections down to these magnitudes. However, no one has been prepared to invest the huge amount of telescope time required to execute an RV survey of these relatively faint stars.

Figure 4.3c shows the distribution of the metallicity for the known host stars. The index used to quantify metallicity generally in astrophysics is a logarithmic indicator relative to the Sun:

$$\left[\frac{\mathrm{Fe}}{\mathrm{H}}\right] = \log_{10}\left(\frac{\mathrm{Fe}}{\mathrm{H}}\right)_* - \log_{10}\left(\frac{\mathrm{Fe}}{\mathrm{H}}\right)_\odot, \qquad (4.2)$$

where the quantities within the round brackets on the right-hand side are the numbers of iron (Fe) and hydrogen (H) atoms present. The units of the quantity [Fe/H] are referred to as **dex**, and the square bracket indicates the logarithmic

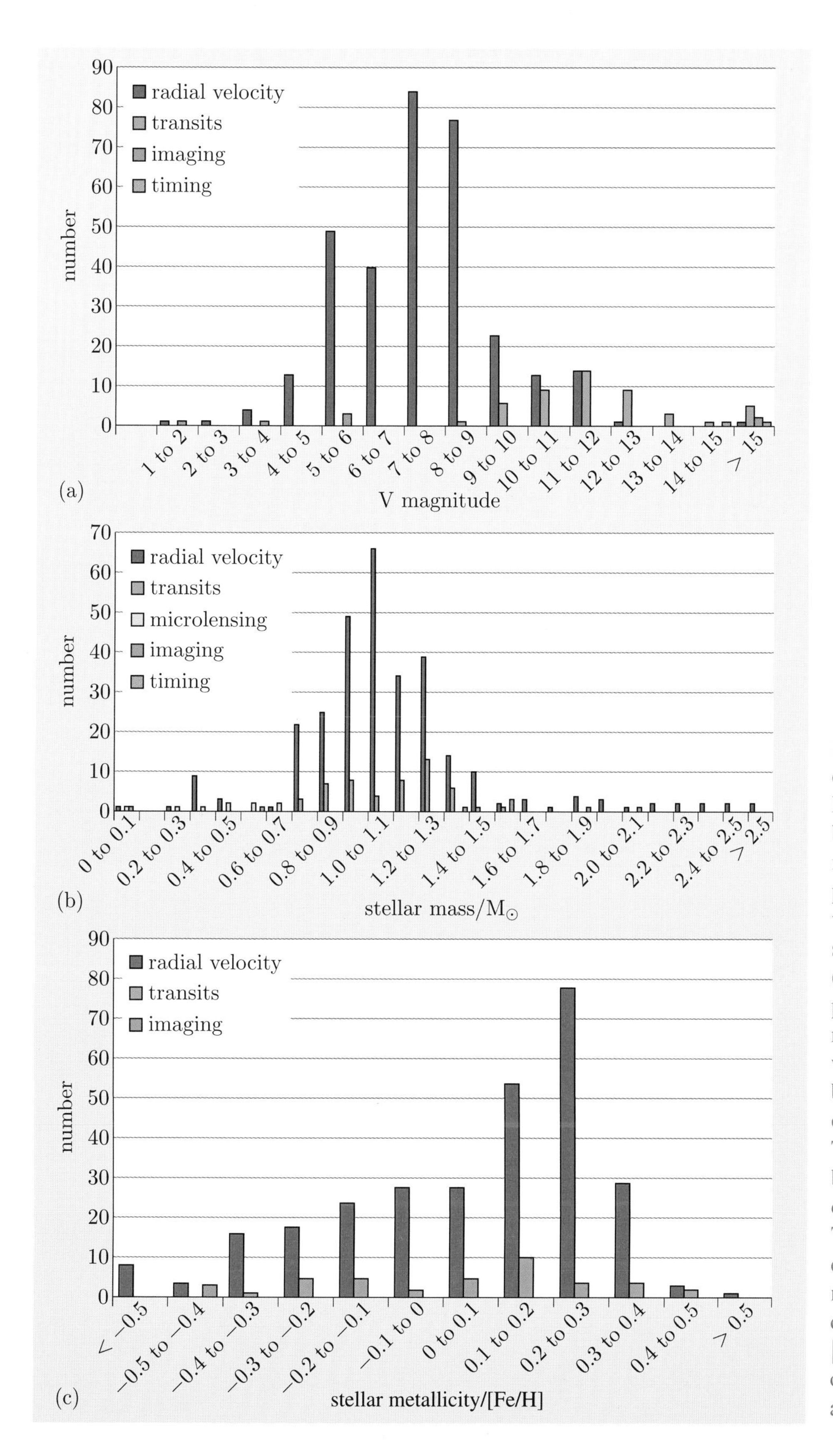

Figure 4.3 Demographics of exoplanet host stars planets known in November 2009, broken down by discovery method. 'Timing' refers to the planets discovered by timing variations of pulsars, pulsating stars or eclipsing binary stars. (a) The V magnitude of known planet hosts. The RV technique requires rather bright stars, while the transit technique can be applied to much fainter objects. (b) The masses M_*. The large peak at $M_* \approx 1\,M_\odot$ is because the RV surveys initially chose Sun-like stars as targets. The peak in the transiting planet distribution is at a slightly higher mass. (c) The stellar metallicity distribution. The quantity [Fe/H] is a logarithmic indicator of the metal content of the stellar atmosphere relative to the Sun.

scale relative to the Sun. Thus a star with [Fe/H] $= 0.3$ has roughly twice the fractional metal content of the Sun, and a star with [Fe/H] $= 0.5$ has roughly three times the fractional metal content of the Sun. Conversely, a star with [Fe/H] $= -0.3$ has roughly half the fractional metal content of the Sun.

It appears that the RV host stars are more metal-rich than the transiting planet host stars. There is no obvious bias in the transit method that might skew the detected host population towards stars of any particular metallicity. We should note, however, that as illustrated in Figure 1.21, the transit search programmes are sampling a bigger volume of the Galaxy than the RV surveys. The effect may arise because the local volume containing the majority of RV host stars is more metal-rich than the larger volume surrounding it. Relatively few stars other than the exoplanet host stars have been studied in sufficient detail to address this. We will return to discuss the metallicity distribution of the RV hosts in Subsection 4.2.2.

- What is the metal content of a star with [Fe/H] $= -0.5$?

- Roughly a third of the fractional metal content of the Sun.

Figure 4.4a shows the distribution of planet masses for the discoveries by the various methods. Figure 4.4b shows the same data, but now they are binned to equal intervals in $\log(M_P/M_J)$. This produces more bins at low masses and fewer bins at high masses. Rather than equal intervals, there are equal factors between adjacent bins. Different aspects of the distribution are highlighted by the two ways of displaying these data.

Note that the RV method gives merely *minimum* planet masses, i.e. $M_P \sin i$, and this contributes to the increasing RV population at very low masses. Figure 4.4b shows the RV population peaks at higher masses than the transit population, which is probably because the RV technique preferentially detects massive planets, while the transit method preferentially detects large planets — these are not necessarily the same planets, as we will see in Section 4.5. The most massive planets are the planets detected by imaging because these have the greatest thermal emission (cf. Subsection 4.3.3) and are hence the most likely planets to have contrast ratios small enough to be directly detectable despite the stellar light.

Figure 4.5 shows that the RV technique has found planets at all separations out to beyond 5 AU with a peak in numbers at about 1 AU. Conversely, the imaging technique has been successful only for large separations, while the transiting technique has been successful only at the very smallest separations.

Finally, Figure 4.6 shows the distribution of orbital periods. The transiting planets are found, as expected, only at the shortest orbital periods.

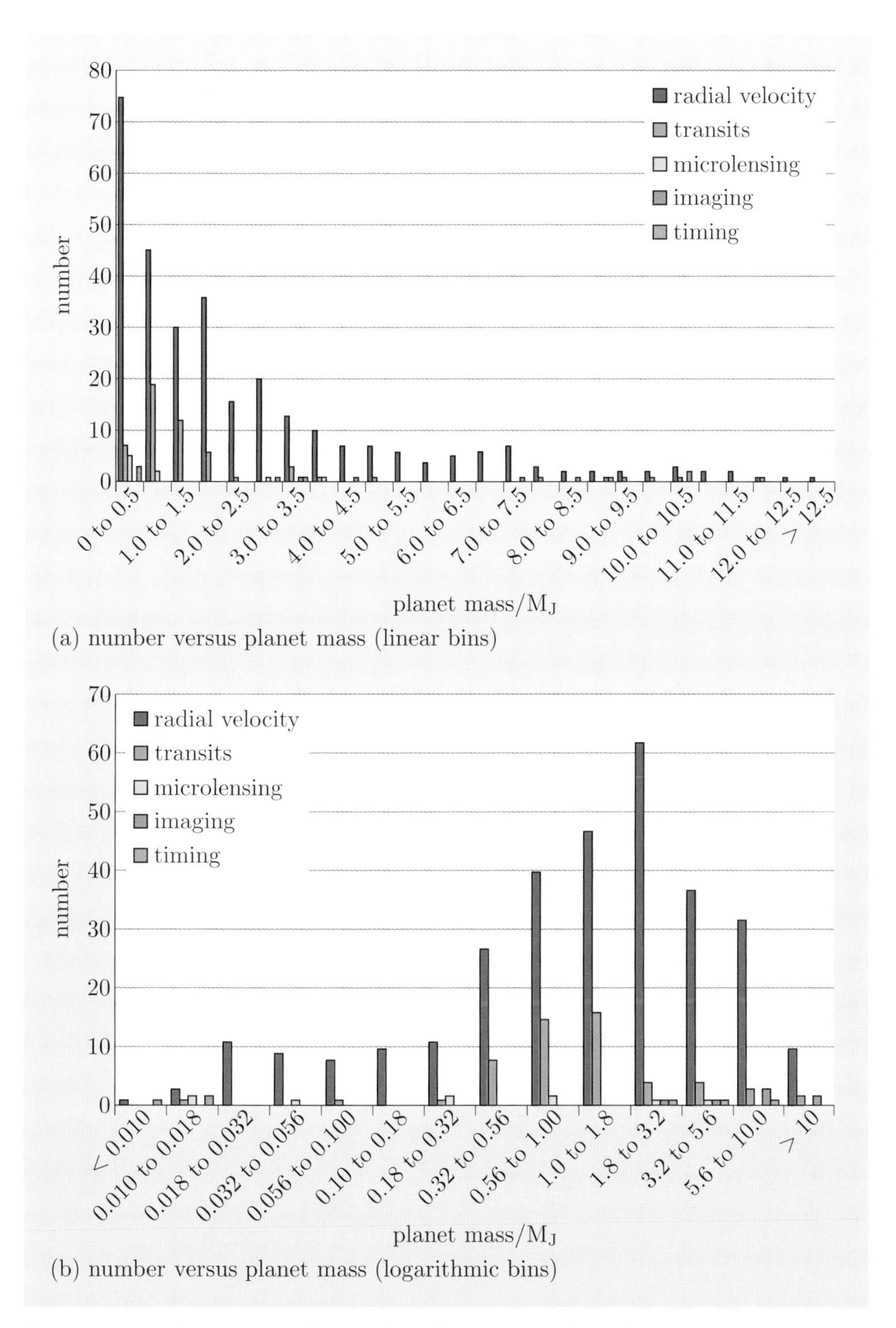

Figure 4.4 Planet mass distribution (November 2009), broken down by discovery method. (a) The number of planets of mass M_P, or minimum masses $M_P \sin i$ for the RV method, in equally spaced mass bins. (b) The same data shown in logarithmically spaced mass bins. There is a factor of $10^{1/4}$ in mass between adjacent bins: between every fourth bin there is a factor of 10 in mass.

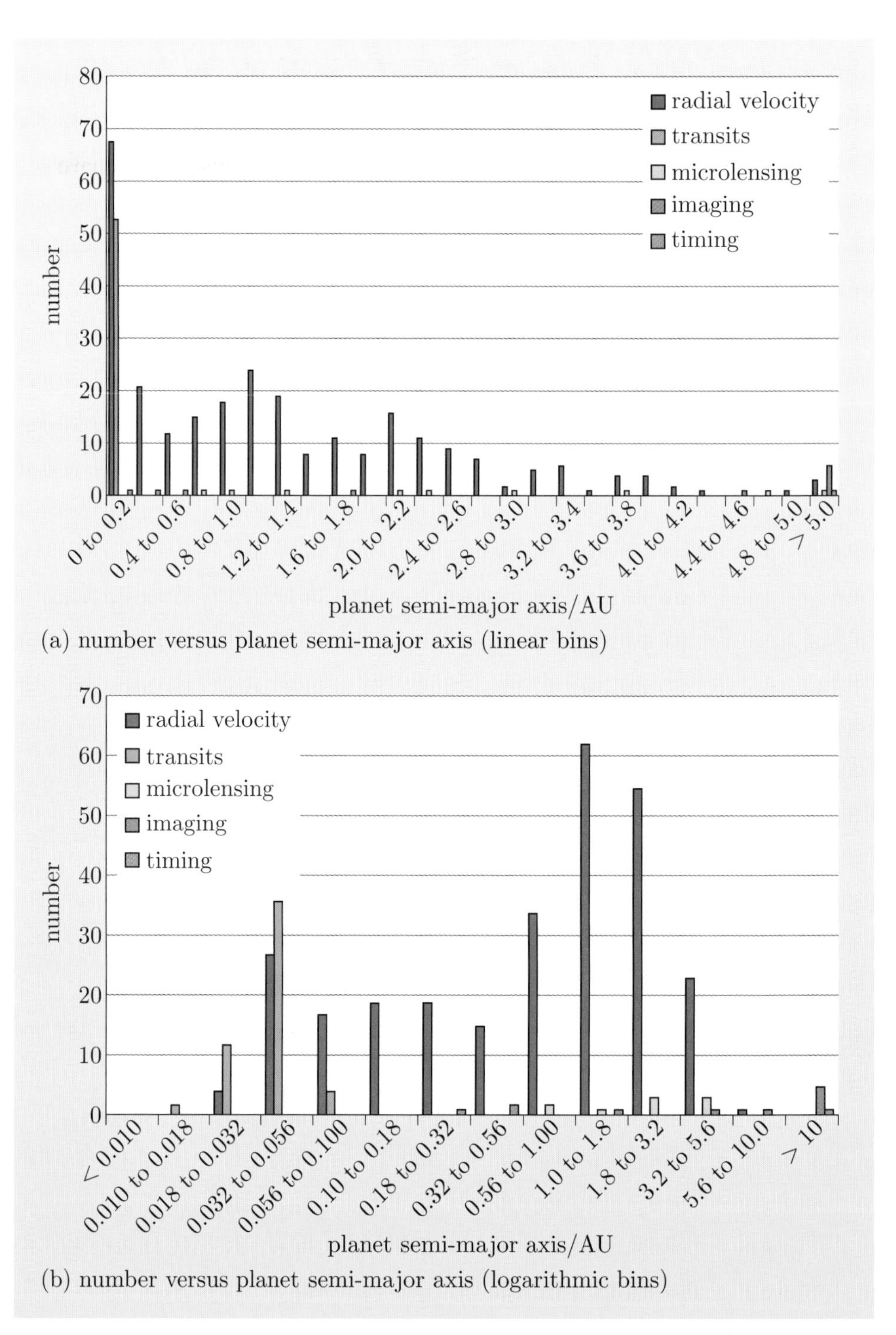

Figure 4.5 The planet orbital semi-major axis distribution (November 2009), broken down by discovery method. (a) The number of planets with semi-major axis a, in equally spaced bins. (b) The same data shown in logarithmically spaced bins. There is a factor of $10^{1/4}$ in orbital semi-major axis between adjacent bins: between every fourth bin there is a factor of 10 in a.

- Why are the planets detected by imaging all in orbits with large semi-major axes?

- Because large separations between the star and planet are needed to resolve the planet from the star, even for nearby stars.

- Why are there no known transiting exoplanets at $a > 0.5$ AU?

❍ The probability of a given planet transiting is strongly dependent on a, with close-in planets being much more likely to transit. Furthermore, it is far easier to detect transits of short-period planets, which of course also tend to have small semi-major axes.

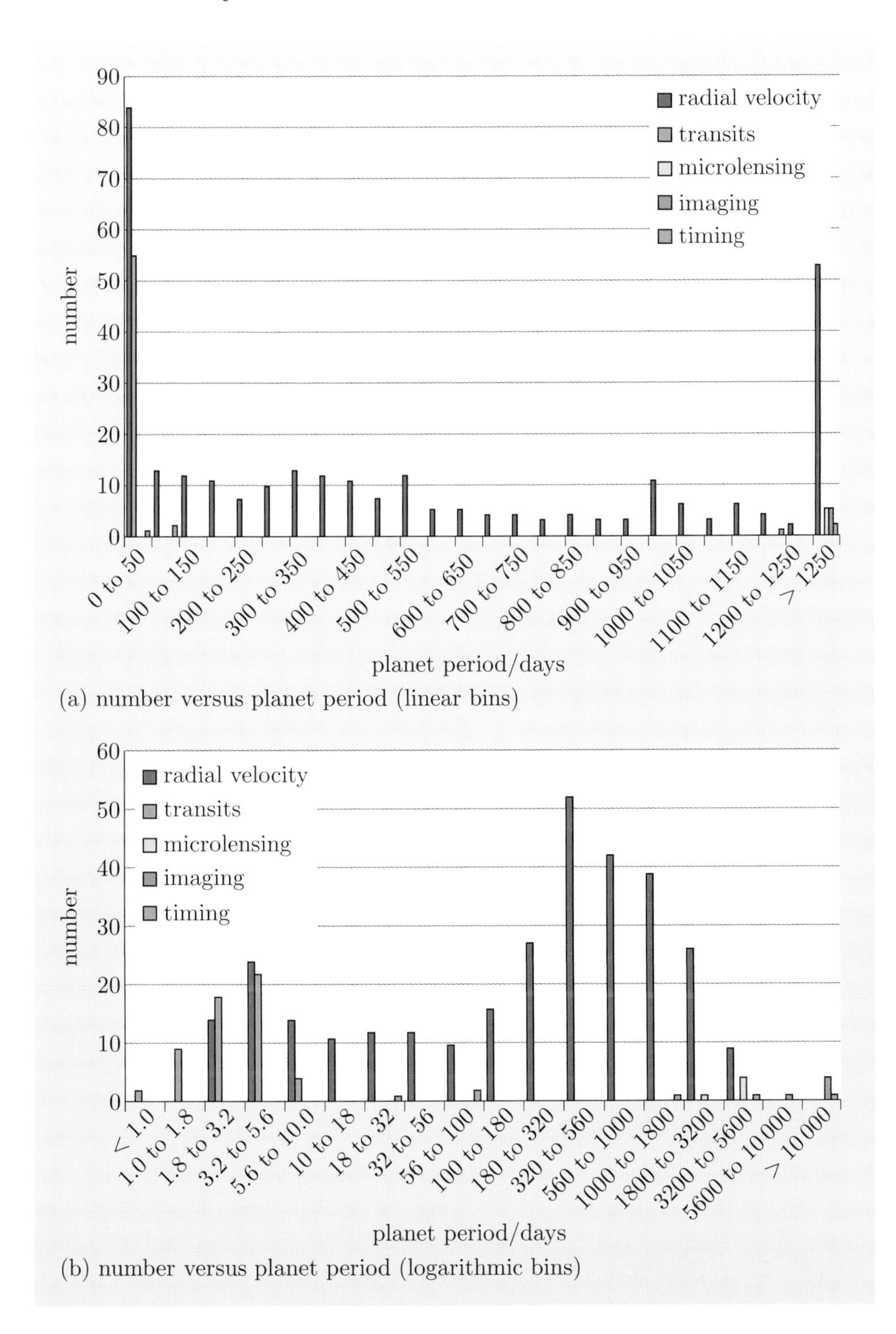

Figure 4.6 The planet orbital period distribution (Nov. 2009), broken down by discovery method. (a) Number of planets with period P, in equally spaced bins. (b) The same data shown in logarithmically spaced bins. There is a factor of $10^{1/4}$ in orbital period between adjacent bins: between every fourth bin there is a factor of 10 in P.

Figure 4.7 illustrates the selection effects in a different way, showing a scatter graph of planet mass, M_P, versus apparent host star magnitude, V. It is obvious in this figure that planets detected by the RV, transiting and imaging methods lie in distinct regions of the graph. As expected, the planets detected by imaging are found primarily around (apparently) bright stars: these are nearby stars, whose known planets have large angular separation from the host star and themselves have large apparent brightness. The planets detected by the RV method are concentrated among host stars with V $\sim$ 9 or brighter: the brighter stars in the skies are chosen for RV studies. The transiting planets' host stars, conversely, are fainter, with V > 8 almost without exception. This is because any given star has a relatively small chance of hosting a transiting planet, and there are a limited number of bright stars in the sky. The wide-field transit surveys have limiting magnitudes V ~ 12, so the known population peaks at around this apparent brightness. Interestingly, as we have already discussed, the planet mass distributions also appear different for the three populations.

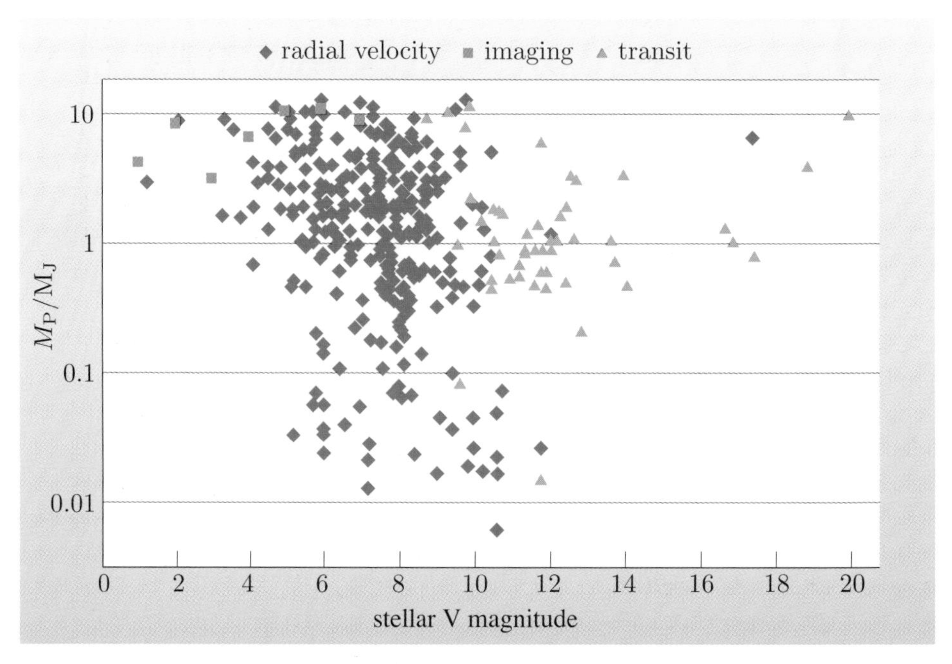

Figure 4.7 Mass of planet versus V magnitude of host star for the RV discovery method, transit and imaging discovery methods. The transiting planets that were first discovered by the RV method belong to the RV population. The RV method has concentrated on brighter host stars, and, on average, finds more massive planets.

Exercise 4.1 (a) According to Figure 4.5, the typical semi-major axis of a planetary orbit among the known exoplanets is $a = 1$ AU. We know, for the Sun, that the orbital period of a planet at $a = 1$ AU is exactly 1 year. Calculate the orbital period of a planet at $a = 1$ AU for stars of mass (i) $M_* = 0.7\,\mathrm{M}_\odot$ and (ii) $M_* = 1.5\,\mathrm{M}_\odot$.

(b) According to Figure 4.6, the typical orbital period among the known exoplanets is $P = 500$ days. Calculate the semi-major axis of the planet's orbit for $P = 500$ days and host star masses of (i) $M_* = 0.7\,\mathrm{M}_\odot$ and (ii) $M_* = 1.5\,\mathrm{M}_\odot$.

(c) If the star were a late M type star with $M_* = 0.2\,M_\odot$, what would be (i) the orbital period of a planet in an orbit with $a = 1$ AU and (ii) the semi-major axis of a planetary orbit with $P = 500$ days?

(d) Based on your answers to parts (a)–(c), how good an indication of the orbital period is a determination of the semi-major axis, and vice versa? ■

4.2.2 The radial velocity planets

As Figures 4.3, 4.4, 4.5 and 4.6 show, the majority of known exoplanets were discovered by the RV technique. With hundreds of exoplanets, this RV population does reveal some facts about the exoplanet population. Thousands of stars have been monitored for reflex RV variations, and consequently it is now clear (October 2008) that $5.9 \pm 1.2\%$ of the main sequence stars in the solar neighbourhood of spectral type F, G or K harbour a giant planet with orbital period shorter than 2000 days and $M_P \sin i > 0.5\,M_J$. Of these giant planets, $9.1 \pm 1.8\%$ have orbital periods shorter than 5 days. Integrating over the distributions, these statistics imply that there is 1 transiting giant planet for every 1350 ± 350 F, G and K dwarf stars.

Figure 4.8 illustrates three properties of the RV population. Figure 4.8a shows that the number of planets in a logarithmically spaced bin rises to a peak for semi-major axes between 1 AU and 2 AU. The decline at larger separations is due to the long orbital periods of planets with these large orbits: few stars have RV measurements with a baseline long enough to detect these planets. Figure 4.8b shows the proportion of stars harbouring RV-detected planets as a function of stellar metallicity. There is a prominent and statistically highly significant rising fraction of detected RV planet hosts as metallicity increases. It seems straightforward to conclude that more metal-rich stars are more likely to harbour planets. An obvious caveat to this conclusion is that the selection effects for the detectability of RV modulations favour metal-rich stars because metal-rich stars have generally stronger photospheric absorption lines, making the RV measurement easier. However, detailed quantitative comparison of the actual measurement uncertainties suggests that the slightly lower RV precision ($\sim$1–2 m s^{-1} poorer precision) for metal-poor stars cannot be the cause of the strong effect seen in Figure 4.8b. The high **planeticity** of metal-rich stars was noticed as early as 1997. Consequently, some researchers sensibly concentrated their planet-hunting effort on metal-rich stars, so these are over-represented in the RV demographics in Figure 4.3. This sort of *literal* selection effect is very difficult to quantify. Figure 4.8b gets around this bias by examining the *fraction* of stars hosting planets rather than the *number* of stars hosting planets.

Planeticity means the abundance of, or prevalence of, or probability for hosting planets.

The final panel of Figure 4.8 indicates that RV planet detection rates are highest for stars with spectral type F5–A5, and decrease as stellar mass decreases. There have been theoretical studies of planet formation that suggest that the density of solids at the mid-plane of a protoplanetary disc is likely to correlate with the mass of the central star, so that planeticity should correlate with stellar mass. Figure 4.8c supports this idea, but we caution that it may be biased by selection effects. We discussed the distribution of planet frequency over semi-major axis and planet mass, $\alpha_P(a, M_P)$, in Chapter 2. Eqn. 2.26 gave the total number of planets per star, η_P, in terms of $\alpha_P(a, M_P)$ and Eqn. 2.28 gave the expected

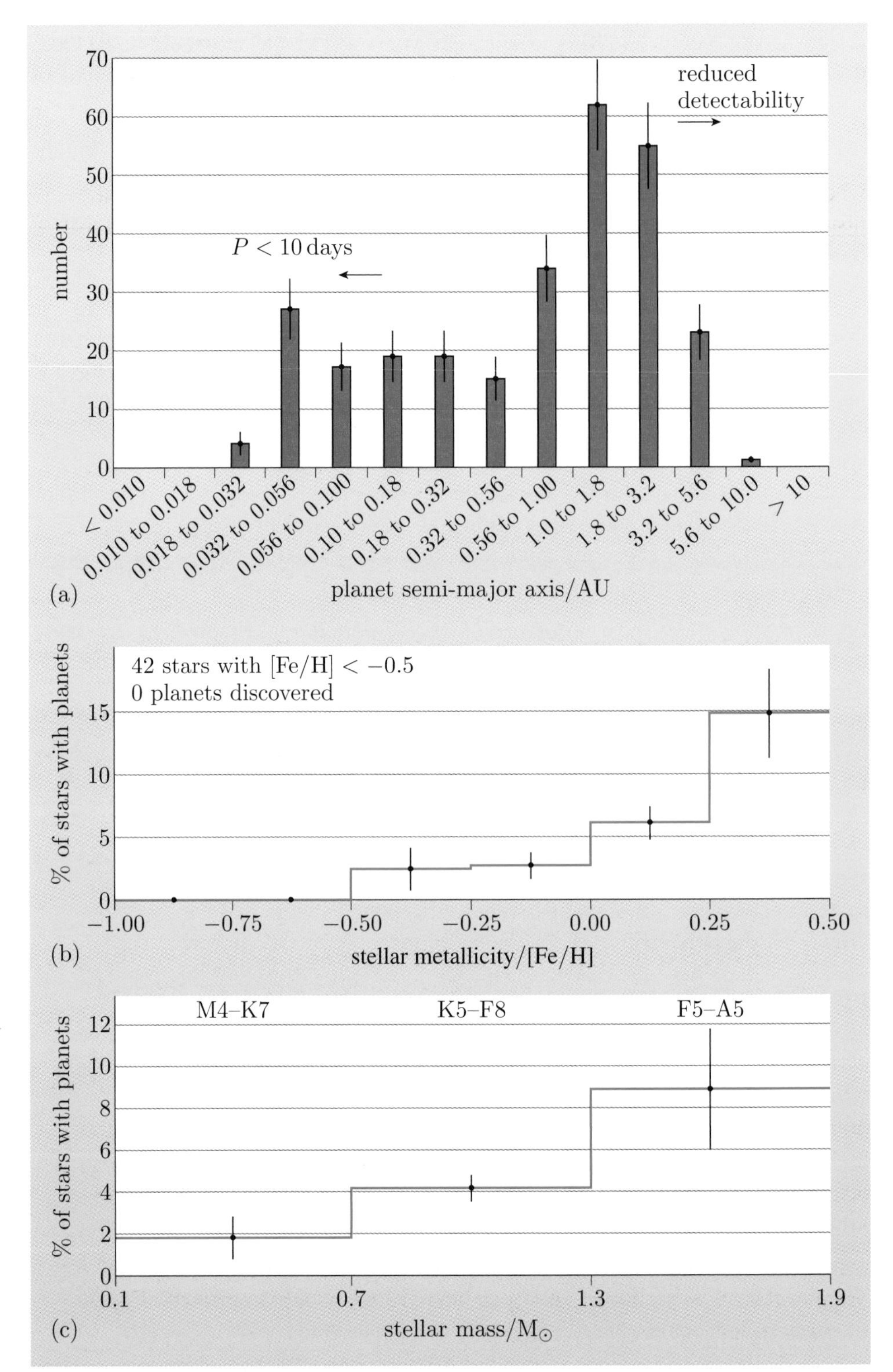

Figure 4.8 (a) The distribution of planets discovered by the Doppler or RV technique as a function of semi-major axis. The horizontal axis is logarithmic. (b) The percentage of stars harbouring RV-detected planets as a function of stellar metallicity. (c) The percentage of stars harbouring RV-detected planets as a function of stellar spectral type.

number of transiting planet discoveries in terms of $\alpha_\text{P}(a, M_\text{P})$, expressing the result as $\text{d}N_\text{P,trans}/\text{d}a\,\text{d}M_\text{P}$. The histograms in Figure 4.8 show conceptually related quantities: Figure 4.8a is a graph of $\Delta N_\text{P,RV-known}/\Delta \log a$, where

$N_{\text{P,RV-known}}$ is the total number of planets discovered by the RV method. The sum over the logarithmically spaced bins gives this total number:

$$N_{\text{P,RV-known}} = \sum \frac{\Delta N_{\text{P,RV-known}}}{\Delta \log a} \Delta \log a, \tag{4.3}$$

where the sum is performed over all the bins, each with width $\Delta \log a$, in the histogram. If we allow the bin size to decrease, this will tend to an integral over a distribution:

$$N_{\text{P,RV-known}} = \int \frac{\mathrm{d}N_{\text{P,RV-known}}}{\mathrm{d}\log a} \,\mathrm{d}\log a. \tag{4.4}$$

We could also consider a two-dimensional distribution, sorting the known RV planets by semi-major axis and by mass, in which case we would have

$$N_{\text{P,RV-known}} = \int_{\log a} \int_{\log M_{\text{P}}} \frac{\mathrm{d}N_{\text{P,RV-known}}}{\mathrm{d}\log a \,\mathrm{d}\log M_{\text{P}}} \,\mathrm{d}\log a \,\mathrm{d}\log M_{\text{P}}, \tag{4.5}$$

just as we had in Equation 2.29 with linearly rather than logarithmically spaced bins. The nomenclature is rather complicated, but the underlying mathematics is straightforward.

- The histogram in Figure 4.8a has an uncertainty ('error') bar on each bin. How do you think these were calculated?

- Each bin gives the number, $\Delta N_{\text{P,RV-known}}$, falling into that bin. These number counts are subject to Poisson statistics, so the uncertainty in each case is given by $\sqrt{\Delta N_{\text{P,RV-known}}}$.

- The histograms in Figure 4.8b,c show the percentage of stars in each bin that have planets detected by RV. How do you think the error bars were calculated in these cases?

- Each bin reports the percentage based on a total number, N_{total}, observed that fall in the bin. Of these, N_{planet} were found to have planets, and the Poisson error on this is $\sqrt{N_{\text{planet}}}$. The uncertainty on the percentage is therefore $(\sqrt{N_{\text{planet}}}/N_{\text{total}}) \times 100$.

Exercise 4.2 Use Figure 4.8c as a starting point, and draw on what you have already learned, to do the following.

(a) Make a quantitative estimate of the fraction of main sequence stars within a distance of 10 parsecs that are likely to host one or more giant planets. Explain your reasoning.

(b) Similarly, make a quantitative estimate of the fraction of main sequence stars brighter than V $\sim$ 10 that are likely to host one or more giant planets. Explain your reasoning.

(c) Are your two estimates the same or different? If they differ, why do they differ?

(d) Compare your answer in part (a) with the results shown in Figure 1.4, and comment on how many undiscovered giant planets are likely to exist within 10 parsecs. ■

Figure 4.9 shows the mass distribution ($M_P \sin i$) of the RV planets. This is the RV data in Figure 4.4, binned differently. The distribution of planet masses rises strongly towards low masses, and the relationship

$$\frac{dN_{P,RV\text{-known}}}{dM_P} \propto M_P^{-1.6} \tag{4.6}$$

provides a good fit to the histogram.

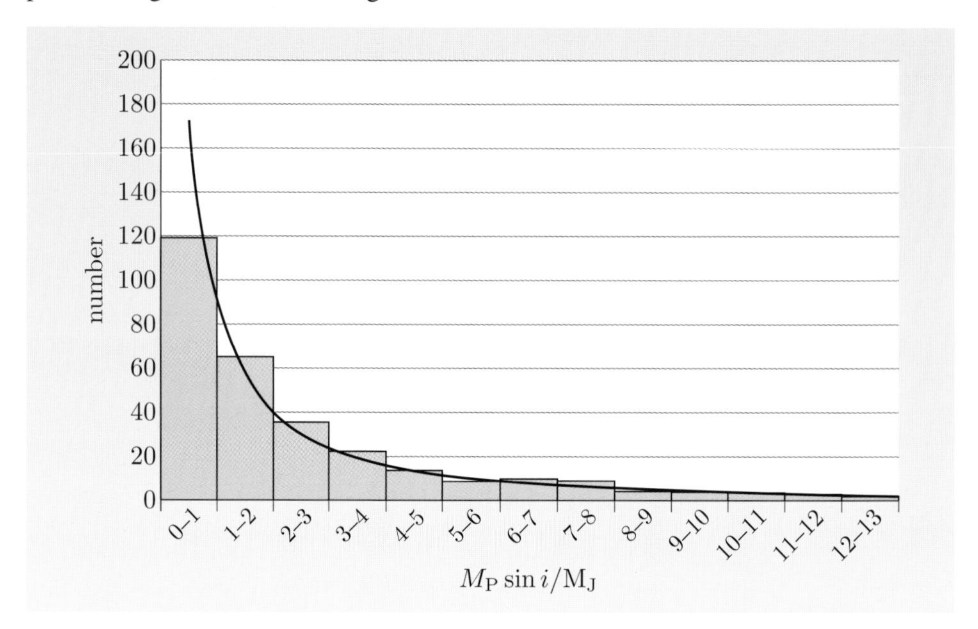

Figure 4.9 The $M_P \sin i$ distribution of planets discovered by the Doppler or RV technique. The distribution rises strongly towards low masses, and is approximated well by Equation 4.6.

- Which planets are easiest to find via the RV technique, low-mass planets or high-mass planets?

- High-mass planets produce a larger reflex RV amplitude, and are therefore easier to find.

- If the actual population of nearby exoplanets contained an equal number of planets in each of the histogram bins in Figure 4.9, how would you expect the detected subset of this population to be distributed?

- The more massive planets would be preferentially detected, so the distribution would rise to higher masses.

- What can we conclude from the distribution shown in Figure 4.9?

- There must be more planets with low values of $M_P \sin i$ than there are with high values of $M_P \sin i$.

Note that the (unknown) factor $\sin i$ does not affect the relative numbers of planets in low-mass and high-mass bins. The orientations of the orbits of all exoplanets are assumed to be random, therefore on average, exoplanets in each mass bin should be subject to the same average value of $\sin i$. The net effect is therefore to reduce the value of $M_P \sin i$ relative to the actual mass M_P in all bins by the same average factor. Consequently, Figure 4.9 implies that the underlying exoplanet mass distribution rises towards low planet masses. Initially, when only a few planets have been detected by the RV method, their masses are subject only to a lower limit

$$M_P > A_{RV} \left(\frac{M_*^2 P}{2\pi G}\right)^{1/3}, \tag{4.7}$$

where we have replaced the unknown value $0.0 \leq \sin i \leq 1.0$ in Equation S2.6 with the inequality. From the empirical results on a handful of systems, the constraint in Equation 4.7 means that it is impossible to say what the mass of any individual planet is unless the value of i can be determined somehow. In fact, it is impossible to say whether any individual RV planet *is* definitely a planet: it could be a brown dwarf in a nearly face-on orbit. Despite this, if we consider the results of a large number of planets and make the assumption that their orbits are randomly oriented, we can make a statistical correction to the mass distribution. In this way, we are sure that the objects detected by the RV surveys are predominantly genuine planets.

A third version of the exoplanet mass distribution is shown in Figure 4.10. This figure uses a logarithmic horizontal axis, like Figure 4.4b, and thus makes clear the drop-off in number of detected exoplanets at very low masses.

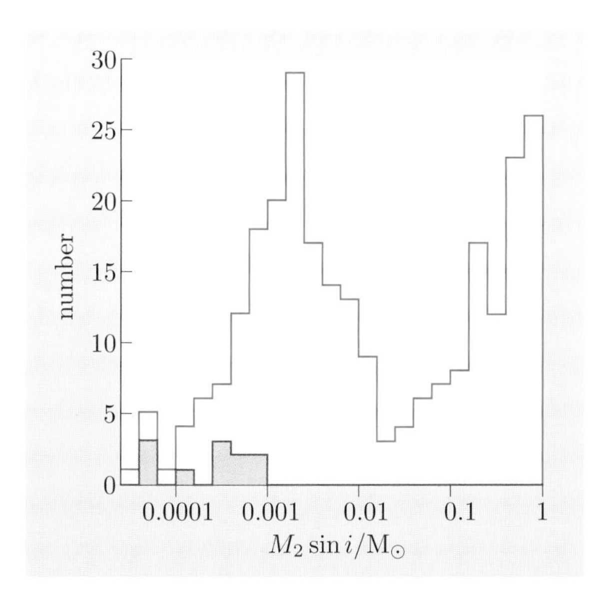

Figure 4.10 The distribution of secondary masses for primaries of spectral type later than F5, plotted on a logarithmic scale in solar-mass units. $1\,\mathrm{M_J} = 0.000\,95\,\mathrm{M}_\odot$, so the maximum just above $0.001\,\mathrm{M}_\odot$ occurs at just above $1\,\mathrm{M_J}$. The graph shows two broad peaks, the first comprised of planets, the second comprised of the secondary stars in stellar binaries. The blue-shaded subset of low-mass planets comprises the planets discovered by the state-of-the-art HARPS spectrograph.

This drop-off is, of course, due to the increasing difficulty of detecting planets as planet mass decreases. Figure 4.10 shows the mass distribution of companions to stars of spectral type F5 or later. Such stars are often referred to as **solar-type** stars, but we will avoid this term as it can be used with a variety of different definitions. The high-mass end of Figure 4.10 is comprised of the stellar companions in binary star systems; at the low-mass end the companions are planets. The small number of low-mass planets that are blue-shaded in the graph are the discoveries by HARPS, an instrument that has a ground-breakingly high RV precision. At the time when Figure 4.10 was made (circa 2007), only a small number of stars had been studied with HARPS; as the number of stars surveyed with this precision increases, we anticipate many more discoveries of exoplanets with mass less than that of Saturn ($\sim 3 \times 10^{-4}\,\mathrm{M}_\odot \approx \frac{1}{3}\,\mathrm{M_J}$). In April 2009 astronomers announced the discovery (using HARPS) of the planet Gliese 581 e, which has a mass only about twice that of the Earth. The measured value of $M_\mathrm{P} \sin i$ is $1.9\,\mathrm{M}_\oplus = 0.0061\,\mathrm{M_J} = 5.8 \times 10^{-6}\,\mathrm{M}_\odot$ for this planet, where the symbol $\mathrm{M}_\oplus$ denotes the mass of the Earth.

Mass units

Astronomers generally adopt conveniently-sized units for the phenomena that they discuss, as SI units are often orders of magnitude too small. We have used the mass of the Sun, $M_\odot$, and the mass of Jupiter, M_J, throughout this book, and have just started discussing exoplanets with masses significantly below that of Jupiter. For convenience we gather in Table 4.1 the conversion factors between these mass units.

Table 4.1 The mass units used in this book, and their conversion factors.

Mass unit	$M_\odot$	M_J	$M_\oplus$	kg
$M_\odot$	$1\,M_\odot$	$1050\,M_J$	$3.33 \times 10^5\,M_\oplus$	1.99×10^{30} k
M_J	$9.55 \times 10^{-4}\,M_\odot$	$1\,M_J$	$318\,M_\oplus$	1.90×10^{27} k
$M_\oplus$	$3.00 \times 10^{-6}\,M_\odot$	$3.15 \times 10^{-3}\,M_J$	$1\,M_\oplus$	5.97×10^{24} k
kg	$5.03 \times 10^{-31}\,M_\odot$	$5.27 \times 10^{-28}\,M_J$	$1.67 \times 10^{-25}\,M_\oplus$	1 kg

The prefixes 'GJ', 'Gl' and 'Gj' have all been used as shorthand for objects in the Gliese catalogue.

Exercise 4.3 The values of the planetary masses, and the instantaneous orbital periods, P_k, semi-major axes, a_k, and orbital eccentricities, e_k, for the planets in the Gliese 581 (usually abbreviated to GJ 581) system are shown in Table 4.2. Use this information to calculate the reflex RV amplitudes of the star due to the contribution of each planet. The mass of GJ 581 itself is $0.31\,M_\odot$. Evaluate the total maximum reflex RV amplitude that these parameters imply for this star. ■

Table 4.2 The properties of the planets in the GJ 581 system.

Planet	$M_P \sin i/M_\oplus$	P_k/days	a_k/AU	e_k
GJ 581 b	15.64	5.369	0.041	0.0
GJ 581 c	5.36	12.929	0.07	0.17
GJ 581 d	7.09	66.8	0.22	0.38
GJ 581 e	1.94	3.149	0.03	0.0

It is not clear whether the low-mass planets (shaded blue in Figure 4.10) have solid surfaces like the Earth or instead have massive envelopes of H and He like Neptune. It seems likely that the extremely-low-mass planets like Gliese 581 e are solid, like the Solar System's terrestrial planets. The only way to really know, however, is by measuring the size of the planet. With a transiting planet, the size can be measured and the planet can be truly labelled a **super-Earth** or a **mini-Neptune**, as the lightest known exoplanets have been termed. We discuss the first transiting super-Earth in the next section.

4.2.3 The transiting exoplanets

This book focuses on the subset of exoplanets that transit because we can obtain far more empirical information for these, as we have seen in Chapter 3. Crucially, we can measure the size for a transiting planet. So far (November 2009) the smallest exoplanet to be announced is CoRoT-7 b.

We saw in Chapter 1 that the probability of a randomly oriented planet executing a transit is given by

$$\text{geometric transit probability} \approx \frac{R_*}{a}. \qquad \text{(Eqn 1.21)}$$

Consequently, the known transiting exoplanet population is strongly biased towards close-in planets. We saw this in Figure 4.5. We also saw in Figure 4.3b,c hints that the host stars of known transiting planets are more massive and less metal-rich than the host stars of the RV-detected exoplanets. It is unclear whether these effects are random statistical flukes, artefacts of the selection effects (most likely on the RV population), or the results of the astrophysical processes leading to a star having a close-in giant planet.

CoRoT-7 b: the first known rocky exoplanet

CoRoT is a French-led ESA satellite that at the time of writing is in the middle of its mission to search for small rocky terrestrial exoplanets. The first of these is CoRoT-7 b, which was announced in February 2009 (Figure 4.11).

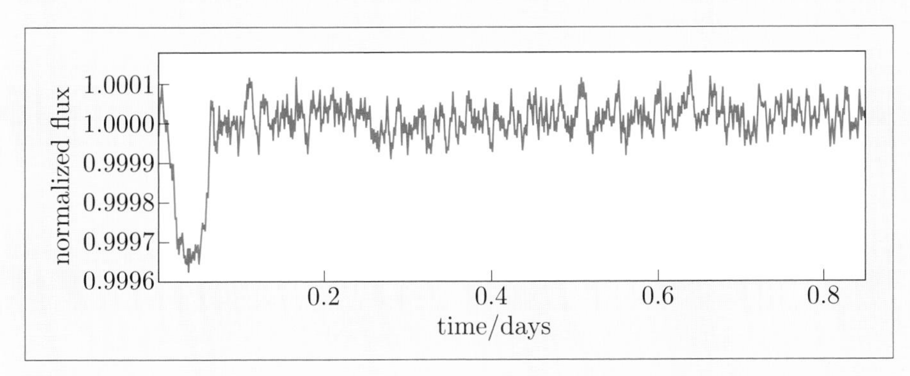

Figure 4.11 CoRoT data showing the transit of CoRoT-7 b.

CoRoT-7 b's orbital period is 0.85 days, and the transit has a depth of only $\Delta F = 0.000\,34\,F$, i.e. only 0.034% of the stellar flux. This tiny transit depth means that the signal-to-noise ratio of the light curve is insufficient to obtain precise parameters, especially as the host star is an active G5V star, so is expected to vary and have star spots. The current (November 2009) best parameters are $R_\mathrm{P} = 1.68 \pm 0.09\,\mathrm{R}_\oplus$. The orbital period is the shortest known, and the radius implies that CoRoT-7 b is a **telluric** planet.

A *telluric* planet is a planet composed primarily of silicate rocks.

We will discuss the radii of the known transiting planets in some detail, but first we will briefly survey the factors that determine the size of a giant planet.

4.3 The structure of giant planets

A full treatment of this topic is far beyond the scope of this book. Here we aim to introduce and describe some of the topic's underpinnings to set the observations of transiting exoplanets in the appropriate context. Our treatment follows parts of the paper 'The interiors of giant planets: models and outstanding questions' (2005, *Annual Review of Earth and Planetary Sciences*, **33**, 493–530) by Tristan Guillot.

4.3.1 The underlying physics

We assert a number of assumptions, the first of which is that we may regard giant planets as being spherical. The remaining assumptions are less obvious, and can be justified by detailed observations and analysis of the giant planets in our Solar System and of the known transiting giant exoplanets. Giant planets are made of a fluid envelope and possibly a dense central core of about $15\,M_\oplus$. The envelope is composed of mostly hydrogen and helium with a small fraction of heavier elements. The core is composed of rocks (**refractory material**), ices (of water and other molecules such as methane, CH_4), or a combination of these. Viscosity, rotation and magnetic fields can all be neglected for the discussion of the gross overall structure. We assume that giant planets form from an extended protoplanetary nebula. The last of the assumptions above is of course intimately related to our picture of star formation.

The basic physics underlying planetary structure is the same as that underlying stellar structure. Planets will adjust on a (fast) **dynamical timescale** to establish **hydrostatic equilibrium**. The **pressure gradient** at any point in the planet must therefore be balanced by the gravitational force per unit volume, and we can write the equation of hydrostatic equilibrium as

In this section, P is pressure, not orbital period.

$$\frac{\mathrm{d}P(r)}{\mathrm{d}r} = -\frac{G\,m(r)\,\rho(r)}{r^2}, \tag{4.8}$$

where $m(r)$ is the mass interior to a sphere of radius r (sometimes referred to as the **enclosed mass** at radius r), and $\rho(r)$ is the density at some radius r.

- If a planet is continually contracting, in what way are we justified in using the equation of hydrostatic equilibrium?

- The contraction is much slower than the dynamical adjustment that would occur if hydrostatic equilibrium were disturbed. At any instant in time, the structure will obey hydrostatic equilibrium to an extremely good approximation.

In a spherical geometry, the mass and density are related by the equation of **mass continuity**, which is the second of the four basic physical principles:

$$\frac{\mathrm{d}m(r)}{\mathrm{d}r} = 4\pi r^2 \rho(r). \tag{4.9}$$

The **temperature gradient** within the planet depends on the processes by which the internal heat is transported through the planet. We will not go into the mathematics of this here.

Finally, **energy conservation** implies that the luminosity increases outwards from the centre of a planet, as each successive shell contributes to the total luminosity. The energy generation equation describes the increase in luminosity, $\mathrm{d}L$, per step in radius, $\mathrm{d}r$, as

$$\frac{\mathrm{d}L(r)}{\mathrm{d}r} = 4\pi r^2 \left(\varepsilon(r) - \rho(r)\,T(r)\,\frac{\partial S(r,t)}{\partial t} \right), \tag{4.10}$$

where $\varepsilon(r)$ is the energy generation rate per unit volume from radioactivity or nuclear reactions. The second term within the parentheses accounts for the energy that is radiated away as a result of **entropy** changes within each annulus. Entropy is a powerful but challenging concept within physics and chemistry, and gives a macroscopic description of energy exchanges related to the microstates accessible to a system at a molecular level. Planets are much more complex than stars in this respect; to remind us of this, we retain the term involving the entropy explicitly in Equation 4.10 rather than incorporating it within ε.

Units of pressure

The SI unit of pressure is the pascal (Pa): $1\,\mathrm{Pa} = 1\,\mathrm{N\,m^{-2}}$. The bar is often used as the unit of pressure in astrophysics; it is related to the pascal by $1\ \mathrm{bar} = 100\,\mathrm{kPa}$, i.e. $1\ \mathrm{bar} = 10^5\ \mathrm{Pa}$. The bar is an intuitively easy unit to interpret because the atmospheric pressure at the surface of the Earth is approximately 1 bar:

$$\begin{aligned} &1\ \text{atmosphere} \\ &= 1.013\,25\ \mathrm{bar} = 1.013\,25 \times 10^5\ \mathrm{Pa} = 1.013\,25 \times 10^5\ \mathrm{N\,m^{-2}}. \end{aligned} \tag{4.11}$$

Worked Example 4.1

(a) Starting from the equations of hydrostatic equilibrium and of mass continuity (Equations 4.8 and 4.9), derive an approximate expression to make an order of magnitude estimate of the pressure, P_c, at the centre of a giant planet.

(b) Evaluate this expression for Jupiter, giving your answer in SI units.

Solution

(a) We have

$$\frac{\mathrm{d}P(r)}{\mathrm{d}r} = -\frac{G\,m(r)\,\rho(r)}{r^2}. \qquad \text{(Eqn 4.8)}$$

To make an order of magnitude estimate, we will perform **dimensional analysis**, by approximating the pressure gradient as linear and writing

$$\frac{P_\mathrm{c}}{R} \approx \frac{GM\rho}{R^2}. \tag{4.12}$$

(Note that the minus sign has vanished because the pressure gradient is negative and we have replaced it with a positive ratio.)

Taking the same approach with the equation of mass continuity, we have

$$\frac{M}{R} = 4\pi R^2 \rho,$$

so

$$\rho \approx \frac{M}{4\pi R^3}. \tag{4.13}$$

The astute reader will have realized, of course, that we could alternatively have made an estimate of the mean density as simply the mass divided by the volume:

$$\bar{\rho} = \frac{3M}{4\pi R^3}. \tag{4.14}$$

Comparing Equations 4.13 and 4.14, we see that our dimensional analysis of the equation of mass continuity gave us a result that is a factor of 3 less than the mean density. This illustrates the power and the limitations of dimensional analysis: it will not give exact answers, but it can provide reliable order of magnitude estimates.

Substituting either of the expressions for ρ into Equation 4.12, we obtain

$$P_{\rm c} \approx \frac{GM^2}{R^4}, \tag{4.15}$$

where we have dropped all the numerical factors of order unity.

(b) Equation 4.15 for Jupiter is

$$P_{\rm c} \approx \frac{G{\rm M}_{\rm J}^2}{{\rm R}_{\rm J}^4},$$

and the numerical values that we require are

$$G \approx 7 \times 10^{-11}\,{\rm N\,m^2\,kg^{-2}}, \quad {\rm M_J} \approx 2 \times 10^{27}\,{\rm kg}, \quad {\rm R_J} \approx 7 \times 10^{7}\,{\rm m}.$$

(Note that we have given these values to single-figure precision because we have already made far more gross approximations in arriving at the expression in Equation 4.15.) So we have

$$\begin{aligned} P_{\rm c} &\approx \frac{7 \times 10^{-11} \times (2 \times 10^{27})^2}{(7 \times 10^7)^4} \, \frac{{\rm N\,m^2\,kg^{-2}\,kg^2}}{{\rm m^4}} \\ &\approx \frac{2.8 \times 10^{44}}{2.4 \times 10^{31}} \, \frac{\rm N}{\rm m^2} \\ &\approx 1 \times 10^{13}\,{\rm N\,m^{-2}}. \end{aligned}$$

Thus our estimate for the pressure at the centre of Jupiter is 1×10^{13} Pa.

4.3.2 The equation of state

In stellar structure the equation of state is simple: the **ideal gas law** is a good approximation to the relationship between pressure, density and temperature throughout a star. The situation is far more complex for giant planets. At the temperatures and pressures characteristic of giant planet interiors, molecules, atoms and ions co-exist. The electrons are **partially degenerate**, which means that their energies are determined by a combination of thermal and quantum effects, i.e. the equations to describe the electron energy distribution are as complex as they could possibly be. The fluid is affected by the electrostatic interactions between the ions, but these interactions do not dominate over the gas pressure provided by the neutral atoms and molecules, which is again the most complicated regime to quantify. The chemical composition is an unknown mixture of many elements, which may interact in ways that have not been characterized empirically.

Degeneracy and the Fermi temperature

Electrons, protons and neutrons are all **fermions**. No two fermions can occupy the same quantum state at the same time. The density of quantum states per unit volume, n_Q, can be derived from the particle's **de Broglie wavelength**. For non-relativistic particles of mass m,

This is derived in the companion book in this series, *Stellar Evolution and Nucleosynthesis*.

$$n_\mathrm{QNR} = \left(\frac{2\pi mkT}{h^2}\right)^{3/2}, \tag{4.16}$$

where we use the symbol n_QNR to denote the quantum concentration in the non-relativistic regime, which applies to planetary interiors. So long as the temperature, T, and number density, n, of identical particles are such that

$$n \ll n_\mathrm{Q},$$

there are plenty of quantum states for all the particles, and the particles will have a **Maxwell–Boltzmann energy distribution**, appropriate to their temperature, T. Conversely, if the density, n, is

$$n \geq n_\mathrm{Q},$$

then there are insufficient quantum states available to particles for them to adopt a Maxwell–Boltzmann energy distribution, and some particles are forced to occupy higher energy states. This situation is what is mean by degeneracy.

These conditions can equivalently be expressed in terms of the temperature. The critical temperature is the **Fermi temperature**, T_F, and a gas is degenerate if

$$T < T_\mathrm{F}, \tag{4.17}$$

where

$$T_\mathrm{F} = \frac{n^{2/3}h^2}{2\pi mk}. \tag{4.18}$$

The condition for whether or not a gas is degenerate can be expressed in terms of the **degeneracy parameter**, θ, where

$$\theta \equiv \frac{T}{T_\mathrm{F}}. \tag{4.19}$$

If $\theta < 1$, the gas is degenerate.

Despite the complications mentioned above, we are confident that the envelopes of giant planets will generally be composed mostly of hydrogen. Figure 4.12 shows the phase diagram for hydrogen, with temperature–pressure profiles calculated for the Solar System giant planets and for HD 209458 b. These models imply that hydrogen is present in molecular form in the outer regions, and as metallic H^+ in the hottest, densest inner regions of the Solar System's giant planets.

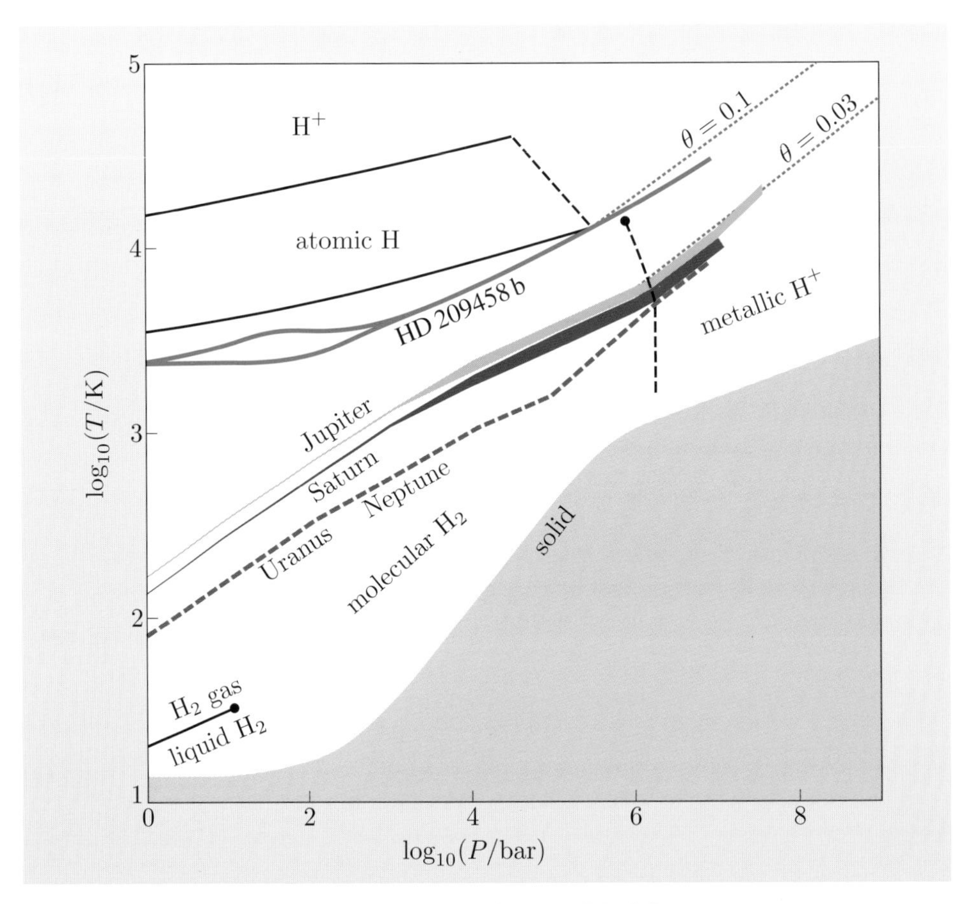

Figure 4.12 The phase diagram for hydrogen. Model temperature–pressure profiles for the Solar System giant planets and for HD 209458 b are indicated. The dashed lines in the upper right indicate the loci of degeneracy parameter, $\theta = 0.1$ and $\theta = 0.03$.

The photospheres of giant planets generally have temperatures in the range $50\,\mathrm{K} < T < 3000\,\mathrm{K}$ and pressures $0.1\,\mathrm{bar} < P < 10\,\mathrm{bar}$, which means that the hydrogen is molecular, and the ideal gas equation is a good approximation. As we go deeper into the interior of the planet, the pressure increases and the ideal gas equation of state becomes a poorer approximation. For Jupiter at $P \sim 10^3$ bar, the pressure exceeds that predicted by the ideal gas equation by about 10%. Deep in the interior, the pressure is $P_\mathrm{c} \sim 1 \times 10^{13}\,\mathrm{Pa} \sim 1 \times 10^8$ bar (as we saw in Worked Example 4.1) and the densities are consequently high. The Fermi temperature is $T_\mathrm{F} > 10^5$ K, and the electrons become degenerate. This is indicated in Figure 4.12 by the degeneracy parameter, θ, with loci of $\theta = 0.1$ and $\theta = 0.03$ shown. These two values are approximately those holding near the centres of the models of HD 209458 b and Jupiter, respectively.

- Why do electrons rather than any other particles become degenerate at the densities prevailing in the cores of giant planets?

- $n_{\rm QNR} \propto m$ so the density of states available is smallest for the lowest-mass particles. The electrons are lighter than any other particle by a factor of $m_{\rm e}/m_{\rm p} \approx 1/1800$.

Exercise 4.4 To answer this question refer to Figure 4.12. Using the information on the model for the temperature–pressure profile of Jupiter and the corresponding value of the degeneracy parameter, θ, deduce the number density of electrons in the central regions of Jupiter. ■

4.3.3 The virial theorem and the contraction and cooling of giant planets

We have mentioned that the thermal emission of planets is powered by their contraction and cooling. Here we examine the processes powering this thermal emission. From Equation 4.8 we can deduce that

$$4\pi r^3 \, {\rm d}P = -4\pi r G \, m(r) \, \rho \, {\rm d}r,$$

and using Equation 4.9 we can substitute for $4\pi r^2 \rho \, {\rm d}r$ to obtain

$$4\pi r^3 \, {\rm d}P = -\frac{G \, m(r)}{r} \, {\rm d}m.$$

If we now integrate this equation over the entire planet we obtain

$$\int_{\rm c}^{\rm s} 4\pi r^3 \, {\rm d}P = -\int_0^{\rm s} \frac{G \, m(r)}{r} \, {\rm d}m$$

or

$$3 \int_{\rm c}^{\rm s} V \, {\rm d}P = -\int_0^{\rm s} \frac{G \, m(r)}{r} \, {\rm d}m,$$

where we have used the symbol V for the volume of the planet, and the letters 'c' and 's' to denote the centre and the surface of the planet, respectively, or alternatively the value zero for the limit at the centre, where appropriate. Integrating the left-hand side by parts gives

$$3 \, [PV]_{\rm c}^{\rm s} - 3 \int_0^{\rm s} P \, {\rm d}V = -\int_0^{\rm s} \frac{G \, m(r)}{r} \, {\rm d}m.$$

Now we use the substitution ${\rm d}V = {\rm d}m/\rho$ and note that the integrated part $[PV]$ vanishes at both the upper and lower limits because $V = 0$ at the centre of the planet and we may neglect the pressure at the surface in comparison with all the other terms in the equation. The quantity on the right-hand side is the **gravitational energy** of the planet, i.e. the energy that has been liberated in assembling the gravitationally bound structure from a tenuous distribution of matter. This should be clear if you consider the work that would need to be done to disassemble the planet by moving each element of mass ${\rm d}m$ to infinity under the gravitational field of the remaining planetary mass. We will give the total gravitational energy of the planet the symbol $E_{\rm GR}$. Consequently, we obtain

$$0 - 3 \int_0^{\rm s} \frac{P}{\rho} \, {\rm d}m = E_{\rm GR}.$$

Thus we have derived the **virial theorem**, which holds generally and has applications throughout astrophysics:

$$3\int_0^{\mathrm{s}} \frac{P}{\rho}\,\mathrm{d}m + E_{\mathrm{GR}} = 0. \qquad (4.20)$$

The virial theorem can be used to relate the gravitational energy and the internal energy of a gravitationally bound object, but to do this we need to know the equation of state of the object. The equation of state governs the value of P/ρ at each point. As we said in the previous subsection, for stars the ideal gas law is a very good approximation to the equation of state, but for planets the equation of state is more complex. Nonetheless we can gain significant insight into the most important trends from analysis of the virial theorem.

We begin by defining the total internal energy,

$$E_{\mathrm{I}} \equiv \int_0^M u\,\mathrm{d}m, \qquad (4.21)$$

where u is the specific internal energy and we have replaced the generic upper limit 's' (denoting the surface of the planet) with the upper limit M since the integration is with respect to the mass element $\mathrm{d}m$.

- What does 'specific' mean in this context?

- It means per unit mass.

We can further define

ζ is the Greek letter zeta, which is usually pronounced squiggle.

$$\zeta \equiv \frac{\int_0^M 3(P/\rho)\,\mathrm{d}m}{\int_0^M u\,\mathrm{d}m} \qquad (4.22)$$

to allow us to conveniently parameterize our ignorance of the equation of state. With these definitions we can rewrite the virial theorem (Equation 4.20) as

$$\zeta E_{\mathrm{I}} + E_{\mathrm{GR}} = 0. \qquad (4.23)$$

By the conservation of energy, we know that the intrinsic luminosity of a planet, L, must be given by

$$\begin{aligned} L &= -\frac{\mathrm{d}}{\mathrm{d}t}(E_{\mathrm{total}}) \\ &= -\frac{\mathrm{d}}{\mathrm{d}t}(E_{\mathrm{GR}} + E_{\mathrm{I}}), \end{aligned} \qquad (4.24)$$

and using Equation 4.23 we have

$$\begin{aligned} L &= -\frac{\mathrm{d}}{\mathrm{d}t}\left(E_{\mathrm{GR}} - \frac{E_{\mathrm{GR}}}{\zeta}\right) = -\frac{\zeta - 1}{\zeta}\frac{\mathrm{d}E_{\mathrm{GR}}}{\mathrm{d}t} \\ &= -\frac{\zeta - 1}{\zeta}\dot{E}_{\mathrm{GR}}, \end{aligned} \qquad (4.25)$$

where we have introduced the standard notation $\dot{x}$ for the derivative of the variable x with respect to time. Here we have implicitly asserted that ζ changes negligibly compared with the change in E_{GR}. This is justified because as the planet contracts, the structure will evolve slowly to maintain hydrostatic equilibrium. ζ changes on the timescale of changes in the equation of state itself.

For example, at the very beginning of its evolution, the planet is a cloud of H_2 gas. For a diatomic ideal gas, $\zeta = 3.2$, and this value of ζ can be used to work out the luminosity generated by the planet's contraction in this earliest stage.

Exercise 4.5 As a planet contracts over a time interval Δt, its gravitational energy will decrease by an amount ΔE_{GR} and it will radiate away an amount of energy E_{rad}.

(a) Derive a general expression for E_{rad} in terms of ΔE_{GR} and ζ.

(b) Evaluate the fraction of ΔE_{GR} that is radiated away for the case of a planet composed entirely of H_2, assuming that the ideal gas law applies.

(c) Explain how conservation of energy is maintained in the collapse.

(d) State what happens to the temperature, pressure and density of the H_2 gas during the collapse, explaining your reasoning. ■

As the planet contracts, the density increases, and as we have already said, except for the very early stages of the planet's evolution, in the interior regions the pressure is provided by degenerate electrons. In these regions the degeneracy removes the equipartition of energy between electrons and ions, and the internal energy is given by

$$E_I = E_e + E_{ion} \tag{4.26}$$

and $E_e \gg E_{ion}$. For a fully degenerate non-relativistic gas, $\zeta = 2$, so Equation 4.25 implies that half of the energy released by contraction is radiated outwards. The other half will go to increasing the internal energy. Dimensional analysis tells us that

$$E_{GR} \propto \frac{1}{R_P} \propto \rho^{1/3}.$$

We can rewrite the proportionality as

$$E_{GR} = A\rho^{1/3}, \tag{4.27}$$

where A is a constant and ρ is the density of the planet. Equation 4.16 implies that the energy of the degenerate electrons is $E_e \propto \rho^{2/3}$, so we can write

We will show that $E_e \propto \rho^{2/3}$ in a worked example below.

$$E_e = B\rho^{2/3}, \tag{4.28}$$

where B is a constant. Combining Equations 4.27 and 4.28 to eliminate ρ, we obtain

$$E_e = B\left(\frac{E_{GR}}{A}\right)^2. \tag{4.29}$$

Differentiating this with respect to time, we obtain

$$\dot{E}_e = \frac{B}{A^2} 2E_{GR}\,\dot{E}_{GR}.$$

We can further manipulate this:

$$\begin{aligned}\dot{E}_e &= 2\frac{B\,E_{GR}^2}{A^2}\frac{1}{E_{GR}}\,\dot{E}_{GR}\\ &= 2\frac{E_e}{E_{GR}}\,\dot{E}_{GR},\end{aligned} \tag{4.30}$$

where we have substituted using Equation 4.29. Equation 4.23 with $\zeta = 2$ tells us that

$$2E_{\rm I} + E_{\rm GR} = 0,$$

so

$$E_{\rm I} = -\frac{E_{\rm GR}}{2}, \tag{4.31}$$

which becomes

$$E_{\rm e} \approx -\frac{E_{\rm GR}}{2} \tag{4.32}$$

using Equation 4.26 and the condition $E_{\rm e} \gg E_{\rm ion}$. Substituting from Equation 4.32 into Equation 4.30, we obtain

$$\begin{aligned} \dot{E}_{\rm e} &\approx -2\,\frac{E_{\rm GR}}{2E_{\rm GR}}\,\dot{E}_{\rm GR} \\ &\approx -\dot{E}_{\rm GR}. \end{aligned} \tag{4.33}$$

From Equation 4.25 with $\zeta = 2$, we have

$$L = \frac{\dot{E}_{\rm GR}}{2}, \tag{4.34}$$

and combining Equation 4.34 with Equation 4.33, we have

$$L \approx -\frac{\dot{E}_{\rm e}}{2}. \tag{4.35}$$

By the conservation of energy (Equation 4.24), we also have

$$\begin{aligned} L &= -\frac{\rm d}{{\rm d}t}(E_{\rm total}) = -(\dot{E}_{\rm GR} + \dot{E}_{\rm e} + \dot{E}_{\rm ion}) \\ &\approx -\dot{E}_{\rm ion}, \end{aligned} \tag{4.36}$$

where we have used Equation 4.33 in the last step. This analysis reveals that the gravitational energy of a contracting giant planet is entirely absorbed by the degenerate electrons (Equation 4.33), while the intrinsic luminosity of the planet is powered by the thermal cooling of the ions (Equation 4.36). Of course, this is a simplification, and we have glossed over myriad complexities, but nonetheless it captures the essential physics.

- Why does $E_{\rm e} \gg E_{\rm ion}$ hold in the degenerate core of a giant planet?

- Because of the Pauli exclusion principle, the electrons are forced into higher energy quantum states than they would occupy at lower densities. This breaks the equipartition of energy between particles, and the electrons have much greater energies than the non-degenerate ions.

- How do the degenerate electrons absorb the gravitational energy as the giant planet contracts?

- As the planet contracts, the density increases. This means that the electrons are forced into higher energy quantum states, and the gravitational energy is required to promote them to these states.

We have analyzed two phases of the evolution. The mature evolution, governed by the equation of state for degenerate non-relativistic electrons, has $\zeta = 2$. The essence of the early evolution is the contraction of H_2 gas, for which $\zeta = 3.2$, and in this case

$$L = -0.69\dot{E}_{GR}, \tag{4.37}$$

as we found in Exercise 4.5.

Worked Example 4.2

Justify the statement that the energy of the degenerate electrons is $E_e \propto \rho^{2/3}$.

Solution

We have

$$n_{QNR} = (2\pi mkT/h^2)^{3/2}, \qquad \text{(Eqn 4.16)}$$

which tells us how the number density of quantum states, n_{QNR}, compares with the temperature, T. If the temperature is high enough, $n \ll n_{QNR}$ and the particles will follow the Maxwell–Boltzmann distribution, and the average energy will be $\sim kT$. If, on the other hand, $n > n_{QNR}$, the particles will be unable to follow a Maxwell–Boltzmann distribution, and will be forced into higher energy states. In this case the energy of the particles is determined not by the temperature, but instead by the energies of the lowest available quantum states.

Since energy is proportional to kT, Equation 4.16 tells us that the number density of quantum states, n_{QNR}, is proportional to energy to the power $3/2$. We could therefore reduce Equation 4.16 to

$$n_{QNR} \propto E^{3/2}. \tag{4.38}$$

For a fully degenerate gas, which we may assume the electrons at the centre of a giant planet to be, the particles will occupy the lowest energy states available, so that the number density, n, will be equal to the critical density corresponding to the energy of the (highest) occupied states:

$$n_e = n_{QNR} \propto E_e^{3/2}. \tag{4.39}$$

Thus the number density of electrons is $\propto E_e^{3/2}$, and as the planet's core is predominantly hydrogen in the metallic state (see Figure 4.12), $n_e \propto \rho$. Using this we obtain

$$\rho \propto E_e^{3/2}$$

or equivalently, making E_e the subject of the equation,

$$E_e \propto \rho^{2/3}, \tag{4.40}$$

which is the proportionality that we desire.

Stepping back from the detail above, the gravitational potential energy still to be liberated from the planet with radius R_P is

$$\Delta E_{GR} \approx \frac{GM_P^2}{R_P}, \tag{4.41}$$

where we have considered the energy required to move the component parts of the planet from its centre to their current positions. If all of this gravitational potential energy were radiated away during a gradual collapse over a time τ, then the average luminosity, L, would be given by

$$L = \frac{\text{energy liberated}}{\text{time taken}} = \frac{\Delta E_{\text{GR}}}{\tau} \approx \frac{GM_{\text{P}}^2}{\tau R_{\text{P}}}.$$

This luminosity is known as the **Kelvin–Helmholtz luminosity**, L_{KH}, and is important in the history of stellar astrophysics:

$$L_{\text{KH}} \approx \frac{GM_{\text{P}}^2}{\tau R_{\text{P}}}. \tag{4.42}$$

As we have already seen, the gravitational energy released by the contraction of a giant planet with an interior supported by degenerate electrons is *not* all radiated away, so the luminosity can be written as

$$L \approx \eta \frac{GM_{\text{P}}^2}{\tau R_{\text{P}}}, \tag{4.43}$$

where τ is the age and η is a factor between 0 and 1 that takes account of the facts that in their mature evolution planets are rather incompressible, and that the bulk of the gravitational potential energy is absorbed by the degenerate electrons. Only the very much smaller energy pool in the thermal energy of the ions is available to power the luminosity. Detailed analysis of this is beyond the scope of this book, so we will simply state that for mature planets $\eta \ll 1$, and with the approximations that we have made in deriving Equations 4.33 and 4.36,

$$\eta \approx \frac{\theta}{\theta + 1}, \tag{4.44}$$

where θ is the degeneracy parameter (cf. Equation 4.19).

4.4 Parameters governing the size of a giant planet

In the previous section we explored the basic physics underlying the structure of giant planets. We derived analytical results with simplifying assumptions. As we stated, the full treatment of all the salient physics is currently challenging researchers, but of course research has progressed quite far beyond the outline that we have examined. Figure 4.13 shows model interiors for the Solar System's giant planets.

4.4.1 Core mass

One factor that we haven't discussed quantitatively is the effect of a planet possessing a core of rocks and ices. Even for the Solar System's giant planets, the mass of these cores is rather uncertain. If there is a core of rock or ice present, this will be denser than hydrogen at similar pressure, so the presence of such a core tends to make the planet smaller for a given total mass, as we can see by comparing the black solid and black dotted lines in Figure 4.14.

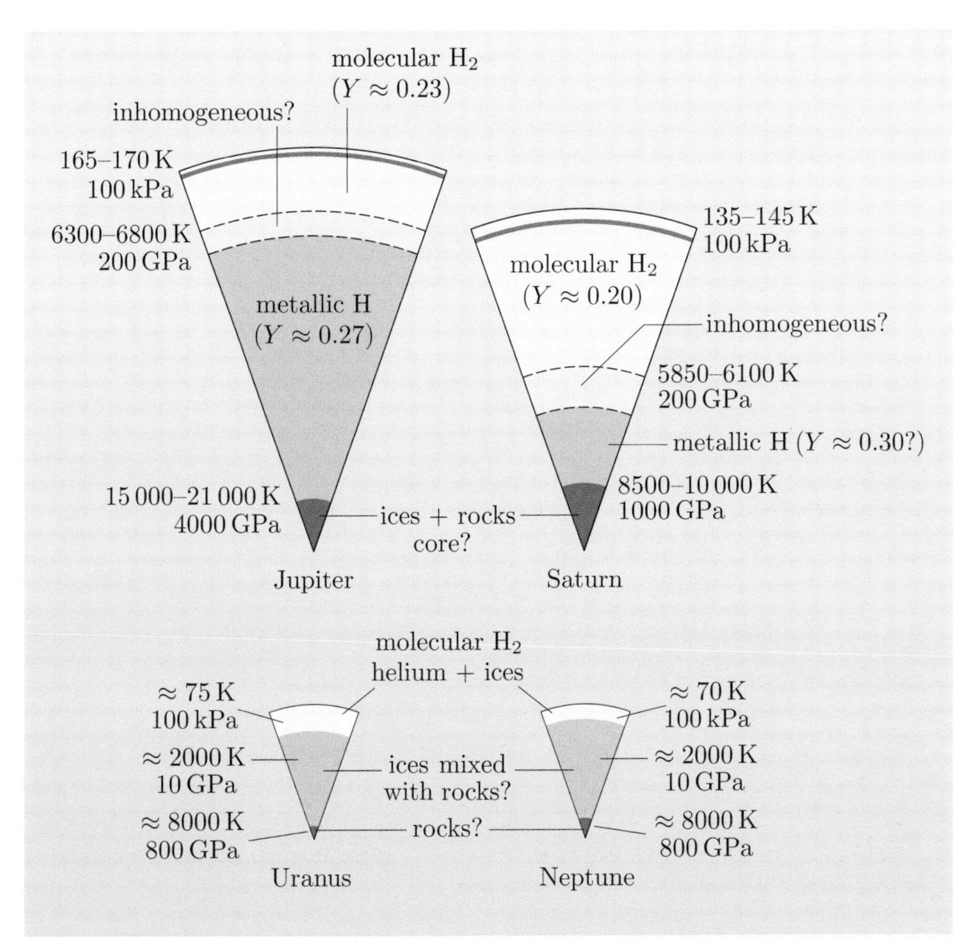

Figure 4.13 Schematic representations of models for the interiors of the Solar System's giant planets. As indicated in the labels, there are significant uncertainties.

Figure 4.14 also considers different values of the tidal heating (see Subsection 4.4.5), as indicated by the different coloured lines, and makes the point that there are many factors that simultaneously affect the planet's radius.

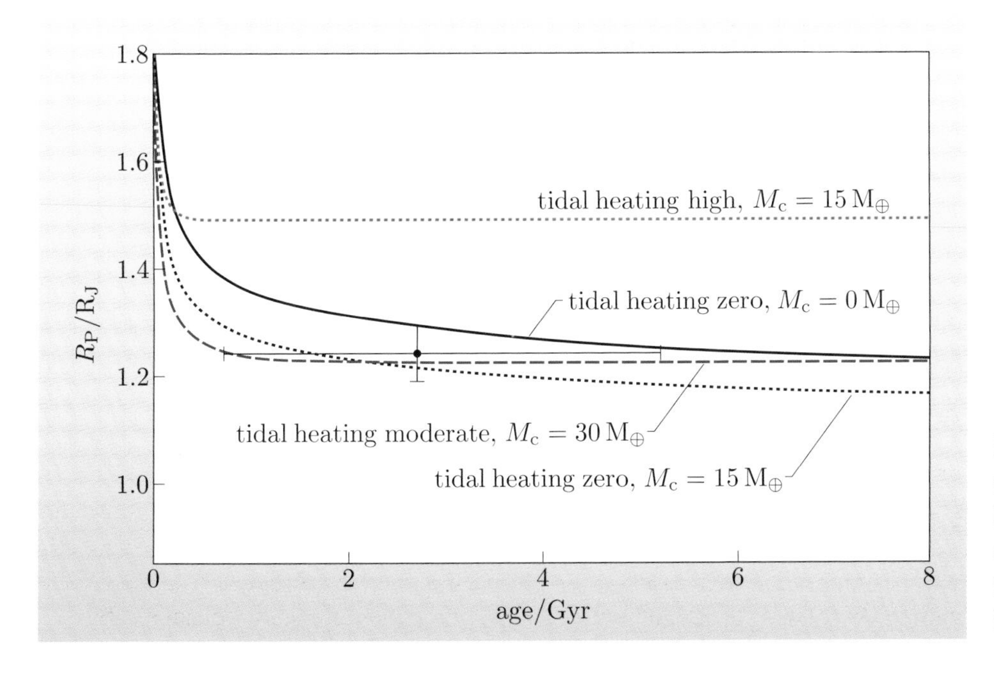

Figure 4.14 The radius evolution of models of the planet HAT-P-1 b. Three different core masses M_c were considered, and three different values of the tidal heating (see Subsection 4.4.5).

4.4.2 Total mass

Figure 4.15 shows the dependence of radius, R_P, on mass, M_P, for five different families of models of planetary structure. For the gaseous planets shown in the three upper curves, masses around $4\,M_J$ produce the largest planets. Planets that are less massive than this have significant non-degenerate regions, and (as naively expected) the radius increases as more mass is added.

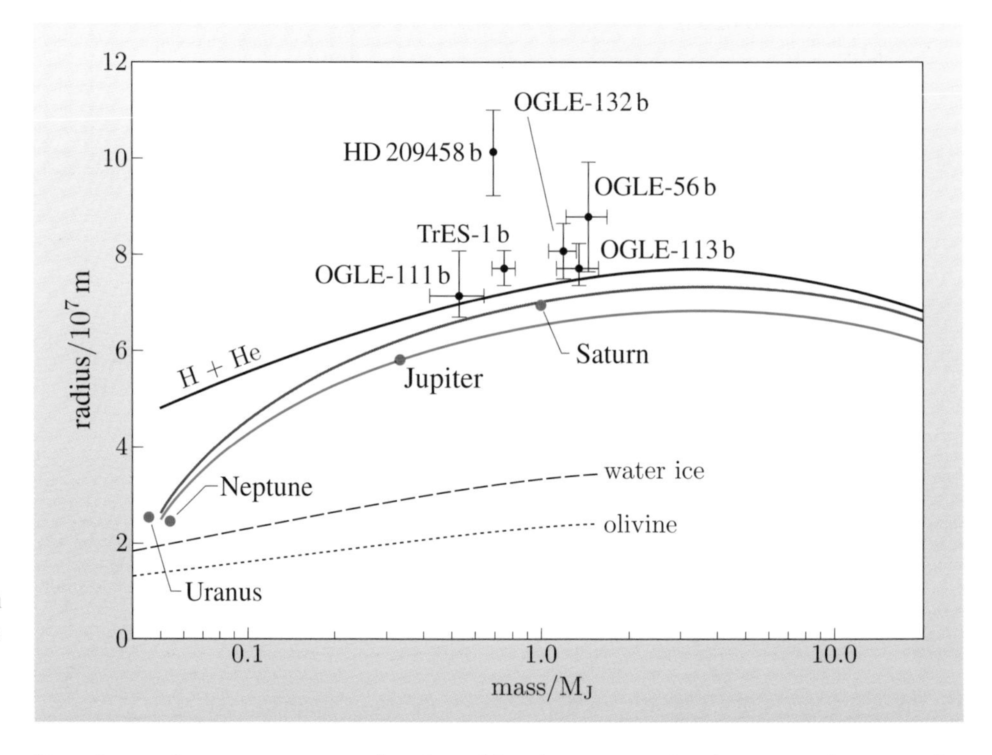

Figure 4.15 Radius versus mass for planetary models of five types. The two lowest curves show approximate mass–radius relationships for pure water ice and pure olivine (one of the most common minerals on Earth). The uppermost curve indicates a composition of 75% H and 25% He (by mass). The second curve from the top is H and He with 30% He by mass, with a $15\,M_\oplus$ core. The third curve from the top differs from the second only in having 36% He by mass; even a small adjustment in abundance makes an appreciable difference in radius.

For planets that are more massive than this, degeneracy dominates, and as mass is added, the radius decreases. This effect is well understood and widely analyzed in the case of white dwarf stars. An appropriate analytic approximation is the **polytropic equation of state**

$$P = K\rho^{(n+1)/n}, \tag{4.45}$$

where K and n are constants. If this holds throughout the planet, then it can be shown that

$$R_P \propto K^{n/(3-n)} M_P^{(1-n)/(3-n)}. \tag{4.46}$$

For a fully degenerate object, $n = \frac{3}{2}$. Real planets are not fully degenerate throughout, but for the interior part of their volume, isolated planets of masses 10, 1 and 0.1 times M_J can be modelled using Equation 4.45 and $n = 1.3$, 1.0 and 0.6, respectively.

4.4.3 Age and irradiation

In Subsection 4.3.3 we discussed the physics underlying the contraction and cooling of isolated giant planets. Of course, the known transiting planets are far from isolated: they are strongly irradiated as a result of their proximity to their host stars. We will discuss this in more detail in Chapter 6. The strong irradiation affects the thermal structure of the planet, and reduces its cooling rate. If the cooling rate is reduced, then the contraction is slowed. Figure 4.16 shows examples of models for three different values of M_P, each with three different irradiation fluxes applied.

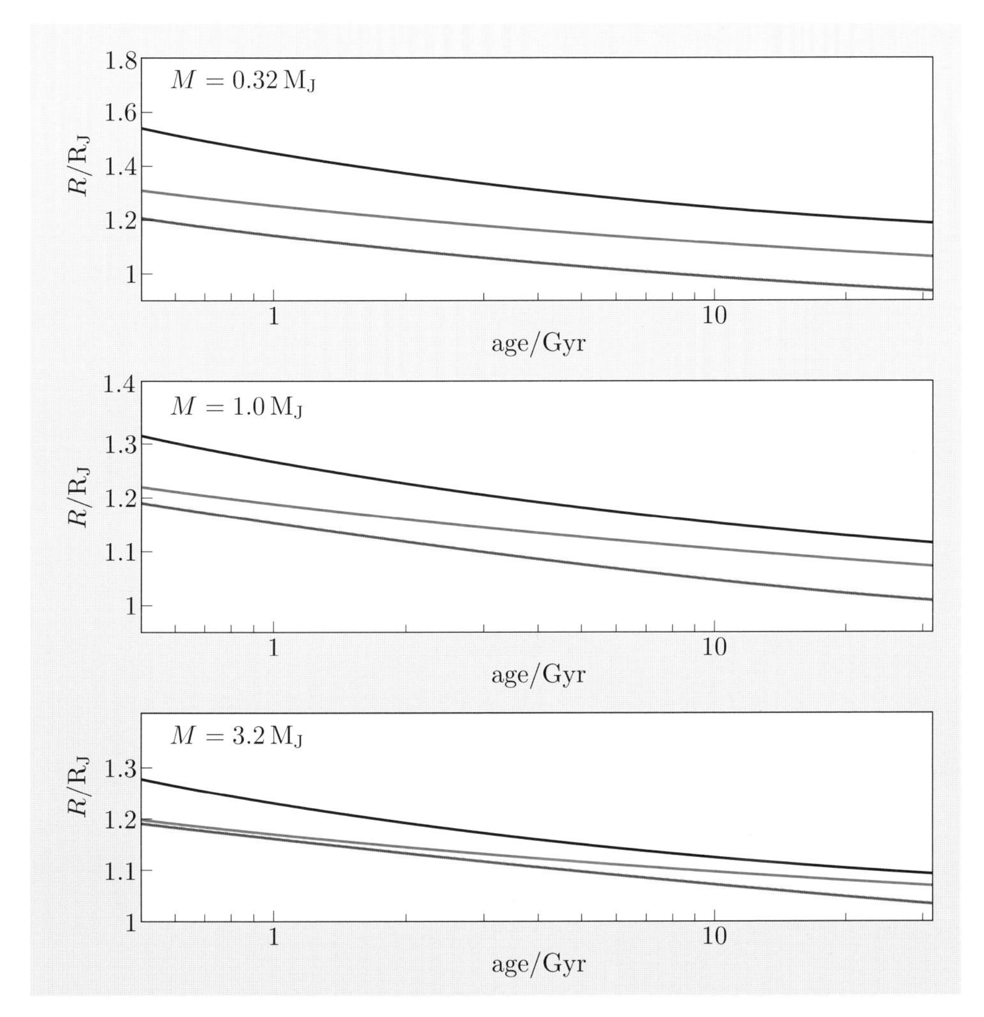

Figure 4.16 Examples of evolutionary models for giant planets; radius versus age for planets of mass $0.32\,M_J$, $1.0\,M_J$ and $3.2\,M_J$. In each case the three solid lines represent three different irradiating fluxes. For any mass and age, the planet is always largest for the highest irradiating flux.

It is important to note that the horizontal axis is logarithmic, so the graph emphasizes the earliest evolution. Initially, the lowest-mass planets are the largest: like stars, higher-mass objects contract more quickly. In each case, the size of the planet at a given age is larger if the irradiating flux is larger.

4.4.4 Metallicity

The contraction and cooling of a giant planet proceeds at a rate that is governed by how quickly energy can be radiated away. If a planet has high metallicity, then the envelope will have a significantly higher opacity than that of a pure hydrogen and helium planet. This is because the heavier chemical elements have far more **atomic energy levels**, and consequently far more **spectral lines**. The mean free path of a photon is therefore much shorter in an envelope of high metallicity, and it takes far longer for radiation to diffuse outwards and escape from the planet. There is evidence (cf. Subsection 4.2.1) that the host stars of known exoplanets tend to be metal-rich, and there is also evidence that the Solar System giant planets are metal-rich compared to the Sun. Consequently, many or most of the known exoplanets are probably metal-rich compared to the Sun. Figure 4.17 illustrates the calculated effect on the radius evolution for a particular planet, TrES-4 b.

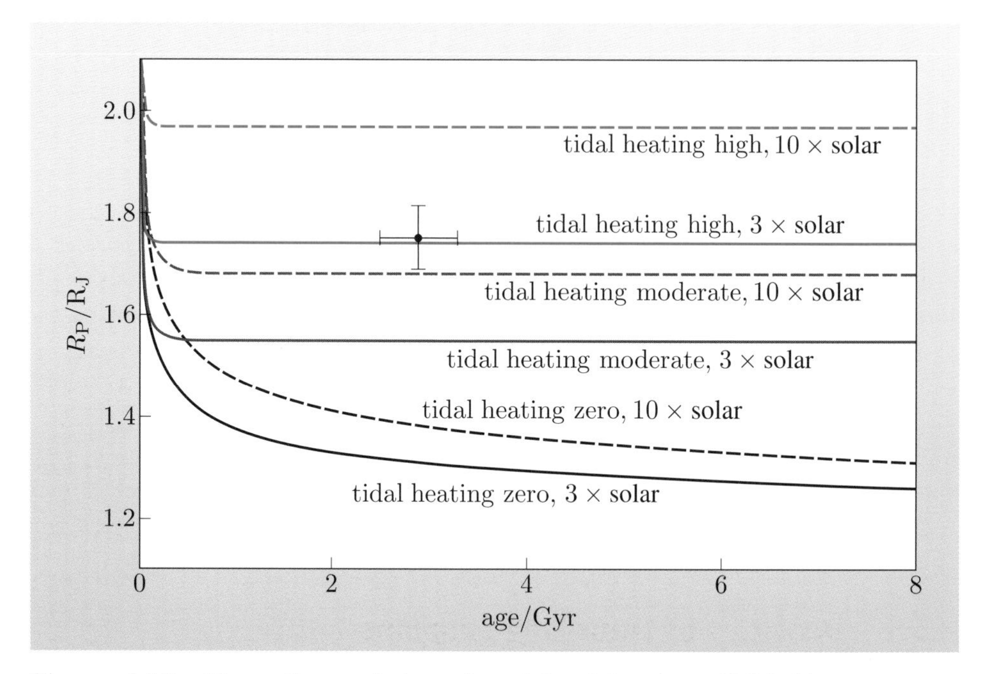

Figure 4.17 The radius evolution of models of the planet TrES-4 b. Metallicities of ten times solar and three times solar are indicated by the dashed and solid lines, respectively. In each case, three different values of tidal heating are also considered, as indicated by the three different colours. In all cases the core mass is fixed at zero.

The figure also shows the effect of three different values of tidal heating, a parameter that we discuss in the next subsection. In all cases higher metallicities, and hence higher opacities, correspond to a larger planet at all stages of the evolution. The effect is particularly marked for non-zero tidal heating: in this case the continual heating of the planet's interior maintains the temperature gradient between the core and the surface.

- Why does the difference in radius between the models with two different metallicity values continuously diminish with time in the case of zero tidal heating?

- As the planet contracts and cools, the heat is gradually lost from the interior. At infinite time the planet will reach absolute zero and the opacity will have no appreciable effect on the radius. As time increases, the temperature gradient and the outwards heat flux continuously diminish, so the two models converge.

4.4.5 Tidal heating

We discuss tidal effects in Chapter 7.

As alluded to in the discussion of Figures 4.14 and 4.17, one of the factors that can potentially influence the radius evolution of a giant planet is tidal heating.The ocean tides on the Earth show us that energy can be dissipated by the influence of tides. In the case of a fluid exoplanet in a slightly eccentric orbit (i.e. $e \sim 0.01$) close to its host star, this tidal heating can be $\sim 1\%$ of the radiation flux. Figures 4.14 and 4.17 show how tidal heating affects the radius evolution. The energy dissipated is

$$\dot{E}_{\text{tide}} \propto \frac{e^2}{Q_{\text{p}}}, \tag{4.47}$$

where Q_{p} is the **tidal dissipation parameter**, which depends on the structure of the planet. As Figures 4.14 and 4.17 show, planets with higher values of $\dot{E}_{\text{tide}}$ are larger throughout their evolution. The models shown calculate the radius evolution assuming that $\dot{E}_{\text{tide}}$ remains constant throughout, but the tidal dissipation will have the effect of reducing the eccentricity and circularizing the orbit. Less importantly, the structure of the planet, and hence its response to tidal forces, might be expected to change as the planet evolves. At least five transiting exoplanets have significantly eccentric orbits, $e > 0.15$, and for these the tidal heating is expected to be significant. These may be relatively young planets, or they may be subject to dynamical interactions with other planets. We will return to this topic in more detail in Chapter 7.

4.5 The mass–radius diagram

4.5.1 Densities of transiting exoplanets

The single most important fact about the transiting exoplanets is that we can measure their radii, and hence make deductions about their structure, composition and evolutionary histories. As we saw in Section 4.4, there are many parameters that jointly govern the radius of a giant planet. Of these, the quantity that we can most accurately and straightforwardly measure is the mass. Consequently, the transiting exoplanet mass–radius plot, Figure 4.18, is one of the keystones of our knowledge about exoplanets.

In Figure 4.18a it is striking that R_{P} is approximately constant for the most massive planets. This is presumably because their size is largely governed by the physics of degenerate matter. As mass increases into the brown dwarf range, size remains approximately constant for this reason, as we mentioned in Chapter 2.

Figure 4.18b immediately makes two facts very clear: there is a very wide range of densities at masses $0.4\,\text{M}_{\text{J}} < M_{\text{P}} < 1.7\,\text{M}_{\text{J}}$; and almost all the transiting exoplanets in this range are less dense than Jupiter. In fact, among exoplanets with $M_{\text{P}} \leq 2\,\text{M}_{\text{J}}$, there is only a handful with a density greater than Jupiter's. This is made even clearer in Figure 4.18c, which, being on a logarithmic scale, emphasizes the low-mass end of the distribution. CoRoT-7 b is the lowest-mass object, and being the only identified terrestrial planet in the sample, it sits alone in its part of the diagram.

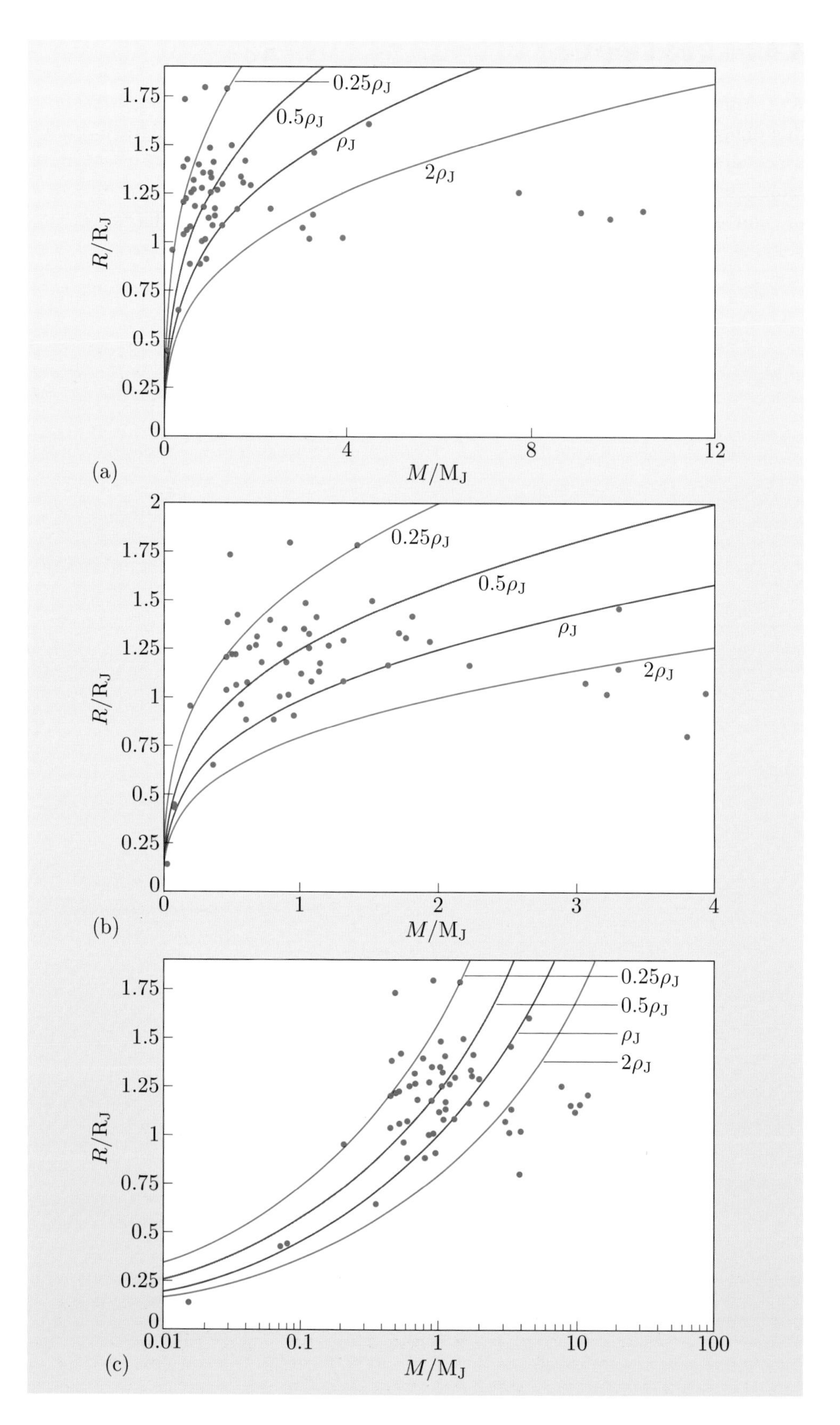

Figure 4.18 The masses and radii of known transiting exoplanets. The curves indicate lines of constant density, with values varying from a quarter of Jupiter's density ($0.25\rho_J$) to twice Jupiter's density ($2\rho_J$). (a) All planets plotted on a linear mass scale. (b) The densely populated region with $M_P < 4\,M_J$. (c) The entire population plotted on a logarithmic mass scale. This emphasizes the lowest-mass planets.

4.5.2 Compositions of transiting exoplanets

As Figure 4.19 illustrates, the measured radius of a planet gives us some idea of that planet's composition. We need to proceed with caution, however: differing proportions of metals, silicates, water and H_2 can give the same total masses and radii. Figure 4.20 shows example model calculations for four planets, including the unusually dense planet HD 149026 b, which is thought to have a large core of heavy elements.

GJ 436 b and HAT-P-11 b are the only known Neptune-sized transiting exoplanets (November 2009). It appears that such planets are intrinsically rare, at least among the close-in planets that have the highest probability of transiting. The discovery of CoRoT-7 b, which has an order of magnitude smaller transit depth, makes it clear that current technology is easily capable of detecting Neptune-sized transiting planets. There could, however, be selection effects at work here. Much

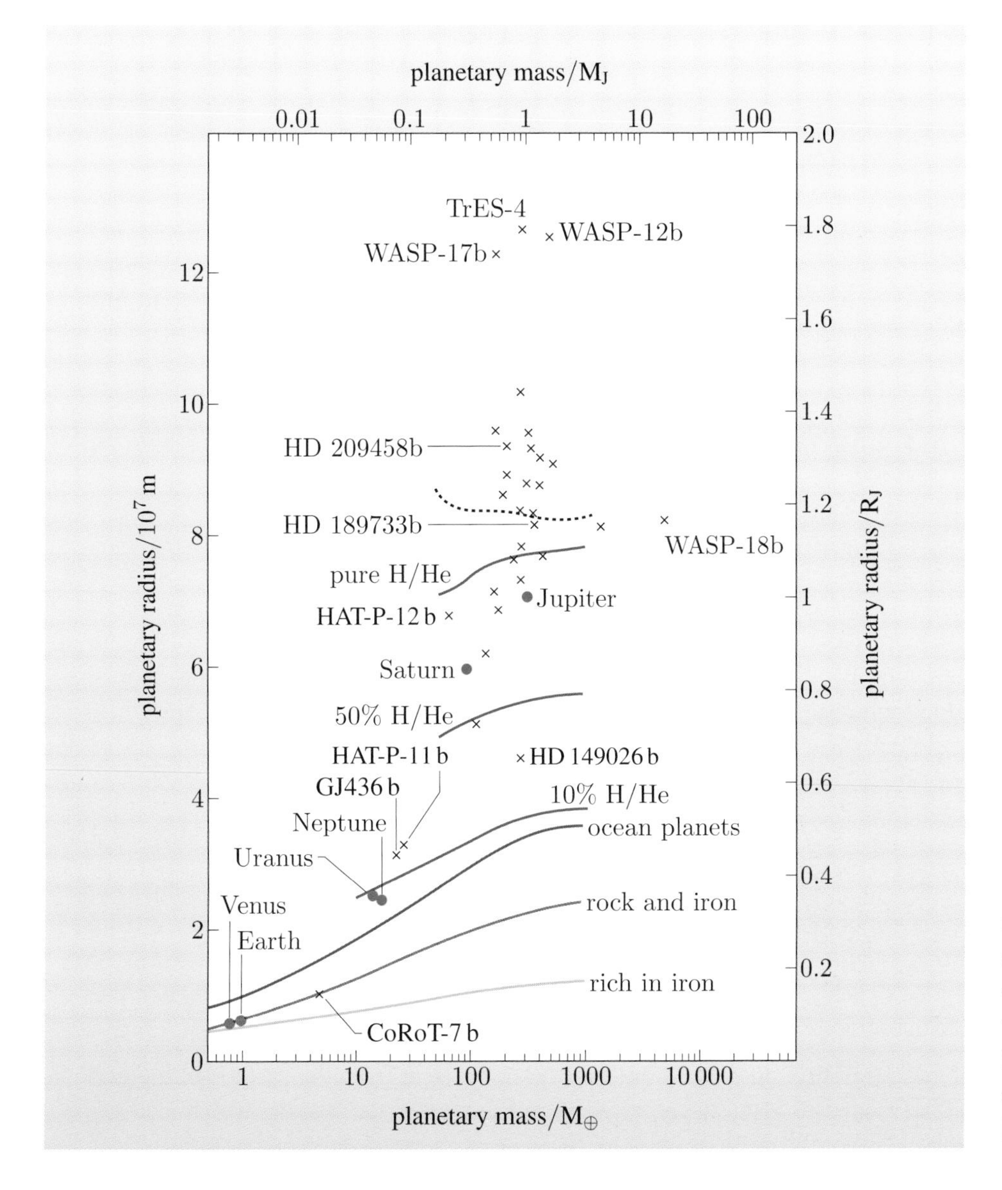

Figure 4.19 The mass–radius relationship for transiting exoplanets and Solar System objects, with expected radii for various simplified compositions indicated. The dotted line is the expected upper limit for pure H/He gas.

interest focused on the discovery of the first terrestrial exoplanet, therefore CoRoT-7 b was a highly-prized discovery. Researchers may not have been motivated to the same degree to vigorously and promptly confirm further exo-Neptunes. This sort of psychological selection effect is almost impossible to quantify. On the bright side, as time passes and the known population grows, it will become clear whether the paucity of exo-Neptunes is real.

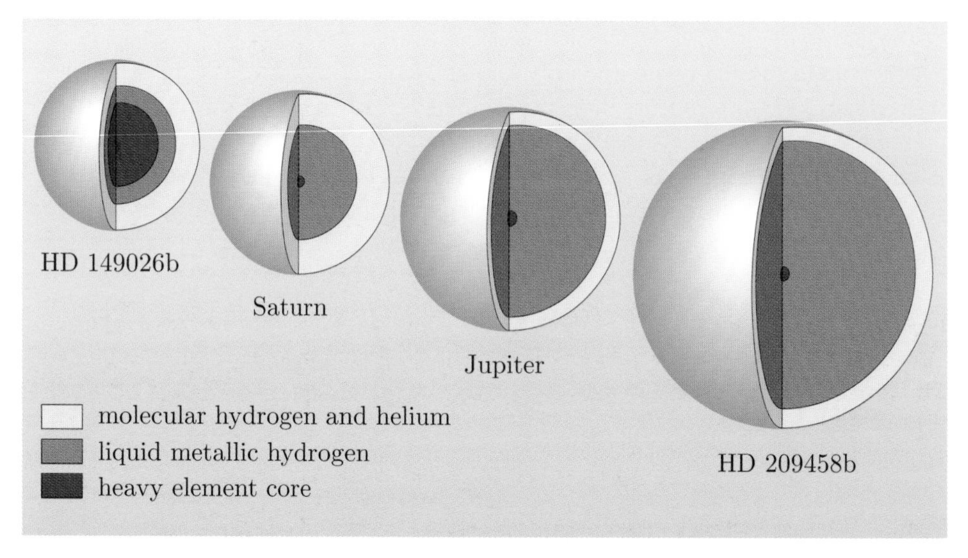

Figure 4.20 Model structures for four giant planets, including the unusually dense HD 149026 b.

4.5.3 Surface gravities of transiting exoplanets

The surface gravity, g, of a planet can be calculated from

$$g = \frac{GM_{\rm P}}{R_{\rm P}^2}. \tag{4.48}$$

It is therefore straightforwardly related to a planet's position in Figure 4.18. Of all the quantities that we might consider for a transiting planet, g has particular importance because it is one of the few things that can be obtained purely empirically. It can be expressed as

$$g = \frac{2\pi}{P} \frac{(1-e^2)^{1/2} A_{\rm RV}}{(R_{\rm P}/a)^2 \sin i}, \tag{4.49}$$

where e and $A_{\rm RV}$ can be obtained from the host star's reflex RV curve, the ratio $R_{\rm P}/a$ can be obtained from the transit light curve, and P can be obtained from either. Figure 4.21 shows the values of g for transiting exoplanets known in 2008 plotted against P. There is a weak but very noticeable correlation between these two quantities.

Just as on Earth, g governs how easy it is to escape from the surface of a planet. The causes underlying the correlation in Figure 4.21 are not known. It is also not clear whether the correlation is an intrinsic property of the Galaxy's exoplanet population or whether it is a selection effect. A plausible explanation may be that short orbital period planets are close to their host stars and their surface layers

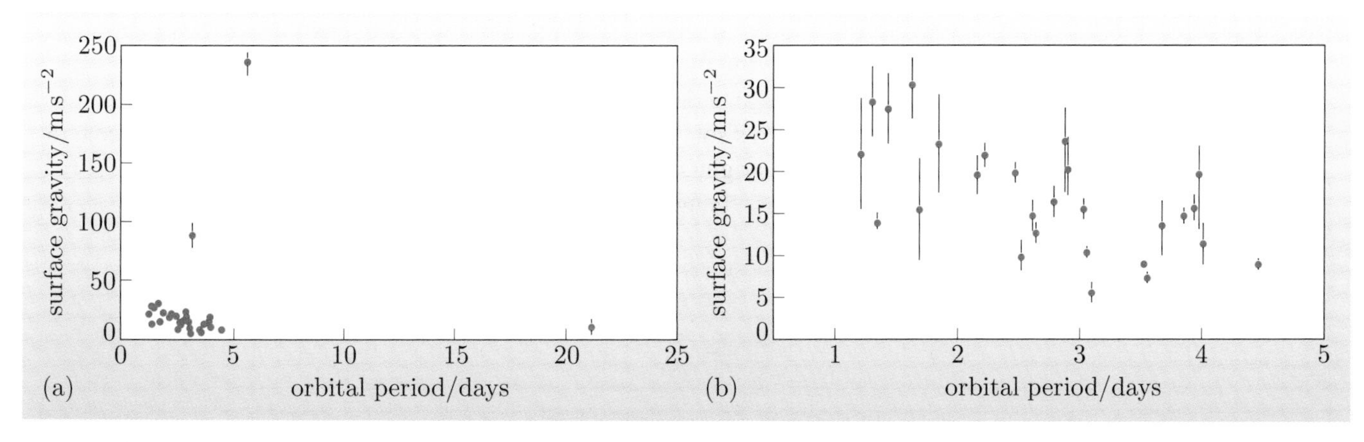

Figure 4.21 (a) Surface gravity plotted against orbital period for transiting exoplanets. (b) A zoom-in on the densely populated lower left corner of panel (a).

may have been photo-evaporated, leaving only a relatively dense former planet core. If this has happened, the closest-in planets, which also have the shortest orbital periods, will have the highest densities and consequently the highest values of g.

4.6 What can we say about the Galaxy's population of planets?

We have been careful to discuss the *known* exoplanet population. Because this population is strongly affected by the selection effects that we have discussed throughout the chapter, it is not straightforward to take these demographics and use them to infer properties of the actual population of planets present in our Galaxy. We have introduced notation to describe the actual population: $\alpha_{\rm P}(a, M_{\rm P})$ is a function describing the distribution of planet frequency in terms of semi-major axis and planet mass. The discovery of the hot Jupiter (or equivalently Pegasid) planets in the mid-1990s implied that $\alpha_{\rm P}(a \approx 0.05\,{\rm AU},\ M_{\rm P} \approx {\rm M_J})$ was much larger than had been previously assumed. In Subsections 2.2.6 and 2.2.7 we showed how we can apply a mathematical description of the selection effects to evaluate the expected number of transit detections as a function of the underlying distribution $\alpha_{\rm P}(a, M_{\rm P})$. Of course, the detectability of a planet's transits depends on its radius, rather than its mass, and as we saw in Section 4.5, there is no straightforward relationship between these two properties. This means that a function $\beta_{\rm P}(a, R_{\rm P})$ describing the underlying population would be more easily related to the known transiting exoplanet population, but as we have seen, $R_{\rm P}$ is related to $M_{\rm P}$ and all of the other properties discussed in Section 4.5, and through this relationship, for each individual planet $\alpha_{\rm P}(a, M_{\rm P})$ is related to $\beta_{\rm P}(a, R_{\rm P})$.

Population synthesis can be used to examine the consistency of the known exoplanet population with various hypotheses about $\alpha_{\rm P}(a, M_{\rm P})$ and about the planet formation processes that lead to the Galaxy's population. Population synthesis has a long history in other areas of astrophysics. In its application to exoplanets, an underlying distribution $\alpha_{\rm P}(a, M_{\rm P})$ is calculated from planet formation models, i.e. a planet population is synthesized. Then a mathematical description of the selection effects and the coverage of one or more discovery surveys is applied to generate a prediction of the population that should have been discovered assuming that the adopted $\alpha_P(a, M_P)$ accurately described the real planet population. Generally, of course, the predictions do not match the actual known population, and this is used to adjust the models so that a new $\alpha_P(a, M_P)$, which provides a better match, is generated. While this process is clearly necessary to determine the properties of the Galaxy's planet population, there are many, many more-or-less free parameters involved. A full discussion of the process is beyond the scope of this book, and would probably be premature, as the known population is growing so rapidly at the time of writing (November 2009).

Nonetheless, there are some empirically constrained properties. Figure 4.8b,c show examples where the statistics of the RV planet host star population have been constrained. This figure strongly suggests that metal-rich stars are more likely to host giant planets, and was generated by carefully accounting for null results as well as discoveries. Because scientists are human, and enjoy the thrill of discovery, null results are often under-reported. This is a shame, as a figure like Figure 4.8b is much more informative than simply plotting the number of discovered planets as a function of host star metallicity. Even here, though, we must be cautious in stating that there are more giant planets around high-metallicity stars than around low-metallicity stars. We cannot completely describe the selection effects that govern the detectability of a planet by the RV technique. As Figure 4.1 shows, stars themselves introduce noise into the velocity curve, and the properties of this noise depend on the detailed structure and composition of each individual star. An absence of a planet detection does not necessarily imply that there is no planet present. Having said all this, Figure 4.8b is highly suggestive, and the straightforward interpretation does make sense in the context of the composition of Solar System objects: the giant planets are metal-rich compared to the Sun, and rocky objects obviously are too.

Summary of Chapter 4

1. All exoplanet detection methods are subject to selection effects. The RV technique, responsible for most discoveries, favours massive, close-in planets with eccentric orbits around low-mass host stars. These produce the largest RV amplitude:

$$A_{RV} = \frac{2\pi a M_P \sin i}{(M_P + M_*)P\sqrt{1 - e^2}}. \quad \text{(Eqn 1.13)}$$

   Detection by the RV technique also depends on the sharpness, strength and stability of the host star's spectral lines. Finally, the RV measurements must span a baseline that exceeds the planet's orbital period.

2. Transit search programmes preferentially find large, close-in planets around small, bright host stars:

$$\text{transit detection probability} \propto R_P^3\, a^{-7/4} L_*^{3/2} R_*^{-5/4}. \quad \text{(Eqn 4.1)}$$

   The population of known transiting exoplanets is consequently strongly biased towards short orbital periods and small orbital separations.

3. The peak in the RV-discovered planet mass distribution is at higher mass than that of the transit-discovered planets. The RV planets have semi-major axes up to 5 AU, with the peak at $\sim$1 AU.

4. Metallicity is measured in dex, a logarithmic scale relative to the Sun. RV planets with $M_\mathrm{P} \sim \mathrm{M_J}$ are preferentially found around main sequence stars of high metallicity, and around stars that are more massive than the Sun. A star of spectral type F5–A5 is 4 times more likely than a star of spectral type M4–K7 to be found to host a Jupiter-mass RV planet. Transit-discovered planet host stars are generally less metal-rich than RV-discovered planet hosts.

5. The mass distribution of known exoplanets rises strongly towards small masses, despite low-mass planets being the most difficult to detect:

$$\frac{\mathrm{d}N_\mathrm{P,RV\text{-}known}}{\mathrm{d}M_\mathrm{P}} \propto M_\mathrm{P}^{-1.6}. \qquad \text{(Eqn 4.6)}$$

HARPS can attain precision of $\sim 20\,\mathrm{cm\,s^{-1}}$ (less than walking speed!), and is now (April 2009) discovering exoplanets with masses as low as $M_\mathrm{P} \sin i = 0.0061\,\mathrm{M_J} \approx 2\,\mathrm{M_\oplus}$. These are super-Earths or mini-Neptunes.

6. The CoRoT satellite has begun to find transits of depth $\sim 0.03\%$ and has found the first transiting terrestrial-sized exoplanet, CoRoT-7 b. Neptune-like planets appear to be rare.

7. Giant planets, like stars, have structures governed by hydrostatic equilibrium, mass continuity and energy conservation:

$$\frac{\mathrm{d}P(r)}{\mathrm{d}r} = -\frac{G\,m(r)\,\rho(r)}{r^2}, \qquad \text{(Eqn 4.8)}$$

$$\frac{\mathrm{d}m(r)}{\mathrm{d}r} = 4\pi r^2 \rho(r), \qquad \text{(Eqn 4.9)}$$

$$\frac{\mathrm{d}L(r)}{\mathrm{d}r} = 4\pi r^2 \left(\varepsilon(r) - \rho(r)\,T(r)\,\frac{\partial S(r,t)}{\partial t}\right) \qquad \text{(Eqn 4.10)}$$

along with the temperature gradient, which depends on the heat transport mechanisms. The central pressure is

$$P_\mathrm{c} \approx \frac{GM^2}{R^4}. \qquad \text{(Eqn 4.15)}$$

This is $\sim 1 \times 10^{13}$ Pa or $\sim 1 \times 10^{8}$ bar for Jupiter.

8. Giant planet interiors are supported by pressure from degenerate non-relativistic electrons. A gas is degenerate if the number density exceeds the quantum concentration, $n > n_\mathrm{Q}$, where

$$n_\mathrm{QNR} = \left(\frac{2\pi mkT}{h^2}\right)^{3/2}, \qquad \text{(Eqn 4.16)}$$

or equivalently if the temperature is less than the Fermi temperature, $T < T_\mathrm{F}$, where

$$T_\mathrm{F} = \frac{n^{2/3}h^2}{2\pi mk}. \qquad \text{(Eqn 4.18)}$$

9. The virial theorem

$$3\int_0^\mathrm{S} \frac{P}{\rho}\,\mathrm{d}m + E_\mathrm{GR} = 0 \qquad \text{(Eqn 4.20)}$$

is widely applied in astrophysics. It relates the gravitational and internal energy of a gravitationally bound object. The equation of state determines P/ρ. The virial theorem can be expressed as

$$\zeta E_{\rm I} + E_{\rm GR} = 0, \qquad \text{(Eqn 4.23)}$$

where

$$\zeta \equiv \frac{\int_0^M 3(P/\rho)\,dm}{\int_0^M u\,dm} \quad \text{and} \quad E_{\rm I} \equiv \int_0^M u\,dm.$$

The luminosity is related to the gravitational contraction by

$$L = -\frac{\zeta - 1}{\zeta}\dot{E}_{\rm GR}. \qquad \text{(Eqn 4.25)}$$

10. The equation of state for giant planets is complex. The essence of the early evolution is the contraction of H_2 gas, for which $\zeta = 3.2$ and

$$L = -0.69\dot{E}_{\rm GR}. \qquad \text{(Eqn 4.37)}$$

Once $n > n_{\rm QNR}$, the gravitational energy released by contraction is entirely absorbed by the degenerate electrons, $\zeta = 2$ and the luminosity is powered by the cooling of the ions:

$$\dot{E}_{\rm e} \approx -\dot{E}_{\rm GR} \qquad \text{(Eqn 4.33)}$$

$$\approx -2L \qquad \text{(Eqn 4.35)}$$

and

$$L \approx -\dot{E}_{\rm ion}. \qquad \text{(Eqn 4.36)}$$

11. The luminosity available from gravitational contraction is the Kelvin–Helmholtz luminosity

$$L_{\rm KH} \approx \frac{GM_{\rm P}^2}{\tau R_{\rm P}}. \qquad \text{(Eqn 4.42)}$$

Mature giant planets are almost incompressible and

$$L \approx \eta\frac{GM_{\rm P}^2}{\tau R_{\rm P}}, \qquad \text{(Eqn 4.43)}$$

where η is given by the degeneracy parameter, θ:

$$\eta \approx \frac{\theta}{\theta + 1}. \qquad \text{(Eqn 4.44)}$$

12. The size of a giant planet is jointly governed by its age, total mass, core mass, metallicity, degree of irradiation and tidal heating. The last of these factors is particularly important for close-in planets in eccentric orbits:

$$\dot{E}_{\rm tide} \propto \frac{e^2}{Q_{\rm p}}, \qquad \text{(Eqn 4.47)}$$

where $Q_{\rm p}$ is the tidal dissipation parameter.

13. The mass–radius diagram is a keystone of comparative planetology. For $M_{\rm P} > 6\,{\rm M_J}$, radius is approximately constant. For $0.4\,{\rm M_J} < M_{\rm P} < 1.7\,{\rm M_J}$, there is a very wide range of densities, but almost all known transiting exoplanets in this mass range are less dense than Jupiter.

14. The surface gravity of a transiting exoplanet can be determined purely empirically:

$$g = \frac{2\pi}{P} \frac{(1 - e^2)^{1/2} A_{\text{RV}}}{(R_{\text{P}}/a)^2 \sin i} \qquad \text{(Eqn 4.49)}$$

and g is correlated with P.

15. The known exoplanet population is the result of the actual Galactic population and strong selection effects, some of which are difficult to quantify. Population synthesis examines this and attempts to deduce $\alpha_{\text{P}}(a, M_{\text{P}})$ and to guide our developing understanding of planet formation.

Chapter 5 Transmission spectroscopy and the Rossiter–McLaughlin effect

Introduction

In Chapter 3 we discussed things that we can learn from analysis of the transit light curve. We considered the planet as an opaque disc occulting a limb darkened stellar disc, and derived expressions for the eclipse area, A_e, and the drop in flux, ΔF, as a function of time. We examined how the light curve allows us to infer precise values for the properties of the star–planet system. It is remarkable that we are able to do this for planets orbiting around stars neither of which we can resolve: all we have to work with is an unresolved point of light and our ingenuity. This chapter examines even more remarkable possibilities that exploit the wavelength-dependence of the transit phenomenon.

In Section 5.2 we will discuss **transmission spectroscopy**, a technique that reveals details about the chemical composition and the extent of the planet's atmosphere. It seems almost science fantasy that we can measure these properties for planets that we can't see, whose presence is only inferred indirectly from observations of the host star. Section 5.3 examines how the transit affects the measured stellar radial velocity: the **Rossiter–McLaughlin effect** is a perturbation to the radial velocity curve that is induced by the planet's disc passing across the disc of a spinning star, and consequently blocking surface elements with different Doppler shifts at different times. Before we consider these two techniques, we first introduce the **equilibrium temperature**, which is one of the most important parameters characterizing a planet.

5.1 The equilibrium temperature

As we discussed in Section 1.1, the closer a planet is to its host star, the more insolation it intercepts per unit area, and hence the hotter it is. We quantify this statement with the equilibrium temperature. We can derive an expression for this temperature based on the simple principle that in equilibrium, the energy received from the host star per unit time is exactly equal to the thermal energy radiated away by the planet per unit time. The planet neither heats up nor cools down, but remains at the equilibrium temperature, T_{eq}. We will assume that the entire planet is characterized by a single temperature. This will be a good approximation if the energy received by the irradiated hemisphere is efficiently transported to the night side (see Figure 5.1).

The host star has luminosity L_*, which is radiated outwards with spherical symmetry. The energy per unit area per unit time at distance d from the star is the flux

Equation 5.1 is identical to Equation 2.2 but uses different notation.

$$F_* = \frac{L_*}{4\pi d^2}, \qquad (5.1)$$

so, as illustrated in Figure 5.1, a planet with radius R_P at a distance a from the star will intercept energy at a rate

$$\dot{E}_{\text{intercept}} = \pi R_P^2 F_*. \qquad (5.2)$$

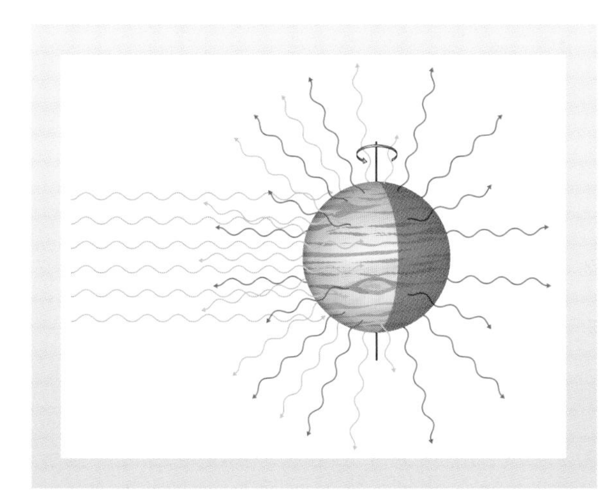

Figure 5.1 A planet receives energy through the insolation of one hemisphere by the host star, reflects some of the incident starlight and emits thermal radiation (indicated in red) from its entire surface area. In general, planets spin, so that each part of the surface spends some time in the irradiated hemisphere, and it is reasonable to assume that the entire surface reaches a single equilibrium temperature.

Note that the area that the planet presents to the radiation from the star is the area of its disc, πR_P^2, rather than its surface area; as shown in Figure 5.1, no irradiation of the night side occurs. The planet reflects a fraction A of the intercepted light, where A is the albedo, consequently energy is absorbed at a rate

$$\dot{E}_\text{abs} = (1 - A)\pi R_\text{P}^2 F_*. \tag{5.3}$$

If we make the simplest possible assumption, that the planet radiates uniformly as a black body at a single temperature, then the thermal flux emitted is

$$\dot{E}_\text{emit} = 4\pi R_\text{P}^2 \sigma T^4. \tag{5.4}$$

In equilibrium, the energy emitted is exactly balanced by the energy absorbed. The equilibrium temperature, T_eq, is thus defined by equating $\dot{E}_\text{emit}$ with $\dot{E}_\text{abs}$:

$$4\pi R_\text{P}^2 \sigma\, T_\text{eq}^4 = (1 - A)\pi R_\text{P}^2 F_*.$$

Making T_eq the subject of the equation, therefore, we obtain

$$\begin{aligned} T_\text{eq} &= \left[\frac{(1 - A)F_*}{4\sigma}\right]^{1/4} \\ &= \frac{1}{2}\left[\frac{(1 - A)L_*}{\sigma\pi a^2}\right]^{1/4}, \end{aligned} \tag{5.5}$$

where we have used Equation 5.1 in the last step, and set $d = a$.

- Why is the area in Equation 5.2 not the surface area of the irradiated hemisphere, i.e. $2\pi R_\text{P}^2$?

- Because the radiation from the star is uniformly spread over a sphere centred on the star. The planet intercepts a fraction of this radiation determined by the fraction of this sphere blocked by the planet. The fraction depends only on the cross-sectional area that the planet presents to the star. Suitably oriented cylinders or ellipsoids or cones with cross-section πR_P^2 would all intercept the same fraction of the star's radiation: the surface area of the intercepting object does not matter, but the cross-sectional area does.

Exercise 5.1 (a) Calculate the equilibrium temperature for the Earth. You may assume that the albedo of the Earth is $A_{\oplus} = 0.30$. (b) Comment on your value.

Exercise 5.2 The exoplanet HD 209458 b is in an orbit with $a = 0.04707$ AU around a star of luminosity $L_* = 1.61 \pm 0.15\,L_{\odot}$. Show that the planet's equilibrium temperature is

$$T_{\mathrm{eq,HD\,209458\,b}} \approx 1400(1 - A)^{1/4}\,\mathrm{K}, \tag{5.6}$$

where A is the (unknown) albedo. ■

5.2 Transmission spectroscopy

Transiting planets give us a remarkable opportunity to make empirical measurements of the properties of their atmospheres. Just as the absorption lines in a stellar spectrum allow us to deduce the chemical composition and physical properties of stellar atmospheres, the technique of transmission spectroscopy allows us to measure the composition and properties of exoplanet atmospheres. This general area of astrophysics began in the early nineteenth century when Wollaston and Fraunhofer noticed the absorption lines in the solar spectrum, and was developed by Lockyer who discovered the chemical element helium in the solar spectrum.

As we derived in Chapter 3, the presence of a transit allows us to measure the radius of the transiting planet. In that discussion we mentioned the dependence of the transit light curve on wavelength as a result of limb darkening; there is another cause of wavelength-dependence of the transit light curve, which we will now discuss. This dependence is potentially extremely informative, as we will see. Figure 5.2 illustrates the basic principle. In general, an exoplanet will possess an atmosphere, so the planet is not in fact an opaque disc with a sharp edge. Instead, the opacity will gradually diminish with height as the atmosphere becomes more tenuous.

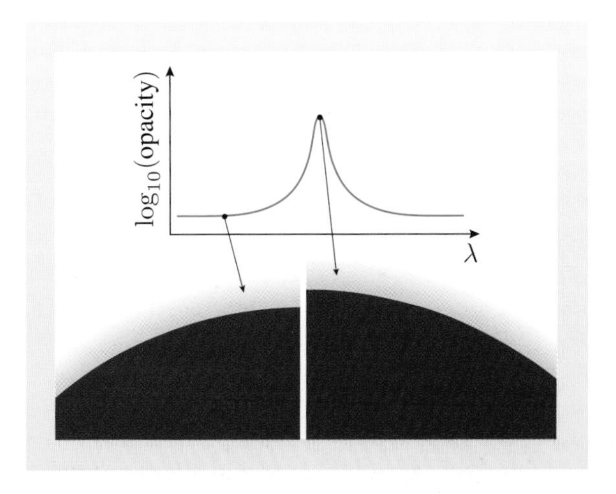

Figure 5.2 A schematic illustration of how the measured radius of a transiting planet depends on wavelength. At the wavelength corresponding to an atomic or molecular spectral line, the opacity is much higher than in the continuum. Therefore the atmosphere at a particular height may be transparent to continuum photons, but opaque to photons at the wavelengths of spectral lines of chemical species present in the atmosphere.

Generally, detecting the wavelength-dependence of the transit depth is a challenging observation, as we are seeking the changes in the depth of a feature that is only ~1% deep. To detect the wavelength-dependence, obviously we must separate the light according to its wavelength. This can be done either using filters that transmit only light of a specific wavelength range, or more commonly using a spectrograph to record the observed spectrum as a function of time.

Worked Example 5.1

The **spectral resolution**, R, of a spectrograph is wavelength, λ, divided by the width of the resolution element, $\delta\lambda$. The exposure time to measure a transit light curve must be no more than about a minute, otherwise the ingress and egress are not resolved. A total of N photons are detected in a spectrum covering wavelength range $\lambda_l < \lambda < \lambda_u$. If these are summed to provide a single measurement of the brightness of the star, the signal-to-noise ratio is SN.

(a) Assuming that the signal-to-noise ratio is determined purely by Poisson statistics, derive an expression for the signal-to-noise ratio, SN_R, for these observations if they are used to measure the brightness of the star at each wavelength in the spectrum with resolution R. State any assumptions that you make.

(b) If the original spectrum covered the range $\lambda_l = 450\,\mathrm{nm} < \lambda < \lambda_u = 850\,\mathrm{nm}$, state the factor by which the signal-to-noise ratio is diminished for observations at resolution R.

Solution

(a) If N photons are detected, and the signal-to-noise ratio is determined purely by Poisson statistics, the uncertainty in the measurement will be $\sqrt{N}$. So

$$\mathrm{SN} = \frac{\mathrm{Signal}}{\mathrm{Noise}} = \frac{N}{\sqrt{N}} = \sqrt{N}.$$

The N photons in the spectrum are spread out over the wavelength range $\lambda_l < \lambda < \lambda_u$. We have no information on the shape of the spectrum, so we will assume that they spread out so each resolution element will have a fraction

$$\frac{\delta\lambda}{\lambda_u - \lambda_l}$$

of the total counts. Thus the number of counts, M, in a single resolution element is given by

$$M = N\frac{\delta\lambda}{\lambda_u - \lambda_l} = NR^{-1}\frac{\lambda}{\delta\lambda}\frac{\delta\lambda}{\lambda_u - \lambda_l} = NR^{-1}\frac{\lambda}{\lambda_u - \lambda_l},$$

where we have introduced the spectral resolution, R, by multiplying through by its definition in terms of λ and $\delta\lambda$, and dividing by R itself; the net effect is to multiply by unity. Applying Poisson statistics, the uncertainty on the counts in a single resolution element is $\sqrt{M}$, so the signal-to-noise ratio is

$$\mathrm{SN}_R = \frac{M}{\sqrt{M}} = \sqrt{M} = \sqrt{NR^{-1}\frac{\lambda}{\lambda_u - \lambda_l}}$$

$$= \mathrm{SN}\,R^{-1/2}\left(\frac{\lambda}{\lambda_u - \lambda_l}\right)^{1/2}.$$

(b) We are given values $\lambda_l = 450\,\text{nm}$, $\lambda_u = 850\,\text{nm}$, so the mid-point of the spectrum is $\lambda = 650\,\text{nm}$. Using these numbers,

$$\text{SN}_R = \text{SN}\,R^{-1/2}\left(\frac{650\,\text{nm}}{850\,\text{nm} - 450\,\text{nm}}\right)^{1/2} = \text{SN}\,R^{-1/2}\left(\frac{650}{400}\right)^{1/2}$$
$$= 1.27\,\text{SN}\,R^{-1/2}.$$

The signal-to-noise ratio at resolution R is diminished by a factor of about $0.8R^{1/2}$ relative to that of the summed spectrum.

As we saw in Worked Example 5.1, the photon-counting signal-to-noise ratio for wavelength-resolved light curves is reduced compared to a broad-band light curve by a factor of about $\sqrt{R}$, where R is the spectral resolution. Even for the brightest transiting exoplanet host stars, it is a challenge to obtain sufficiently high signal-to-noise ratios, and consequently the results so far focus on a few particularly strong spectral features observed in transits of the brightest known host stars. As we discussed in Chapter 2, transit light curves are often dominated by systematic errors arising from imperfect correction for the effects of the Earth's atmosphere. For this reason, most of the transmission spectroscopy results have come from observations made with the Hubble Space Telescope. Before we study some of these stunning observational results, we first consider what determines the position of the edge of the planet's disc at continuum wavelengths.

Basic atmospheric structure

If we assume an ideal gas equation of state, then the pressure in a planet's atmosphere is given by

$$P = nkT, \tag{5.7}$$

where n is the number density of particles, and k is Boltzmann's constant. In general the atmosphere will be composed of many different species, so $n = \sum n_i$, and the pressure, P, is the sum of the partial pressures due to each species. The mass density, ρ, is of course related to the number density, n:

$$\rho = n\mu u,$$

where μ is the **mean molecular mass**, i.e. the average value for the mass of a particle in the atmosphere in units of the atomic mass unit, u. Consequently, we can express the pressure as

$$P = \frac{\rho kT}{\mu u}. \tag{5.8}$$

If the atmosphere is in hydrostatic equilibrium, then the downwards force of gravity must be balanced by the pressure gradient. This is illustrated in Figure 5.3, where the **atmospheric scale-height**, H, has been assumed to be negligibly small compared with the radius of the planet, so we can treat the geometry as though the atmosphere is plane-parallel. The element of

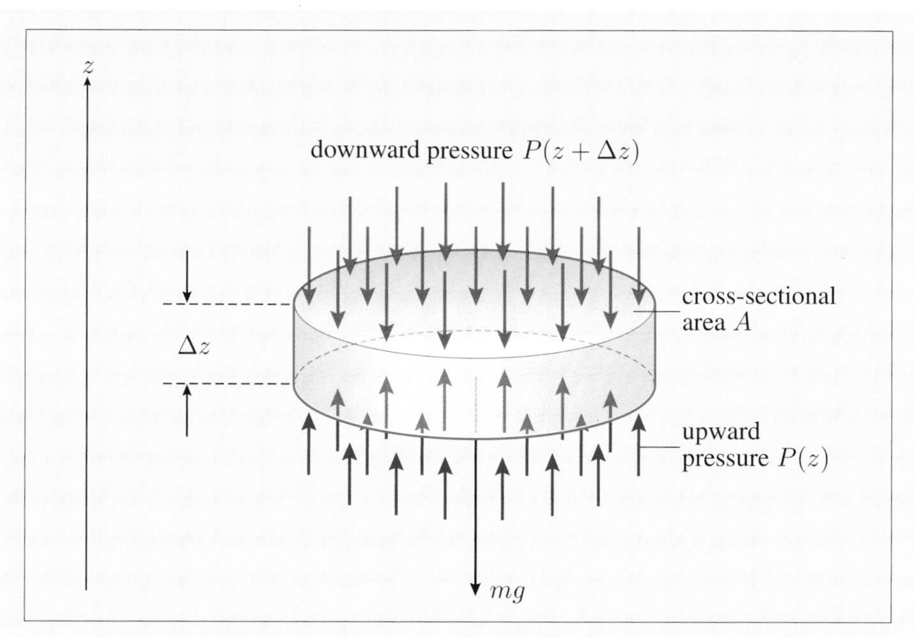

Figure 5.3 A cartoon showing the balance of forces on a plane-parallel atmosphere.

atmosphere shown in Figure 5.3 has cross-sectional area A, and consequently experiences a downwards pressure force of $A\,P(z + \Delta z)$ on its upper surface that is balanced by the difference between the upwards pressure force on the lower surface and the downwards force of gravity. Since the mass, m, of the element is given by

$$m = \rho V = \rho A\,\Delta z,$$

then the balance of forces is

$$A\,(P(z + \Delta z) - P(z)) = \rho A\,\Delta z\,g,$$

where g is the acceleration due to gravity at height z in the planet's atmosphere. We implicitly assume in our treatment here that the variation in g with z may be neglected. This is akin to our explicit assumption that we may adopt a plane-parallel geometry: both assumptions are effectively that $H \ll R_{\rm P}$. Cancelling A, dividing through by Δz and taking the limit $\Delta z \to 0$, we obtain

$$\frac{\mathrm{d}P}{\mathrm{d}z} = -\rho g = -\frac{P\mu u}{kT}g.$$

If we assume that the variation in μ and T with z may be neglected, this is one of the simplest forms of differential equation, and has the solution

$$P = P_0 \exp(-z/H), \tag{5.9}$$

where H is the atmospheric scale-height and is given by

$$H = \frac{kT}{g\mu u}. \tag{5.10}$$

5.2.1 Atmospheric transparency and clouds

In a typical planetary atmosphere, energy is transported outwards predominantly by convection until the column through the layers above becomes optically thin. This level is called the **tropopause**. Above the tropopause, radiation is the primary means by which energy is transported outwards. The pressure at which the transition occurs depends on the atmospheric composition, the gravity, and temperature. Gas giant planets tend to have radii that are more or less independent of mass (cf. Figure 4.18) so all else being equal, higher-mass planets have higher gravity.

The column density, N, above a particular height, z, is given by

$$N = \int_z^\infty n\,\mathrm{d}z'. \tag{5.11}$$

Using Equations 5.7 and 5.9, the column density is therefore

$$\begin{aligned} N &= \int_z^\infty \frac{P(z')}{kT(z')}\,\mathrm{d}z' \\ &= \int_z^\infty \frac{P_0}{kT(z')} \exp(-z'/H)\,\mathrm{d}z'. \end{aligned} \tag{5.12}$$

If we make the approximation that the temperature variation with z may be neglected, then

$$\begin{aligned} N &\approx \frac{1}{kT} \int_z^\infty P_0 \exp(-z'/H)\,\mathrm{d}z' \\ &\approx \frac{1}{kT} \left[-HP_0 \exp(-z'/H)\right]_z^\infty \\ &\approx \frac{HP_0}{kT} \exp(-z/H) \\ &\approx \frac{H\,P(z)}{kT} \\ &\approx \frac{P(z)}{g\mu u}, \end{aligned} \tag{5.13}$$

where the definite integral vanished at the upper limit, so we were left with the quantity at the lower limit and thus cancelled the minus sign. In the final step we substituted for H from Equation 5.10.

The column density above a particular level determines the optical depth, and hence the location of the tropopause. If the composition and hence μ are held constant, then Equation 5.13 tells us that the column density above a particular pressure is inversely proportional to the planet's gravity. Thus in general higher-mass giant planets should have a lower column density above any specified pressure level, and consequently their atmospheres in general should be more transparent at any particular pressure level.

- Two otherwise identical gas giant planets have differing mass due to the presence of a large rocky core in one of them. Which one's tropopause occurs at the higher pressure?

- The tropopause occurs at the level where the atmosphere above becomes transparent, so the higher-gravity planet will have the tropopause at the higher pressure level (cf. Equation 5.13). Since the radii are identical, the higher-mass planet will have the higher gravity. Thus the higher-mass planet has the tropopause at the higher pressure level.

- Under what circumstances could the approximation in Equation 5.13 be justifiably written as an equality?

○ If the temperature was isothermal, i.e. non-varying, so taking $T(z)$ outside the integral was unequivocally justified. We also require $H \ll R_P$ as used in the derivation of Equation 5.9.

The visible disc of Jupiter is composed of ammonium clouds, with pollutants, that reflect the incident sunlight. Clouds are generally expected to form in planetary atmospheres, and have a vital role in governing the thermal structure and the visible disc. Clouds are, unfortunately, rather difficult to reduce to a simple analytical description that can produce robust general predictions. The dominant source of uncertainty in Earth climate modelling is the behaviour of clouds as a function of temperature. Figure 5.4 shows the sort of cloud structure that is expected for Jupiter-like planets at three different temperatures.

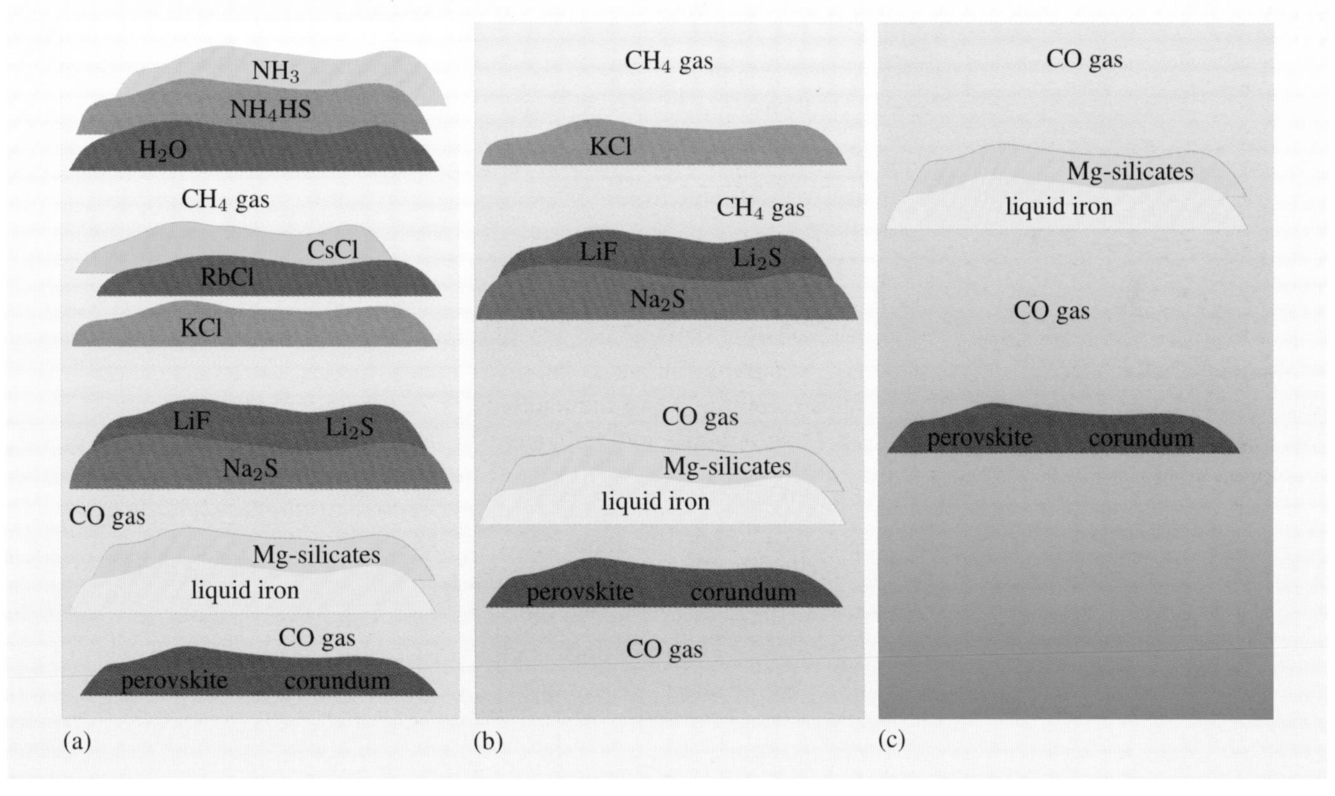

Figure 5.4 A schematic illustration of the cloud structure expected for (a) Jupiter's atmosphere, (b) a Jupiter-like planet too warm for water clouds to form, and (c) a typical hot Jupiter found in transit searches. The overall cloud structure is similar in each panel, but cloud levels move upwards as the atmosphere is warmed. The clouds are labelled with the condensates of which they are formed: perovskite is $CaTiO_3$; corundum is Al_2O_3, which is also the mineral in rubies and sapphires.

If we move up through a planet's atmosphere, the first clouds will be encountered when the temperature falls sufficiently for a species to condense at the ambient pressure and density. Thus the lowest clouds in each of the three panels are the same, but they occur at larger pressures and densities (i.e. deeper within the atmosphere) for the cooler planets. As the temperature is increased from Jupiter's, the uppermost cloud layers are progressively missing: hotter atmospheres do not condense the more volatile condensates. Thus a hot Jupiter typical of the transiting exoplanets (Figure 5.4c) may possess only the three deepest level cloud layers shown in Figure 5.4a.

Since the influence of clouds on the observable properties is likely to be profound, and the theoretical possibilities are so uncertain, the prediction of exoplanet properties is fraught with uncertainty. Even where observations exist, the framework for interpretation is nascent, and the current (August 2009) consensus may evolve significantly as new and better observations and models become available. In our discussion we will use examples that include the most reliable observational results and illustrate the relevant fundamental concepts.

5.2.2 The first detection of an exoplanet's atmosphere

Figure 5.5 shows the stunning first detection of the atmosphere of a planet around another star, taken from the paper 'Detection of an extrasolar planet atmosphere' (2002, *Astrophysical Journal*, **568**, 377–84) by David Charbonneau, Timothy Brown, Robert Noyes and Ronald Gilliland. These observations were made using the Hubble Space Telescope to perform time-series spectroscopy, carefully measuring the spectrum of HD 209458 during the transit of its planet HD 209458 b. The instrument was adjusted to observe the region of the spectrum that includes the strong sodium D lines. These are the spectral lines responsible for the orange light in the low-pressure sodium street lights that are widely used in the UK. Charbonneau and collaborators performed an extremely careful data analysis, revealing that the photometric dimming during transit in a bandpass centred on the sodium features is deeper by $(2.32 \pm 0.57) \times 10^{-4}$ relative to simultaneous observations of the transit in adjacent bands. In other words, the transit is about 1.0002 times deeper in the sodium D lines than in the nearby continuum: the depth is $\Delta F = 0.0164\,F$ in the continuum and $\Delta F \approx (0.0164 \times 1.0002)\,F$ in the sodium D lines. This is illustrated schematically in Figure 5.5a, and the data themselves are shown in Figure 5.5b,c.

One of the discovery team said of sodium: 'It's the spectral equivalent of a skunk. You don't need very much in the air before you notice it.'

The observational result in Figure 5.5 is consistent with expectations. As we saw in Exercise 5.2, the equilibrium temperature of HD 209458 b is $T_{\text{eq}} \approx 1400(1 - A)^{1/4}$ K, and this is within the range of effective temperatures of brown dwarfs. Brown dwarf spectra show very strong absorption lines from alkali metals, such as sodium, and therefore it is reasonable to expect that HD 209458 b's atmosphere might also harbour the species producing these lines. Model calculations unanimously predicted that HD 209458 b's transmission spectrum would show strong sodium D line features, and examples of such model calculations are shown in Figure 5.6.

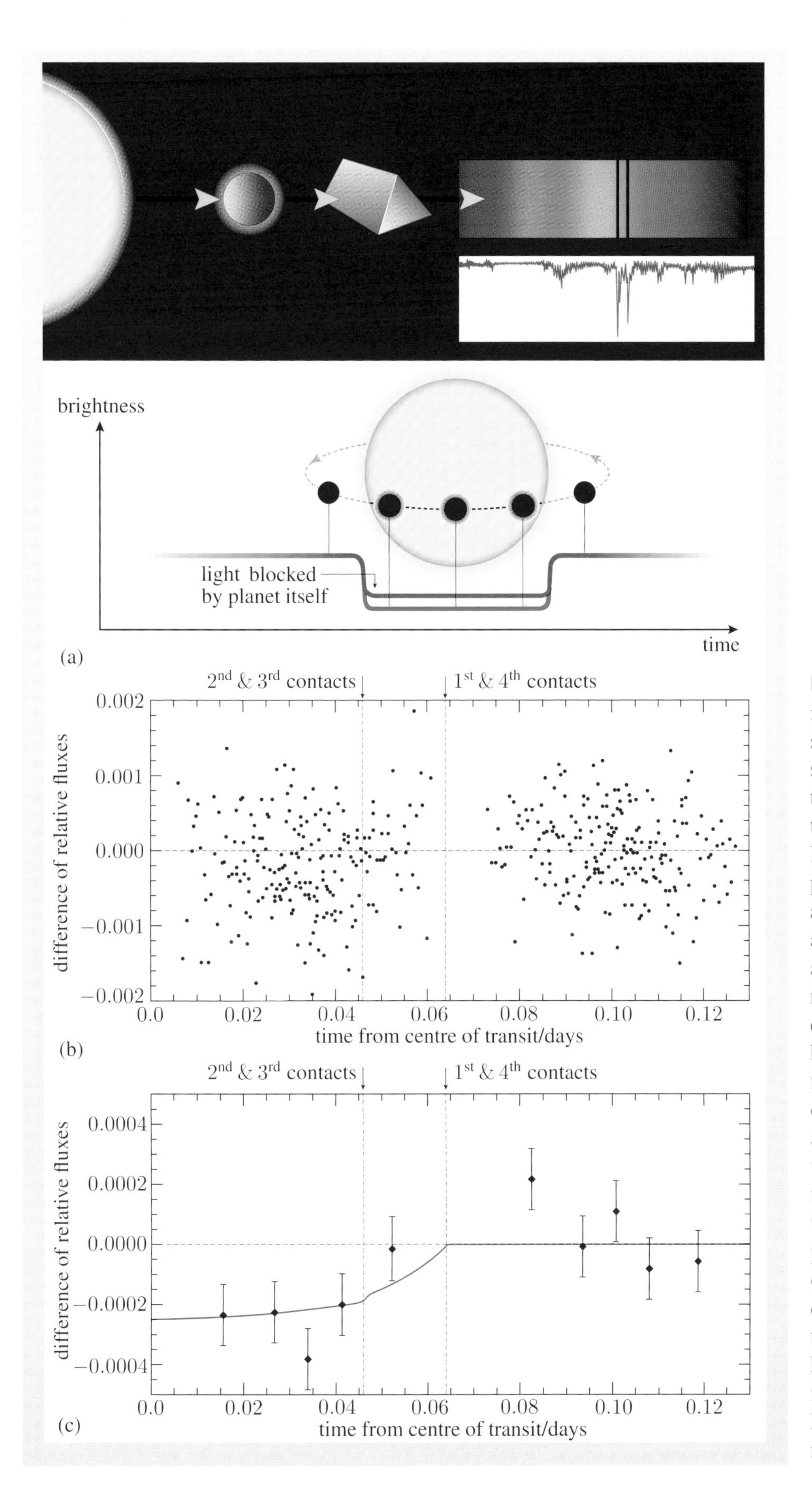

Figure 5.5 (a) A schematic illustration of transmission spectroscopy. The whole spectrum gets dimmer during the transit, but at the wavelengths of the strong sodium D absorption lines, the dimming is more pronounced. This is due to additional blocking of light by absorption in the planet's atmosphere. (b) The data revealing the first detection of an exoplanet atmosphere. The horizontal axis measures time from mid-transit, i.e. the light curve has been 'folded' on its axis of symmetry, and the vertical axis shows the flux in the sodium D lines relative to the nearby continuum. This panel shows the data at the original time resolution. (c) The data averaged into longer time-interval bins. The first four here clearly show that the transit is slightly deeper in the sodium D lines. The solid curve is a model.

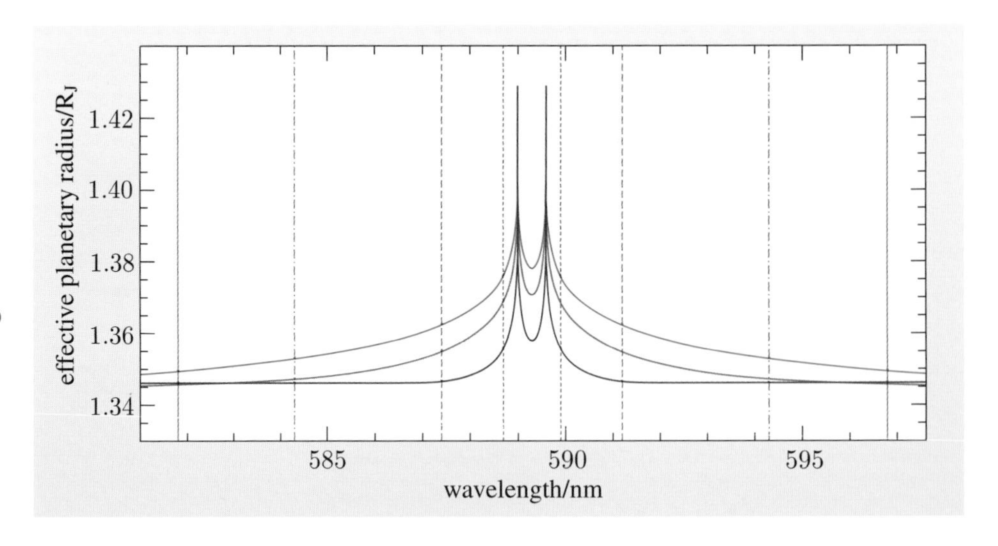

Figure 5.6 Examples of model calculations of the effective planetary radius as a function of wavelength. In all the models shown, the presence of atomic sodium in the atmosphere results in a larger opaque planetary disc at the wavelengths of the sodium D lines. All the models shown are qualitatively similar, with a $\sim 7\%$ enhancement in the effective planetary radius at the central wavelengths of the two spectral lines.

5.2.3 HD 209458 b's transit in Lyman α

Once the results shown in Figure 5.5 appeared, it seemed likely that other spectral lines might also give measurable transmission spectroscopy signals. Lyman α is an obvious spectral line to try, and observations of HD 209458 b's transit in Lyman α were published in 'An extended upper atmosphere around the extrasolar planet HD 209458 b' (2003, *Nature*, **422**, 143) by Alfred Vidal-Madjar and colleagues.

- Why is Lyman α an obvious spectral line to observe?

○ It is a spectral line of hydrogen, and giant planets are assumed to be composed predominantly of hydrogen. It is a transition from the ground state, and therefore is likely to be strongly present in absorption from a gas containing hydrogen atoms. It is the longest wavelength line in the Lyman series, and the other lines are further in the ultraviolet, where observations are increasingly difficult to make.

Exercise 5.3 (a) Referring to Figure 5.7, how deep is the transit of HD 209458 b at the wavelength of Lyman α?

(b) The radius of HD 209458 is $1.146\,R_\odot$. What radius, R_P, for HD 209458 b is implied by the transit depth of HD 209458 b at the wavelength of Lyman α? Comment on your answer, comparing it to the radius expected for a giant planet, and the currently accepted value (February 2009) for the radius of this planet, which is $R_P = 1.359 \pm 0.015\,R_J$.

(c) Can you think of any explanation for your answers in parts (a) and (b)? ■

Figure 5.8 illustrates the findings that we examined in Exercise 5.3. The transit depth at the wavelength of Lyman α implies that the planet HD 209458 b is surrounded by a huge **exosphere** containing neutral hydrogen. This cloud could be formed as a result of evaporation of the atmosphere caused by the intense irradiation from the host star, and this explanation was the most obvious hypothesis. However, an alternative interpretation has been put forward, and we will discuss this in Subsection 5.2.5 below.

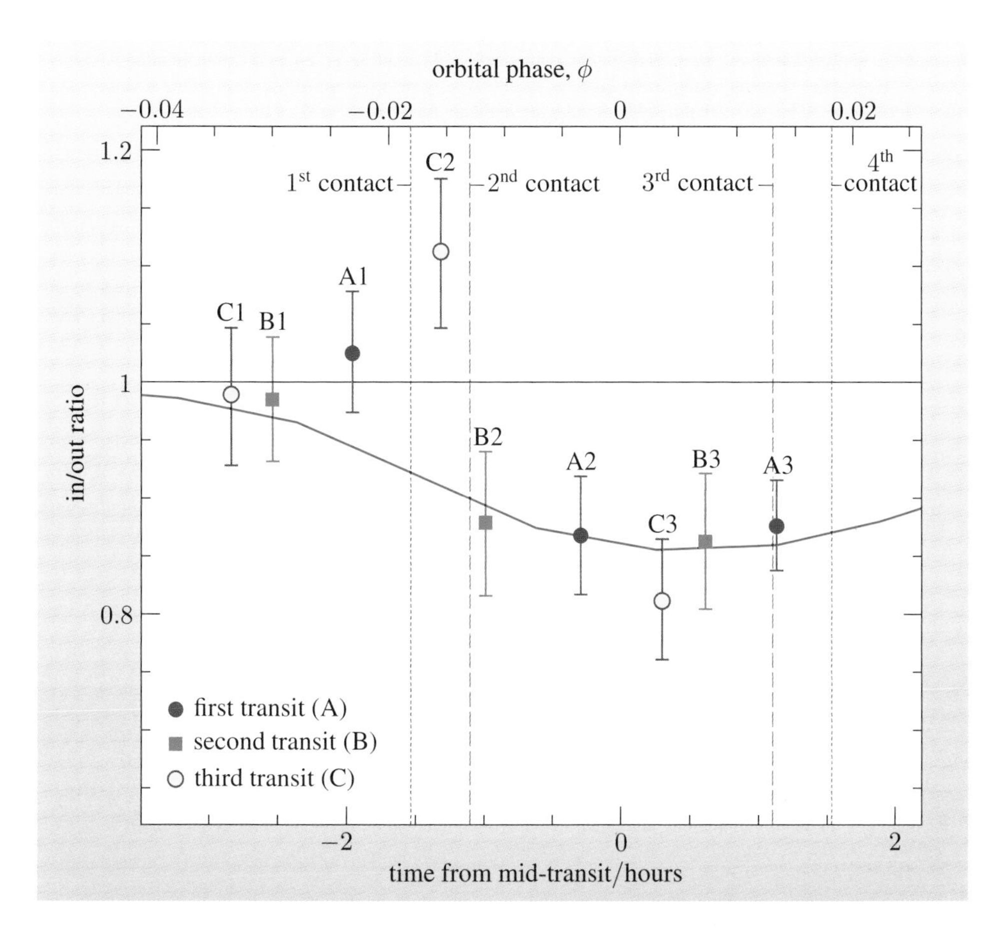

Figure 5.7 The transit of HD 209458 b observed at the wavelength of Lyman α. The different symbols correspond to data taken in different orbits and phase-folded. The dashed vertical lines denote the four contacts as deduced from the optical continuum ephemeris. The solid line is the result of a model calculation.

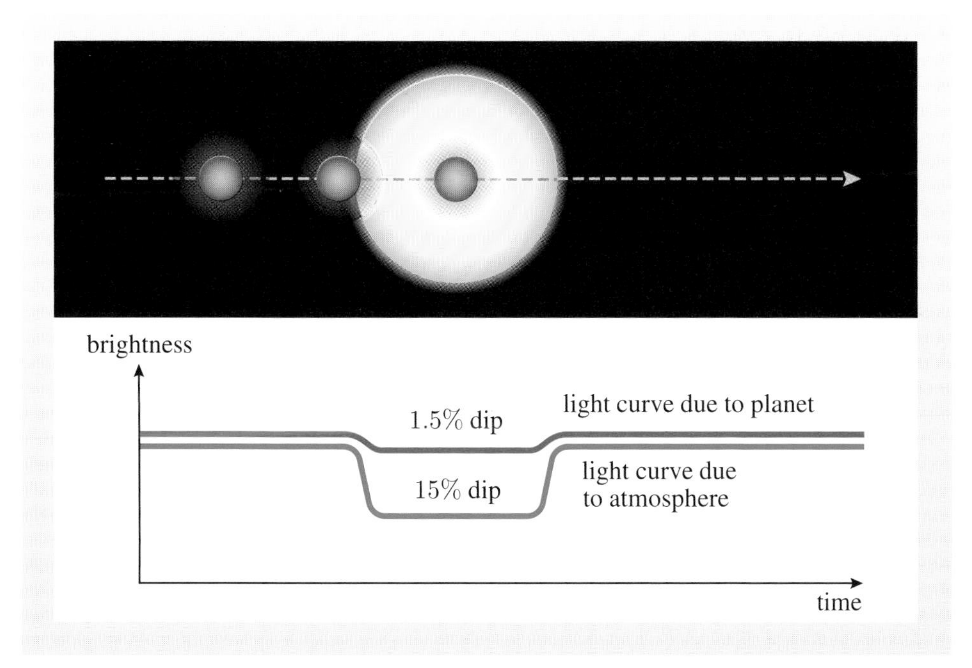

Figure 5.8 A schematic illustration of the transit of HD 209458 b. Red indicates the opaque disc of the planet itself, and blue indicates the larger diffuse cloud of hydrogen that absorbs in Lyman α. The upper part of the figure illustrates the inferred situation, while the lower part gives a simplified depiction of the continuum and Lyman α transits.

5.2.4 O I and C II absorption during HD 209458 b's transit

The transit depth in O I and C II is also ∼10%, as Figure 5.9 shows. This has been interpreted as the entrainment of these species (neutral oxygen atoms and singly ionized carbon) in the photo-evaporated outflow of neutral hydrogen. Figure 5.9a is equivalent to Figure 5.7, but with fresh data. These data indicate a 5% depth, significantly shallower than that of Figure 5.7. It suggests that the transit depth in Lyman α may be variable, though both sets of data are noisy. The lack of a signal in the Si IV line suggests that this high-ionization species is not present in any significant abundance in HD 209458 b's exosphere.

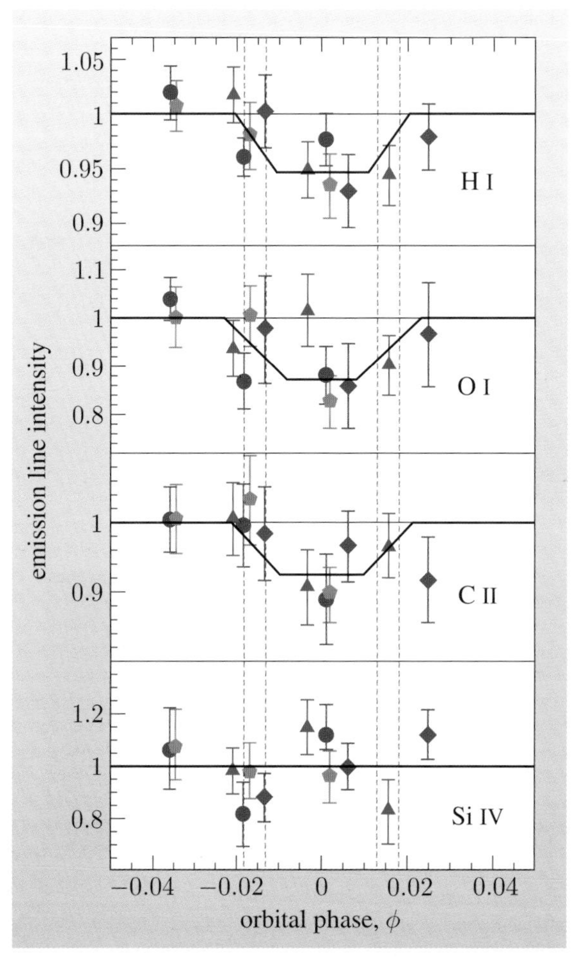

Figure 5.9 Four distinct transits, colour-coded, at the wavelengths of (a) Lyman α, (b) O I, (c) C II and (d) Si IV. Panel (a) shows fresh data repeating the observations of Figure 5.7, and shows a reduced, but still notably large, Lyman α transit depth. Panels (b) and (c) show that neutral oxygen and singly ionized carbon also have transit depths ∼10%. The higher-ionization species, Si IV, shows no transit signal.

The ultraviolet transmission spectroscopy discussed in Subsection 5.2.3 and here differs in one important respect from the sodium D results in Subsection 5.2.2. This is illustrated in Figure 5.10: the spectral lines being discussed are coronal emission lines, rather than photospheric absorption lines. This stellar line emission is patchy and time-variable, and can extend beyond the limb of the star, so potentially hugely complicates the interpretation of the transit light curve.

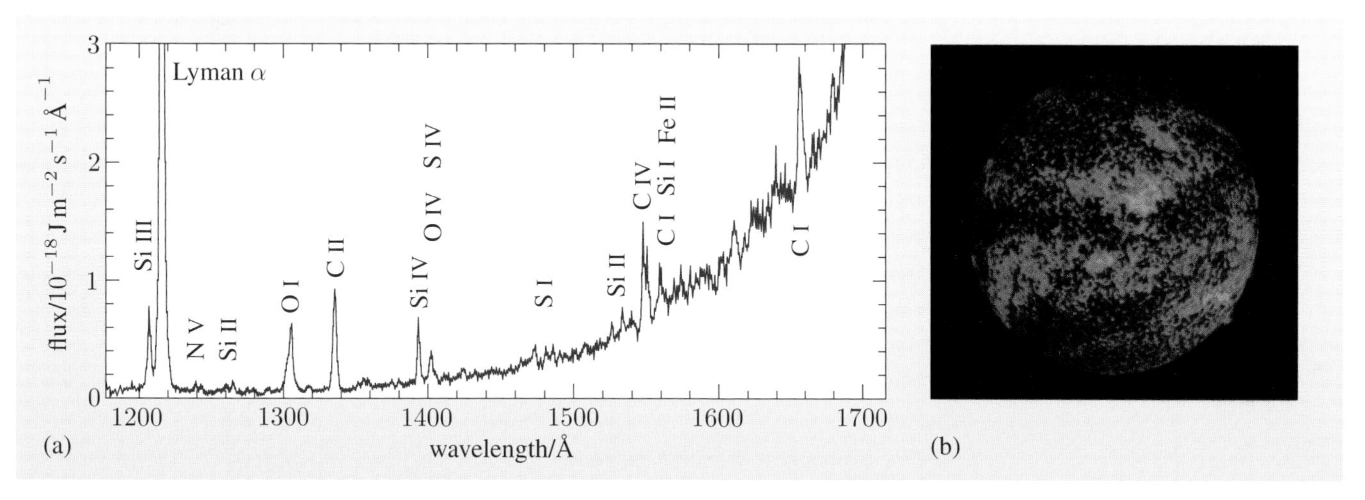

Figure 5.10 (a) The ultraviolet spectrum of HD 209458 b, showing that the large transit depths occur in emission lines. (b) An image of the Sun in Lyman α light. The emission is coronal and is patchy and time-variable.

- What sort of light curve would result from the transit of a planet across the image shown in Figure 5.10b?

- The light curve is likely to be variable and unpredictable, with depths that overestimate the planet's size where an intense region of emission is occulted.

5.2.5 An alternative explanation of the Lyman α absorption

The idea that the hot Jupiters' atmospheres are being evaporated is an obvious and attractive explanation of the results that we examined in Exercise 5.3. It may also underlie the correlation between surface gravity and orbital period that is shown in Figure 4.21. There is, however, an alternative possible explanation. Mars and Venus also have extended hydrogen coronae, which were first detected by their Lyman α signature. These coronae were also initially interpreted as the uppermost layers of an escaping exosphere, and temperatures of $\sim$700 K were inferred. When spacecraft subsequently visited Mars and Venus, the measured temperatures of their exospheres were about three times lower than required for the coronae to be photo-evaporated. The energetic neutral atoms observed around Mars and Venus are now understood to be produced in charge exchange between energetic solar wind protons and the planetary exosphere. It is possible that a similar mechanism explains the extended Lyman α cloud surrounding HD 209458 b; however, it is also clear that HD 209458 b is far more irradiated than either Mars or Venus.

For several years there has been no telescope capable of performing the time-series ultraviolet spectroscopy shown in Figure 5.7. Consequently, no observations exist of the transits of other exoplanets at Lyman α, and it has not been possible to obtain the higher signal-to-noise ratio data that might discriminate between the hydrodynamic 'blow-off' scenario and the energetic neutral atom hypothesis. A new instrument was installed on the Hubble Space Telescope (HST) in 2009, and several observations will address this issue. The very highly irradiated planet WASP-12 b shows similar enhanced transit depths in Mg II and other ultraviolet lines.

5.2.6 Rayleigh scattering in HD 189733 b — the sky is blue

So far all the transmission spectroscopy results that we have shown have been for HD 209458 b. As we saw in Exercise 5.1, transmission spectroscopy requires the collection of many photons in order to reach the required signal-to-noise ratios for detection of spectral features. Thus it comes as no surprise that the first results and the most numerous results are for this, the brightest of the known transiting exoplanet host stars. We now move on to discuss the results for the second brightest known (August 2009) host star: HD 189733.

There have been claims of detection of absorption from water vapour in the atmosphere of this exoplanet, which may prove to be correct. Currently, however, Occam's razor probably favours the interpretation of the HST transmission spectroscopy suggested by Figure 5.11. This figure compiles the wavelength-dependent radius determinations from a variety of sources, and shows that they are more or less consistent with a simple physically motivated curve.

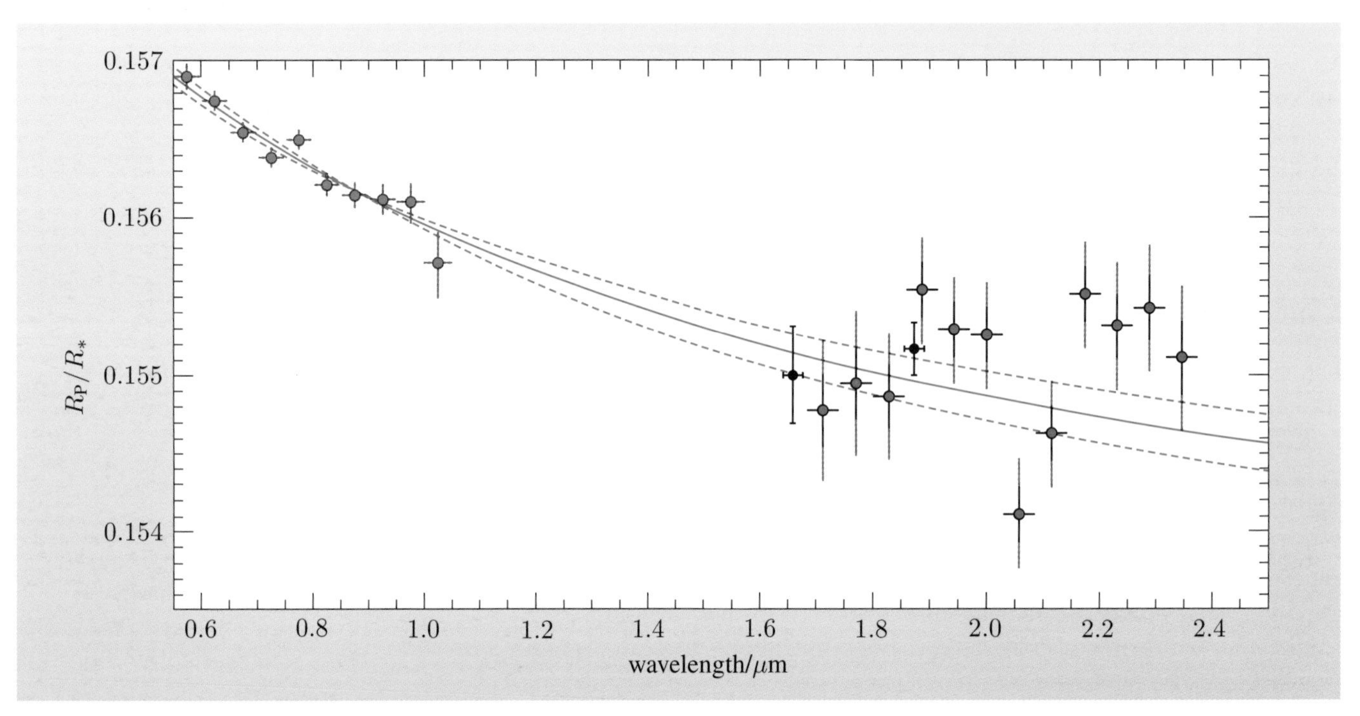

Figure 5.11 The radius of HD 189733 b as a function of wavelength. The left-hand side is in the optical while the right-hand side is in the infrared. All the data came from the HST, with differing instrumental configurations colour-coded. The solid red line is the curve that would be expected from Rayleigh scattering in the atmosphere, with the dashed red lines indicating the uncertainties in this prediction.

If radiation is passed through a gas composed of particles that are much smaller than the wavelength of radiation, the scattering cross-section, σ_R, presented by each particle is proportional to λ^{-4}. This scattering of radiation by particles with size much less than λ is called **Rayleigh scattering**. A familiar example of this is the blue daylight sky: the sky's brightness is sunlight that has been scattered by the Earth's atmosphere, and since short wavelength radiation is subject to the highest scattering cross-section, it is the most likely to be scattered. This is why the diffuse, scattered light appears blue, while direct sunlight has a yellow colour: the short wavelength sunlight is preferentially scattered.

The red curve in Figure 5.11 shows the trend in the inferred value of R_P caused by the host star's light being Rayleigh scattered in the transiting exoplanet's atmosphere. For the shortest wavelengths, where the scattering is most effective, the atmosphere is effectively opaque at altitudes where it is transparent to longer-wavelength radiation. The smooth dependence of σ_R on wavelength results in a smooth variation of the inferred R_P with wavelength. The smooth curve is clearly an excellent match to the general trend in these data. While some points appear significantly above or below the curve, perhaps indicating the effect of spectral lines on the atmospheric opacity, it is important to remember that one in three Gaussian-distributed data points is expected to differ from the model by more than one standard deviation, σ. To make things worse, the systematic errors (not included within the error bars) are large in the context of the extremely high precision measurements shown.

We have too few letters: we have used σ_R for the Rayleigh scattering cross-section, and σ for the standard deviation. These are two completely distinct quantities, not to be confused.

5.2.7 CO and H_2O with Spitzer

Atoms generally produce sharp spectral lines, while molecules generally produce broader features known as **molecular bands**. These bands are predominantly in the infrared. The Spitzer Space Telescope is an observatory optimized for the infrared and has been extensively used to study exoplanets, as we will see in Chapter 6. Spitzer results for HD 189733 b are compiled in Figure 5.12, and may indicate the presence of water and carbon monoxide.

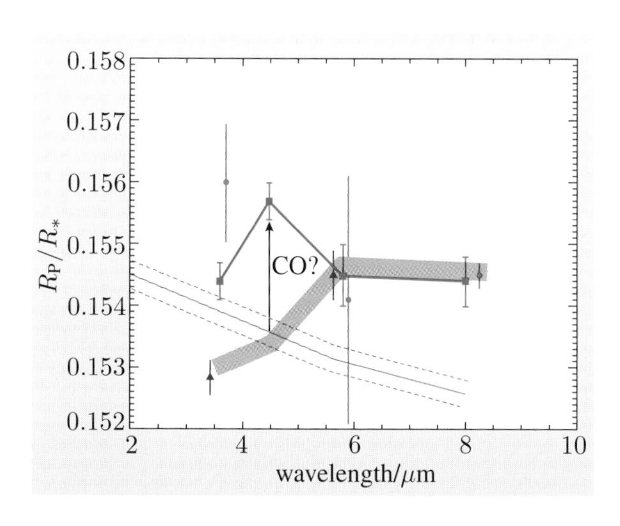

Figure 5.12 The radius of HD 189733 b as a function of wavelength. All the data, plotted in red, green and purple, were taken with Spitzer and analyzed by different researchers. The solid and dashed grey lines are the Rayleigh scattering curves, and the turquoise band shows the values predicted by a model that has water as the only absorbing molecule. The radius measurement at 4.5 μm may indicate the presence of CO.

5.2.8 Na D absorption from HD 189733 b: the first ground-based atmosphere detection

The atmospheric chemistry of HD 189733 b has been unequivocally detected in the data shown in Figure 5.13, which is also the first such detection made from the ground. The telescope used was the Hobby–Eberly telescope at the McDonald Observatory in west Texas. This observatory is one of the darkest sites in North America, so the skies are free from the light pollution that would make this work extremely challenging in most of Europe.

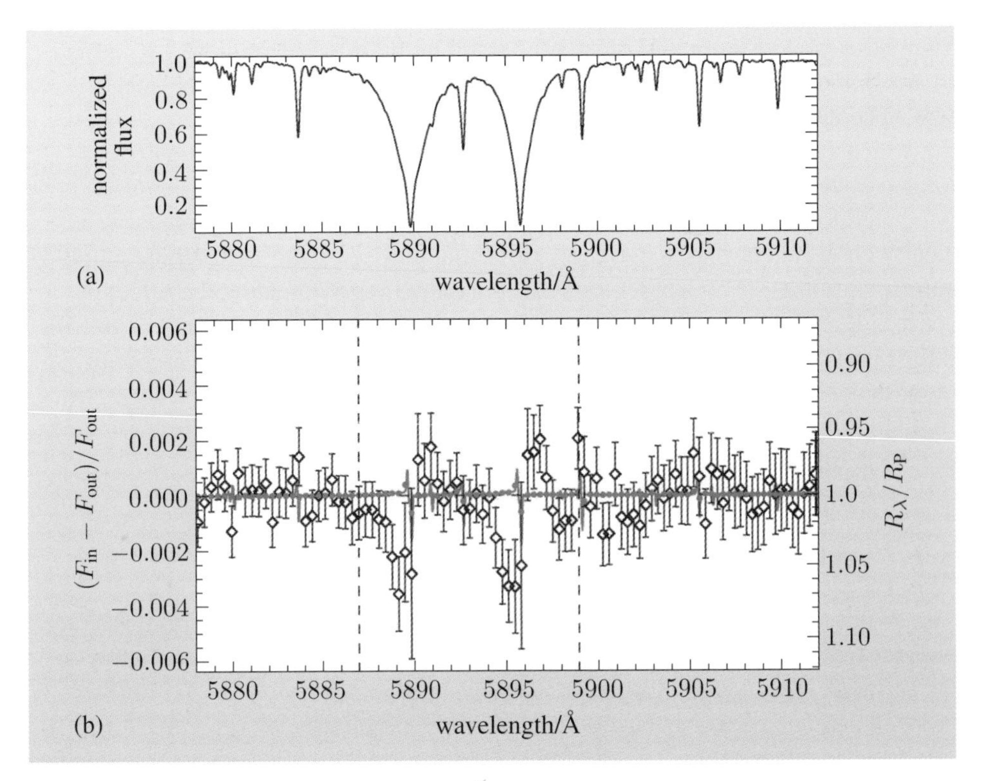

Figure 5.13 (a) The spectrum of HD 189733 showing the prominent Na D absorption lines. (b) Absorption by Na D in the atmosphere of HD 189733 b is clearly detected. The left axis gives values of the fractional excess absorption in the atmosphere, while the right axis is calibrated to give the corresponding values of $R_{\rm P}$.

5.2.9 What next?

In this section we have given a few illustrative results. Since there are now (November 2009) a dozen known exoplanets transiting host stars brighter than $V = 10$, we expect many more atmospheric detections in the near future. While the atmospheric detections are likely to remain limited to species with particularly strong spectral signatures, the technique offers an invaluable opportunity to quantitatively examine exoplanet atmospheric physics and chemistry. As Figure 5.6 illustrates, the results can be compared to model calculations and consequently can lead to inferences about atmospheric temperature and pressure profiles, and the height and/or pressure at which cloud decks form, as well as the more obvious inferences about the abundances of detected species. This technique will play an important role in the general goal of setting the Solar System planets in context.

5.3 The Rossiter–McLaughlin effect

In Chapter 1 we introduced the radial velocity method of planet discovery. The Doppler shift of the light from the host star reveals the component of the host star's reflex orbital motion in a direction directly towards or away from us. In general, the host star will also be spinning on an axis. Unless this axis points exactly towards the observer, the stellar spin will cause some parts of the star's disc to move towards the observer while other parts move away from the observer. The second row in Figure 5.14 illustrates this schematically: red and blue regions respectively represent redshifted and blueshifted regions of the star's disc. This motion occurs while the star as a whole follows its reflex orbital motion.

Generally, of course, we cannot resolve the disc of a star, and our measurements of the stellar radial velocity are an average, where the light from the entire visible disc is effectively integrated to produce a single spectrum.

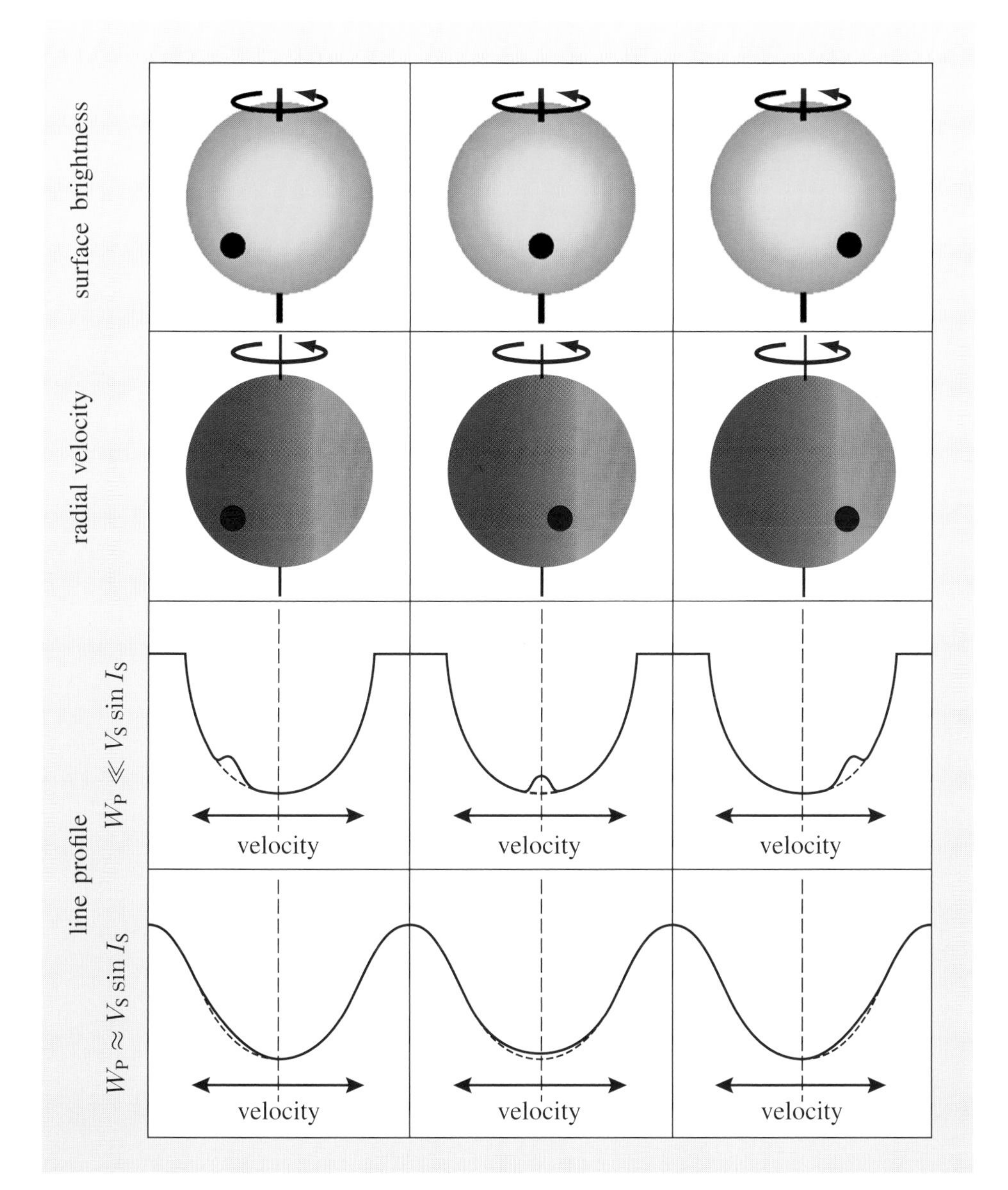

Figure 5.14 Generally, stars spin. Consequently, light from one half of the star will be blueshifted, while light from the other half is redshifted. If a planet transits the star, it will block regions of continuously varying Doppler shift, and this will cause an observable signature in the star's radial velocity curve. The three columns in this figure correspond to three instants during a transit. The top row indicates stellar surface brightness, the second row indicates the surface radial velocity, the third row illustrates the line profile that would be observed if the stellar spin is the dominant source of line broadening. The bottom row shows the line profile in the case where other line-broadening mechanisms are important.

- ● What effect will the stellar spin have on the observed spectral lines?
- ❍ The lines will be broadened as some of the light will be redshifted while other light is blueshifted. The width of the spectral lines will be determined by the cumulative effects of the intrinsic line width, broadening due to the thermal

motions of the atoms, broadening due to bulk motions in the stellar photosphere, broadening due to the spin of the star, and broadening due to the finite resolution of the instrument. The stellar spin can be one of the largest contributors to this broadening.

The third row of Figure 5.14 illustrates how the transit of a planet affects the observed spectral lines in the case of the stellar spin being the dominant line-broadening mechanism; the fourth row shows the line profiles produced when other line-broadening mechanisms are equally important.

The radial velocity of a star is, of course, measured using the photospheric absorption lines. During a transit, as Figure 5.14 shows, this radial velocity measurement will be contaminated: some of the light from the stellar disc is missing, affecting the line profile. If the planet blocks blueshifted light, as in the first column of Figure 5.14, then the measured radial velocity is shifted redwards compared to the true orbital radial velocity. These shifts due to the transiting planet are known as the **Rossiter–McLaughlin effect**. Our discussion follows that of the paper 'Prospects for the characterization and confirmation of transiting exoplanets via the Rossiter–McLaughlin effect' (2007, *Astrophysical Journal*, **655**, 550–63) by Scott Gaudi and Joshua Winn.

This effect was first described by Rossiter and McLaughlin in connection with eclipsing binary stars, where the same effect is seen.

- If the planet blocks redshifted light from the host star, how will this affect the measured radial velocity?
- The measured radial velocity is shifted bluewards compared to the true orbital radial velocity.

5.3.1 Spin–orbit angle and the projected spin–orbit angle

In Figure 5.14 the projections on the sky of the stellar spin and the orbital angular momentum vectors are aligned. It is clear that in this case the perturbation to the orbital radial velocity is antisymmetric about the time of mid-transit. This is illustrated in Figure 5.15.

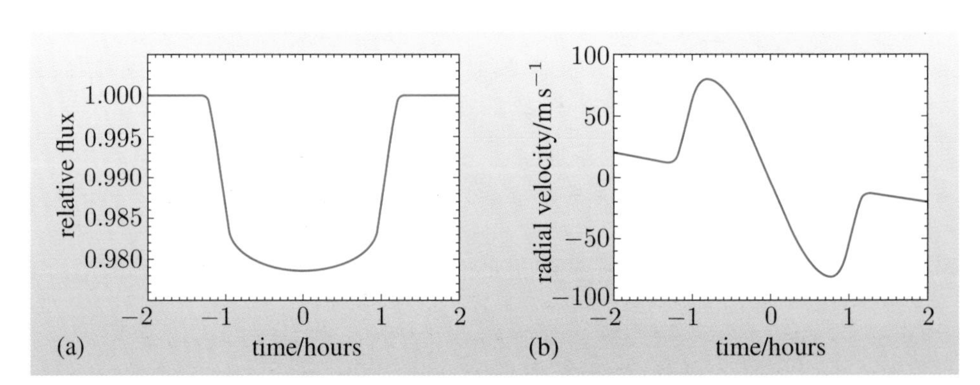

Figure 5.15 Simulation of the signatures of a giant planet transiting its host star. The parameters have been chosen to approximate those of TrES-1. (a) The familiar photometric transit. (b) The corresponding effect on the radial velocity measurements of the host star. The smoothly declining radial velocity curve during the out-of-transit measurements before transit ingress and after transit egress is interrupted by the Rossiter–McLaughlin effect perturbation, which appears as an antisymmetric blip in this example.

While it is likely from angular momentum considerations that the spin and orbital angular momentum vectors will be roughly aligned, they are unlikely to be exactly aligned. In our Solar System the alignment is within about $10°$, with the Sun's rotation axis being at an angle $\psi_\odot = 7.15°$ to the normal to the ecliptic. If the projections on the sky of the two vectors are misaligned, the Rossiter–McLaughlin effect will no longer produce an antisymmetrical perturbation. This is illustrated in Figure 5.16 for $b/R_* = 0.5$ and three different angles, β, between the projected stellar spin and the projected orbital angular momentum vectors.

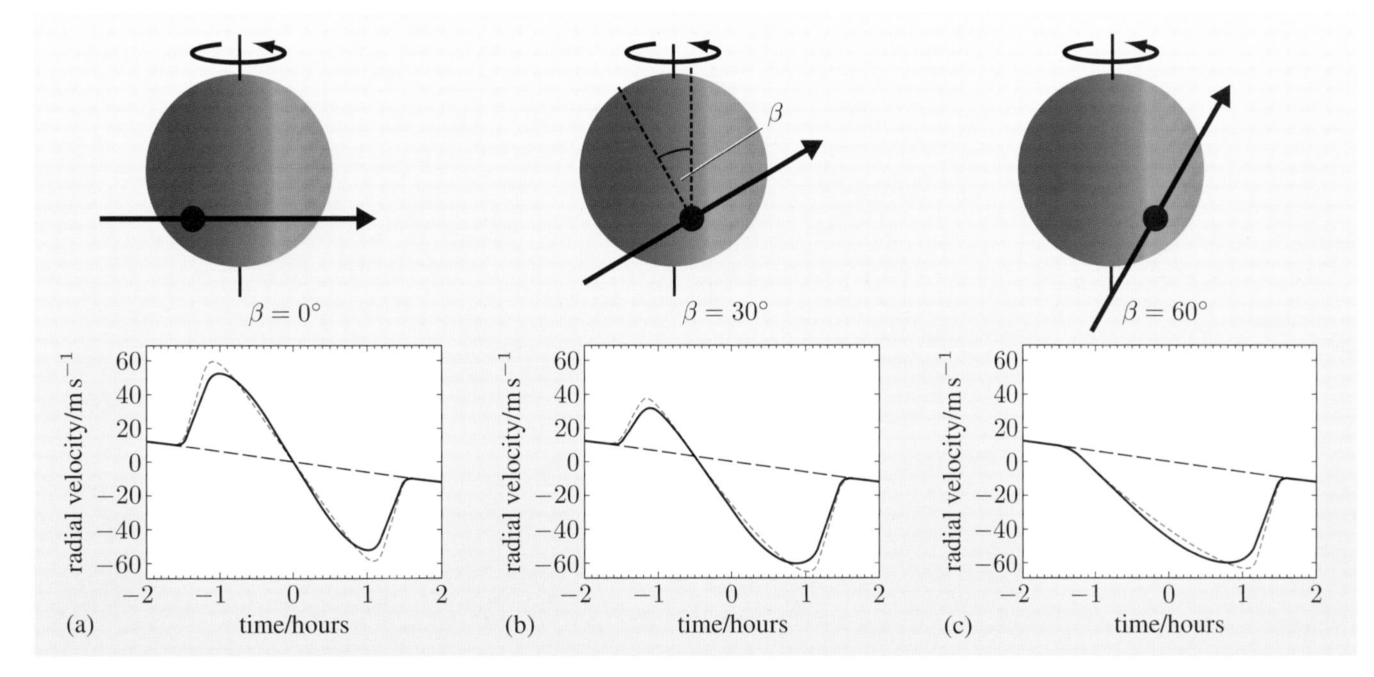

Figure 5.16 Three simulations showing Rossiter–McLaughlin radial velocity signatures produced by differing relative orientations of the stellar spin and orbital angular momentum vectors. All three panels show transits with otherwise identical parameters, and would produce identical photometric transits. The solid lines are calculated with a simple limb darkening law, while the red dashed line shows the calculated effect neglecting limb darkening. The projected spin–orbit angle, β, is indicated in each case.

The **spin–orbit angle**, ψ, between the stellar spin and the orbital angular momentum vectors, is a fundamental property of the planetary orbit, and must be specified along with the semi-major axis, a, and the eccentricity, e, to fully define the motions in the system. ψ has particular importance in constraining the evolutionary history of the planet: if the planet has migrated to its current orbit via tidal interactions with a protoplanetary disc, then we would expect $\psi \approx 0°$. In contrast, if the planet has undergone scattering with other planets or the orbit has been shrunk by tidal interactions with a stellar binary companion, we might expect occasional large misalignments. Generally, however, we have no information about the component of the stellar spin angular momentum vector, $\mathbf{\Omega}_\mathrm{S}$, along our line of sight. For this reason, we can only empirically constrain the **projected spin–orbit angle**, β.

5.3.2 Analysis of the Rossiter–McLaughlin effect

The radial velocity variation of the star due to the reflex orbital motion is (from Equation 1.12)

$$\Delta V(t) = \frac{2\pi a M_\mathrm{P} \sin i}{(M_\mathrm{P} + M_*)P\sqrt{1-e^2}} \left(\cos(\theta(t) + \omega) + e\cos\omega\right). \tag{5.14}$$

If we use the symbol $ER(t)$ to denote the empirical radial velocity variation, this will be the sum of $\Delta V(t)$ and $\Delta V_\mathrm{RM}(t)$, where the latter is the Rossiter–McLaughlin radial velocity variation caused by the planet's disc occulting part of the spinning stellar surface:

$$ER(t) = \Delta V(t) + \Delta V_\mathrm{RM}(t). \tag{5.15}$$

- How does the quantity $\Delta V(t)$ differ from the quantity $V(t)$ defined in Equation 1.12?

- $V(t)$ is the sum of $\Delta V(t)$ and the constant $V_{0,z}$, which is the radial velocity of the stellar system with respect to the Solar System.

Exercise 5.4 (a) Show that the amplitude of the orbital radial velocity variation, A_RV, as given in Equation 1.13, can be expressed as

$$A_\mathrm{RV} = \left(\frac{2\pi G}{P}\right)^{1/3} \frac{M_\mathrm{P} \sin i}{(M_\mathrm{P} + M_*)^{2/3}} \left(1 - e^2\right)^{-1/2}. \tag{5.16}$$

(b) Consequently, show that if we assume that $M_\mathrm{P} \ll M_*$ and $e \approx 0.0$, then

$$A_\mathrm{RV} \approx 8.9\,\mathrm{cm\,s^{-1}} \left(\frac{P}{\mathrm{yr}}\right)^{-1/3} \left(\frac{M_\mathrm{P}\sin i}{\mathrm{M}_\oplus}\right) \left(\frac{M_*}{\mathrm{M}_\odot}\right)^{-2/3}. \tag{5.17}$$

■

To calculate $\Delta V_\mathrm{RM}(t)$ we need to quantify the radial velocity values corresponding to each element on the spinning stellar surface. Figure 5.17 illustrates the orientation of the spinning star with respect to the observer.

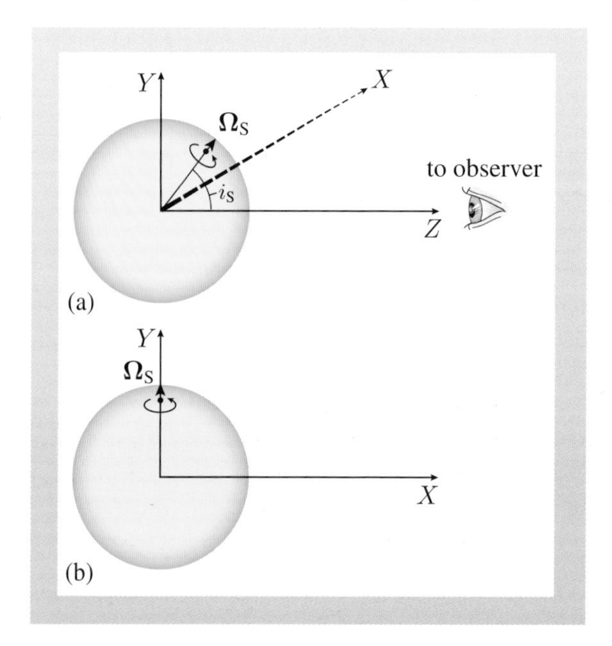

Figure 5.17 The (X, Y, Z) coordinates used to analyze the stellar spin. (a) The observer lies in the $+Z$-direction, and the spin vector $\boldsymbol{\Omega}_\mathrm{S}$ lies in the YZ-plane. The inclination of the stellar spin axis to the Z-axis is i_S. (b) On the plane of the sky, the projected stellar spin axis coincides with the Y-axis.

To analyze the spin, we adopt Cartesian coordinates (X, Y, Z) centred on the star, with the observer located in the $+Z$-direction. The Y-axis is chosen so that the spin vector, $\boldsymbol{\Omega}_\mathrm{S}$, is in the YZ-plane; this means that the X-axis is orthogonal to $\boldsymbol{\Omega}_\mathrm{S}$. We normalize this coordinate system such that the surface of the (assumed spherical) star is at $X^2 + Y^2 + Z^2 = 1$, i.e. R_* is the unit length in our coordinate system. Note that this Cartesian system differs from the (x, y, z) coordinates that we adopted in Chapter 1 to analyze the orbital motion. The orientation of the spin axis with respect to the observer is measured by the angle i_S.

- Describe the orientation of the stellar spin axis if $i_\mathrm{S} = 0.0$.
- For $i_\mathrm{S} = 0.0$, the stellar spin axis lies exactly in the direction of the observer.
- Where on each part of Figure 5.17 would the stellar spin pole be for $i_\mathrm{S} = 90°$?
- The spin axis would be along the Y-axis. In both parts of the figure, the pole would be on the limb of the star at the top where the Y-axis crosses the limb.
- Estimate the value of i_S shown in Figure 5.14.
- The projection of the pole on the plane of the sky lies slightly within the disc. This probably corresponds to $i_\mathrm{S} \approx 80°$.

With our adopted coordinates, the angular velocity of the star is

$$\boldsymbol{\Omega}_\mathrm{S} = (0, \Omega_\mathrm{S} \sin i_\mathrm{S}, \Omega_\mathrm{S} \cos i_\mathrm{S}). \tag{5.18}$$

The velocity of any point on the stellar surface, $\boldsymbol{v}$, is the vector product of $\boldsymbol{\Omega}_\mathrm{S}$ and the position vector $\boldsymbol{R} = R_*(X, Y, Z)$. Here and in the subsequent analysis we have multiplied the normalized (X, Y, Z) coordinates by the unit length R_* to obtain the physical quantity, in this case the position of the stellar surface in physical units, required to derive the velocity $\boldsymbol{v}$:

$$\begin{aligned} \boldsymbol{v} &= \boldsymbol{\Omega}_\mathrm{S} \times \boldsymbol{R} \\ &= \Omega_\mathrm{S} R_* (0, \sin i_\mathrm{S}, \cos i_\mathrm{S}) \times (X, Y, Z) \\ &= \Omega_\mathrm{S} R_* (Z \sin i_\mathrm{S} - Y \cos i_\mathrm{S}, X \cos i_\mathrm{S}, -X \sin i_\mathrm{S}). \end{aligned} \tag{5.19}$$

$\Omega_\mathrm{S} R_*$ is of course simply the speed of the stellar surface at the equator, which we will term V_S. The components of $\boldsymbol{v}$ in the X- and Y-directions are in the plane of the sky; it is only the Z-component of $\boldsymbol{v}$ that produces a radial velocity. The Z-axis points towards the observer, while the convention for radial velocities is that motions away from the observer are positive. Thus, for any small element on the surface of the star, the radial velocity due to the stellar spin is

$$v_\mathrm{el} = -v_Z = \Omega_\mathrm{S} R_* X \sin i_\mathrm{S} = V_\mathrm{S} X \sin i_\mathrm{S}. \tag{5.20}$$

The observed radial velocity is given by integrating the radial velocity contributions of the elements on the visible stellar disc. For each element, the contribution should be weighted by the emergent intensity, $I(X, Y)$ (cf. the top row of Figure 5.14). In the case of an unocculted symmetrical stellar disc, each element at a positive value of X will produce a radial velocity contribution that is equal and opposite to the contribution from the corresponding element at $-X$. If the radial velocity is measured during a transit, however, some of the surface is occulted, leading to unbalanced emission and the Rossiter–McLaughlin radial velocity perturbation

$$\begin{aligned}\Delta V_{\mathrm{RM}}(t) &= \frac{\int_{\text{visible disc}} v_{\mathrm{el}}\, I(X,Y)\,\mathrm{d}A}{\int_{\text{visible disc}} I(X,Y)\,\mathrm{d}A} \\ &= V_{\mathrm{S}} \sin i_{\mathrm{S}} \frac{\iint X\, I(X,Y)\, R_*^2\,\mathrm{d}X\,\mathrm{d}Y}{\iint I(X,Y)\, R_*^2\,\mathrm{d}X\,\mathrm{d}Y}. \end{aligned} \tag{5.21}$$

The numerator here is simply the velocity multiplied by the emergent intensity integrated over the visible disc, while the denominator is the emergent intensity alone integrated over the disc to normalize the result.

- Why did the factor R_*^2 appear within the two integrals in the second line of Equation 5.21?

○ Because we need to integrate over the area of the stellar disc, and the area element is $\mathrm{d}A = R_*\,\mathrm{d}X \times R_*\,\mathrm{d}Y$.

In the simplest possible case where we neglect limb darkening and assume that the intensity is uniform across the stellar disc,

$$I(X,Y) = \begin{cases} I_0 & \text{if } X^2 + Y^2 \leq 1 \text{ and } (X - X_{\mathrm{P}})^2 + (Y - Y_{\mathrm{P}})^2 \geq R_{\mathrm{P}}^2/R_*^2, \\ 0 & \text{otherwise.} \end{cases} \tag{5.22}$$

Here, $(X_{\mathrm{P}}, Y_{\mathrm{P}})$ is the position of the centre of the planet in the coordinates defined in Figure 5.17. When the planet's disc lies completely within the star's disc, the two double integrals in Equation 5.21 are straightforward. The integral in the denominator is simply the visible area of the stellar disc multiplied by the intensity I_0:

$$R_*^2 \iint I(X,Y)\,\mathrm{d}X\,\mathrm{d}Y = \pi\left(R_*^2 - R_{\mathrm{P}}^2\right) I_0,$$

where the visible area of the stellar disc is simply the area of the stellar disc minus the area of the planet's disc. The integral in the numerator is the first moment about the Y-axis, and evaluates to

$$R_*^2 \iint X\, I(X,Y)\,\mathrm{d}X\,\mathrm{d}Y = -X_{\mathrm{P}}\,\pi R_{\mathrm{P}}^2 I_0.$$

Here again the result is as expected: the missing flux is the intensity, I_0, multiplied by the area of the planet, and its mean location is at the centre of the planet $(X_{\mathrm{P}}, Y_{\mathrm{P}})$, so the first moment about the Y-axis is $-X_{\mathrm{P}}$ times the missing flux. Taking these two results and substituting them into Equation 5.21, we obtain

$$\Delta V_{\mathrm{RM}}(t) = -V_{\mathrm{S}} \sin i_{\mathrm{S}} \left(\frac{R_{\mathrm{P}}^2}{R_*^2 - R_{\mathrm{P}}^2}\right) X_{\mathrm{P}}. \tag{5.23}$$

If we imagine a situation such as that shown in Figure 5.16a, the planet is at a negative value of X_{P}, and the occulted stellar surface has a spin motion moving it towards the observer (out of the page), meaning that the missing flux would have made a negative contribution to the radial velocity. Thus this situation would lead to a positive value of ΔV_{RM}. Consequently, we can see that the minus sign in Equation 5.23 is correct: for negative X_{P} the resulting value of ΔV_{RM} is positive, and of course vice versa.

- The reasoning above verifying the sign of $\Delta V_{\rm RM}$ in Equation 5.23 implicitly assumed that the spin vector was oriented upwards in Figure 5.16. Mathematically, what would happen in the case of a star–planet system that was oriented with the spin vector, $\mathbf{\Omega}_{\rm S}$, the other way up?

- The (X, Y, Z) coordinate system was defined relative to the orientation of $\mathbf{\Omega}_{\rm S}$, so $\mathbf{\Omega}_{\rm S} \cdot \widehat{\mathbf{Y}}$ is always positive, where $\widehat{\mathbf{Y}}$ is the unit vector in the Y-direction.

All that remains to calculate the Rossiter–McLaughlin velocity perturbation in this simplest case is to deduce an expression for $X_{\rm P}$ in terms of more fundamental parameters. If we approximate the locus of the planet across the disc of the star as a straight line, then Figure 5.18 shows the geometry.

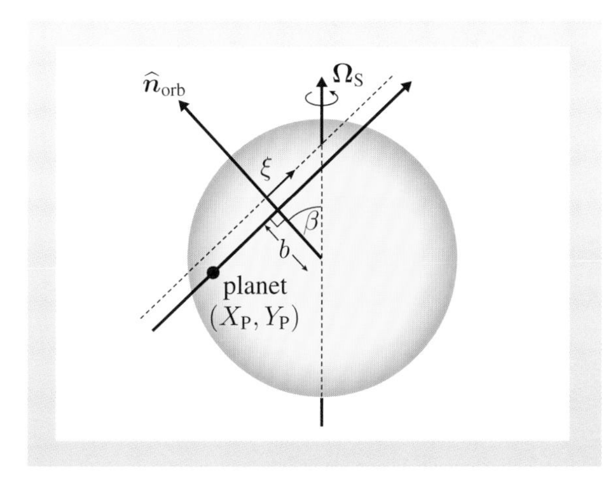

Figure 5.18 The locus of the transit across the disc of the star determines $X_{\rm P}$.

We used ξ to represent a different quantity in Chapter 2.

We introduce the displacement ξ, which is measured from the point of mid-transit along the transit chord. The motion along the chord is approximately at a constant speed, so long as $a \gg R_*$, which must be the case if we can consider the locus to be a straight line. If we assume therefore that $\dot{\xi}$ is constant, then

$$\xi = \frac{2l(t - T_{\rm c})}{T_{\rm dur}}, \tag{5.24}$$

where $T_{\rm c}$ is the time of mid-transit, and we measure t with respect to $T_{\rm c}$; l is half the length of the transit chord, as shown in Figure 3.3 and given in Equation 3.3; $T_{\rm dur}$ is the transit duration, as given in Equation 3.4 for a circular orbit. From Figure 5.18 we can see that

$$X_{\rm P} = \frac{\xi \cos\beta - b \sin\beta}{R_*}. \tag{5.25}$$

Equations 5.23, 5.24, 5.25, 3.3 and 3.4 allow us to calculate the Rossiter–McLaughlin radial velocity perturbation in terms of the fundamental system parameters in this simplest possible case.

In general, of course, the calculation is more complex. The complications include: the ingress and egress phases where the planet's disc is not completely within the star's disc; elliptical orbits; limb darkening; **differential rotation** of the star, where the angular velocity at the surface is latitude-dependent; and the case of small a. These can all be treated by generalizing the approach, but the details of the derivations are beyond the scope of this book.

We can break down the Rossiter–McLaughlin radial velocity perturbation into a dimensionless function, $g(t)$, and an overall amplitude, A_S, where the subscript indicates the origin of the effect in the stellar spin:

$$\Delta V_{RM}(t) = A_S\, g\left(t, X_P, Y_P, \frac{R_P}{R_*}, u, \ldots\right). \tag{5.26}$$

Here u indicates the limb darkening coefficient(s), Y_P comes into play if there is differential rotation, and the ellipsis indicates dependence on parameters that we will not explore in detail here. The overall amplitude is

$$A_S = \Omega_S R_* \sin i_S \left(\frac{R_P^2}{R_*^2 - R_P^2}\right) = V_S \sin i_S \left(\frac{R_P^2}{R_*^2 - R_P^2}\right), \tag{5.27}$$

and in the simplest case that we developed in Equation 5.23,

$$g(t) = \begin{cases} -X_P & \text{in transit,} \\ 0 & \text{otherwise.} \end{cases} \tag{5.28}$$

The normalization of the (X, Y, Z) coordinate system means that $X_P < 1$ throughout the transit, so the dimensionless function $g(t)$ has a range $-1 < g(t) < 1$ in the simplest case. For the more complicated situations indicated by the general expression in Equation 5.26, the function $g(t)$ remains of order unity, so A_S is always a good indicator of the amplitude of the effect.

- How can we be sure that $g(t)$ is invariably of order unity?

- The Rossiter–McLaughlin effect arises because some of the stellar surface is blocked by the planet. The fraction blocked is approximately the part of the right-hand side of Equation 5.27 in parentheses, and the maximum radial velocity of this part is given by the part of the right-hand side of Equation 5.27 outside the parentheses. It is impossible to generate an effect significantly larger than this.

5.3.3 Comparison of the amplitudes of the orbital radial velocity and the Rossiter–McLaughlin effect

It is interesting to consider how large an effect on the orbital radial velocity curve we expect from the analysis that we have performed. Combining Equations 5.16 and 5.27, we obtain

$$\frac{A_S}{A_{RV}} = \frac{V_S R_P^2 P^{1/3} (M_P + M_*)^{2/3} \sqrt{1 - e^2} \sin i_S}{(2\pi G)^{1/3} M_P (R_*^2 - R_P^2) \sin i}.$$

If we assume that $M_P \ll M_*$ and $R_P^2 \ll R_*^2$, this reduces to

$$\begin{aligned} \frac{A_S}{A_{RV}} &\approx \left(\frac{P}{2\pi G}\right)^{1/3} \frac{V_S \sqrt{1 - e^2} \sin i_S}{M_P^{1/3} \sin i} \left(\frac{M_*}{R_*^3}\right)^{2/3} \left(\frac{R_P^3}{M_P}\right)^{2/3} \\ &\approx \left(\frac{P}{2\pi G M_P}\right)^{1/3} \frac{V_S \sqrt{1 - e^2} \sin i_S}{\sin i} \left(\frac{\rho_*}{\rho_P}\right)^{2/3}, \end{aligned} \tag{5.29}$$

where we have introduced the densities ρ_* and ρ_P of the star and planet, respectively. We retained the eccentricity in the input expression for A_{RV} as it can have a profound effect on the orbital radial velocity amplitude. In contrast, we have explicitly set $e = 0$ in the analysis that we performed on the Rossiter–McLaughlin effect, but as we argued at the end of Subsection 5.3.2, the eccentricity cannot have a significant effect on the amplitude of the Rossiter–McLaughlin perturbation.

Exercise 5.5 (a) Show that the amplitude of the Rossiter–McLaughlin effect for an exo-Earth is

$$A_S \approx 0.42\,\mathrm{m\,s^{-1}} \times \left(\frac{V_S \sin i_S}{5\,\mathrm{km\,s^{-1}}}\right)\left(\frac{R_P}{\mathrm{R_\oplus}}\right)^2\left(\frac{R_*}{\mathrm{R_\odot}}\right)^{-2}. \tag{5.30}$$

(b) Use your expression and a result from Exercise 5.4 to show that the ratio A_S/A_{RV} for an exo-Earth is

$$\frac{A_S}{A_{RV}} \approx 4.7\left(\frac{V_S \sin i_S}{5\,\mathrm{km\,s^{-1}}}\right)\left(\frac{R_P}{\mathrm{R_\oplus}}\right)^2\left(\frac{M_P \sin i}{\mathrm{M_\oplus}}\right)^{-1}\left(\frac{P}{\mathrm{yr}}\right)^{1/3}\left(\frac{\rho_*}{\rho_\odot}\right)^{2/3}. \tag{5.31}$$

(c) Comment on the prospects for the detection of exo-Earths by their transit light curve, their reflex orbital radial velocity variation, and their Rossiter–McLaughlin effect. ■

Equation 5.31 reveals that the Rossiter–McLaughlin effect becomes increasingly prominent compared to the orbital radial velocity as the period increases and the mass of the planet decreases. For long-period, low-mass planets, the amplitude of the Rossiter–McLaughlin perturbation can exceed that of the orbital radial velocity variation. Thus the Rossiter–McLaughlin effect may be used in the future to detect or confirm transiting Earth-like exoplanets.

5.3.4 Results from Rossiter–McLaughlin observations

Figure 5.19 shows radial velocity curves exhibiting the Rossiter–McLaughlin effect. These can be modelled using the equations outlined in Subsection 5.3.2 above to yield values for the projected spin–orbit angle, β. The results of these analyses to date (May 2009) are summarized in Table 5.1.

As we said in Subsection 5.3.1, the spin–orbit angle, ψ, is of fundamental importance, but we are only able to measure its projection on the sky, β. This situation is similar to the planetary mass constraints yielded by the radial velocity measurements: we want to know the value of M_P, but the measurements only tell us the value of $M_P \sin i$. Despite this, by assuming that exoplanet orbits are randomly oriented, we can make a statistical correction to the mass distribution. Consequently, we are sure that the planets detected by radial velocity methods are predominantly genuine planets. In a similar way, without knowing the value of i_S for an individual system, it is impossible to learn the value of ψ from the measured value of β. Nonetheless, if a large number of measurements of β are made for an ensemble of different systems, it is possible to say whether or not the results are consistent with various assumptions about the distribution of the unmeasurable angle ψ.

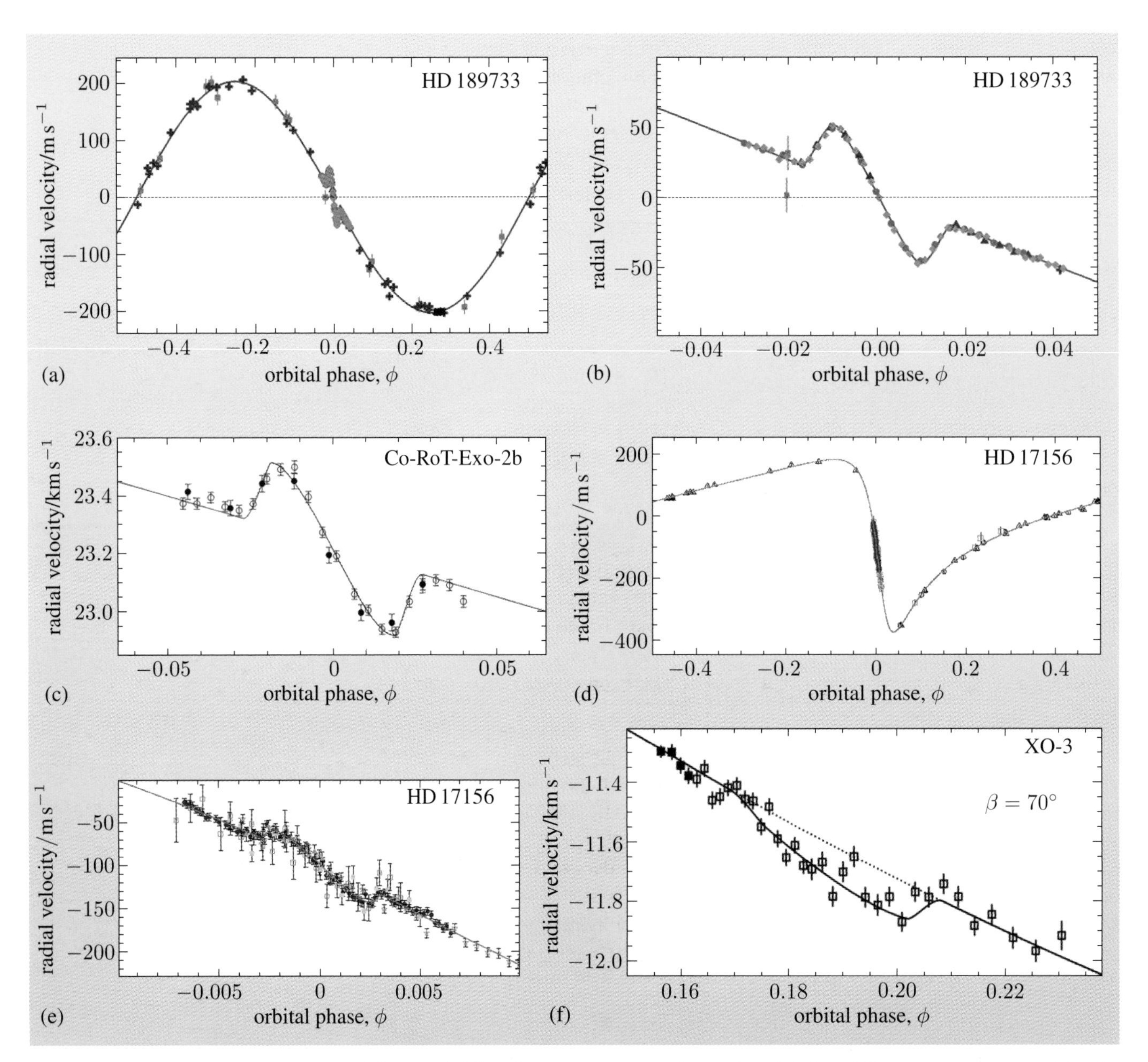

Figure 5.19 Some examples of Rossiter–McLaughlin measurements. (a) The radial velocity measurements for HD 189733 as a function of orbital phase for the planet HD 189733 b. (b) A zoom-in on the HD 189733 radial velocities around orbital phase 0, showing the Rossiter–McLaughlin perturbation. (c) The measured Rossiter–McLaughlin perturbation for CoRoT-2 b. (d) The radial velocity measurements for HD 17156, which has a planet in a highly eccentric orbit. (e) A zoom-in on the HD 17156 radial velocities around orbital phase 0, showing the Rossiter–McLaughlin perturbation. (f) The measured Rossiter–McLaughlin perturbation for XO-3. The planet XO-3 b is in an orbit that is misaligned with the stellar spin: $\beta \approx 70°$, producing a negative velocity perturbation throughout the transit.

Table 5.1 Rossiter–McLaughlin effect results for the projected spin–orbit angle, β (May 2009). Except for XO-3 b, the angles are consistent within 2σ of perfect spin–orbit alignment.

Exoplanet	Projected spin–orbit angle β/deg
HD 189733 b	-1.4 ± 1.1
HD 209458 b	0.1 ± 2.4
HAT-P-1 b	3.7 ± 2.1
CoRoT-Exo-2 b	7.2 ± 4.5
HD 149026 b	1.9 ± 6.1
HD 17156 b	9.4 ± 9.3
TrES-2 b	-9.0 ± 12.0
HAT-P-2 b	1.2 ± 13.4
XO-3 b	70.0 ± 15.0
WASP-14 b	-14.0 ± 17.0
TrES-1 b	30.0 ± 21.0

The results in Table 5.1 are just about plentiful enough to begin statistically analyzing the properties of the ensemble, and this has been done. The statistical constraints are more complex than the treatment of the simple $\sin i$ projection factor that features in the analysis of the orbital radial velocity planet mass determinations. Figure 5.20 summarizes the constraints on ψ that can be made given a particular measured value of β.

The statistical analysis of the 11 systems in Table 5.1 suggests that the sample might be composed of representatives from two distinct populations: the first being systems that are almost perfectly aligned, so $\psi \approx 0°$; the second having the spin angular momentum vector randomly oriented relative to the orbital angular momentum vector. A cursory inspection of the data in Table 5.1 is already suggestive of this, with XO-3 b showing the only strong evidence for misalignment. However, XO-3 b may not be the only one of the 11 systems drawn from an isotropic distribution; several of the other systems may also be representatives of this second population.

Measurements of the Rossiter–McLaughlin effect for newly discovered transiting planets are currently being made. By the time that this book is in print, there will probably be a significant number of additions to the results tabulated in Table 5.1, and consequently the statistical constraints on ψ will be stronger. Models for exoplanet formation and evolution are capable of predicting the probability distribution for ψ for the various scenarios. As the number of measurements increases, the ensemble of results should be able to distinguish between possible evolutionary histories for the transiting exoplanet population. Thus the Rossiter–McLaughlin effect should ultimately constrain the migration history of the close-in giant planets, and consequently illuminate our general picture of the formation and evolution of planets.

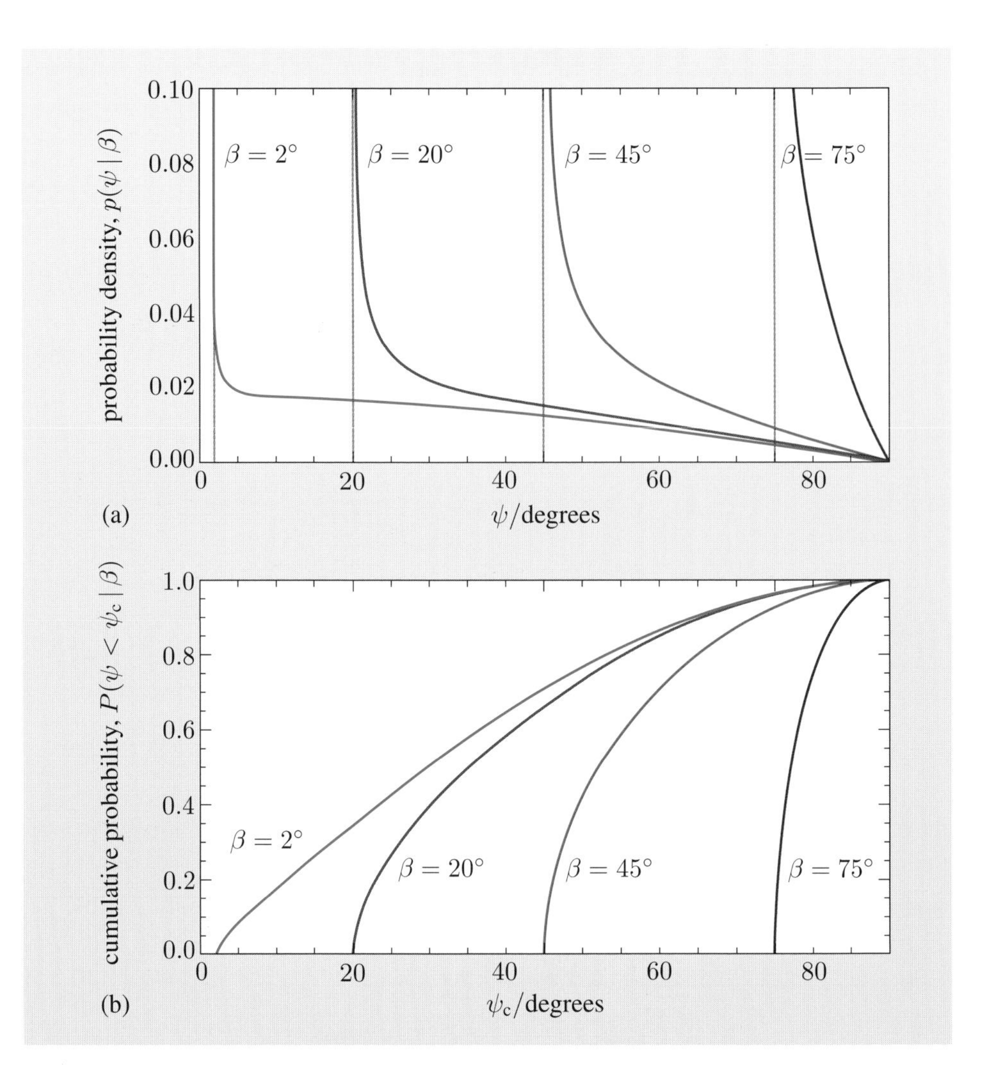

Figure 5.20 The statistical inferences about the spin–orbit angle, ψ, that can be made from measuring the projected spin–orbit angle, β. (a) The probability density for each value of ψ given a particular value of β. The vertical lines indicate $\psi = \beta$ in each case, and the probability is highest at $\psi = \beta$ but there is a significant probability of ψ having any value up to the maximum possible, $90°$. (b) The same probability information presented in a different way: in this case the curves show the probability that ψ has a value less than any particular value. In each case, obviously, the minimum possible value is $\psi = \beta$.

Summary of Chapter 5

1. The equilibrium temperature of a planet is given by

$$T_{\text{eq}} = \frac{1}{2}\left[\frac{(1-A)L_*}{\sigma\pi a^2}\right]^{1/4} \qquad \text{(Eqn 5.5)}$$

and provides the simplest estimate of the temperature of the planet's surface.

2. Transmission spectroscopy examines the wavelength-dependence of the transit depth, thus inferring the presence of chemical species in the atmosphere. It is sensitive to species in the exoplanet's upper atmosphere if the atmosphere is optically thin to continuum radiation.

3. A plane-parallel ideal gas atmosphere in hydrostatic equilibrium has an exponentially falling pressure

$$P = P_0 \exp(-z/H), \qquad \text{(Eqn 5.9)}$$

where z is the height in the atmosphere and H is the scale-height:

$$H = \frac{kT}{g\mu u}. \qquad \text{(Eqn 5.10)}$$

4. The column density, N, above a particular height, z, is given by

$$N \approx \frac{P(z)}{g\mu u}. \tag{Eqn 5.13}$$

The tropopause occurs where the layers above become transparent. Higher-gravity planets have higher-pressure tropopauses.

5. A typical hot Jupiter is too hot for most of the cloud layers present in Jupiter to form. Only perovskite/corundum, metallic liquid iron and Mg-silicate cloud layers are expected. The properties of clouds are, however, notoriously difficult to predict.

6. Na has been detected in the atmospheres of both HD 209458 b and HD 189733 b. HD 209458 b is surrounded by a large cloud that absorbs in Lyman α, C II and O I, possibly arising from the planet's atmosphere. The ultraviolet–optical–infrared opacity in HD 189733 b's atmosphere suggests Rayleigh scattering with absorption due to water and CO in the infrared.

7. The Rossiter–McLaughlin effect is the perturbation on the orbital radial velocity curve caused by the transit of a planet across the disc of a spinning star. The perturbation is

$$\Delta V_{\mathrm{RM}}(t) = A_{\mathrm{S}}\, g\left(t, X_{\mathrm{P}}, Y_{\mathrm{P}}, \frac{R_{\mathrm{P}}}{R_*}, u, \ldots\right), \tag{Eqn 5.26}$$

where $g(t)$ is a dimensionless function of order unity and the amplitude is

$$A_{\mathrm{S}} = \Omega_{\mathrm{S}} R_* \sin i_{\mathrm{S}} \left(\frac{R_{\mathrm{P}}^2}{R_*^2 - R_{\mathrm{P}}^2}\right) = V_{\mathrm{S}} \sin i_{\mathrm{S}} \left(\frac{R_{\mathrm{P}}^2}{R_*^2 - R_{\mathrm{P}}^2}\right). \tag{Eqn 5.27}$$

In the simplest case with no limb darkening and neglecting the ingress and egress phases,

$$g(t) = \begin{cases} -X_{\mathrm{P}} & \text{in transit,} \\ 0 & \text{otherwise.} \end{cases} \tag{Eqn 5.28}$$

8. The ratio of the amplitudes of the Rossiter–McLaughlin effect and the orbital reflex radial velocity curve is

$$\frac{A_{\mathrm{S}}}{A_{\mathrm{RV}}} \approx 4.7 \left(\frac{V_{\mathrm{S}} \sin i_{\mathrm{S}}}{5\,\mathrm{km\,s^{-1}}}\right) \left(\frac{R_{\mathrm{P}}}{R_\oplus}\right)^2 \left(\frac{M_{\mathrm{P}} \sin i}{M_\oplus}\right)^{-1} \left(\frac{P}{\mathrm{yr}}\right)^{1/3} \left(\frac{\rho_*}{\rho_\odot}\right)^{2/3}. \tag{Eqn 5.31}$$

For long-period, low-mass planets, $A_{\mathrm{S}} < A_{\mathrm{RV}}$. Earth-like transiting exoplanets may be detected or confirmed using the Rossiter–McLaughlin effect.

9. The Rossiter–McLaughlin effect measures the projected spin–orbit angle, β. The spin–orbit angle, ψ, is a fundamental property, and can be used to constrain the evolutionary history of the system. The stellar spin orientation is generally unknown or very poorly known, so inferences about ψ can only be made through statistical analysis. This suggests that there are two populations: the first with the stellar spin and orbital angular momentum axes closely aligned, the second being effectively isotropic. These two populations reflect different evolutionary paths.

Chapter 6 Secondary eclipses and phase variations

Introduction

In Chapter 5 we saw how the planetary atmospheric chemistry and the stellar spin orientation can be measured for transiting planets. Both these techniques analyzed the starlight blocked by the planet during the transit. In this chapter we consider instead the light from the planet itself. The previous sentence seems a little like science fantasy: a few years ago the idea of being able to detect and analyze light from planets in close orbits around stars other than our Sun seemed unlikely. As we stressed in Chapter 1, stars are generally MUCH MUCH brighter than planets, so detecting an exoplanet's light is not easy. Figure 6.1 illustrates the configurations during transit and at the beginning of **secondary eclipse**: at the opposite side of the orbit from the transit, the planet passes behind the host star and its heated face is eclipsed.

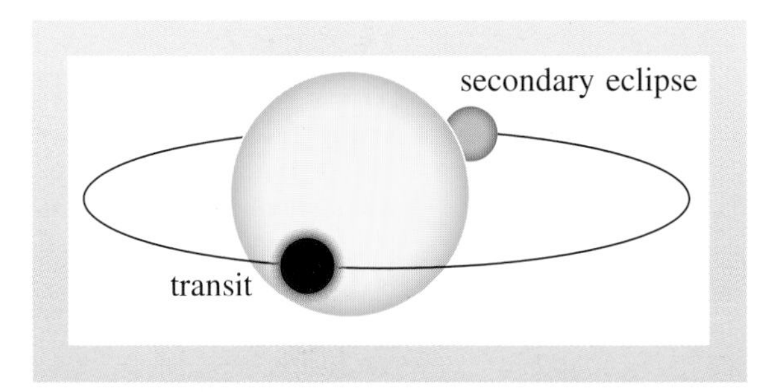

Figure 6.1 A schematic diagram showing a planet transiting its host star, with light being transmitted through the planet's atmosphere, which is represented by the blue annulus. The transit occurs at inferior conjunction of the planet; at superior conjunction of the planet, the planet passes behind the star and a secondary eclipse occurs. Between the transit and the secondary eclipse a continuously varying amount of the heated face of the planet is visible: this causes a smooth variation in the amount of light detected from the planet.

We will always refer to transits and secondary eclipses, but some authors also use the term **primary eclipse** to refer to the transit. At the secondary eclipse, the thermal emission from the heated face of the planet is occulted, and a (very!) small sharp drop in the total light from the system can, in principle, be detected. This drop allows us to distinguish the light from the planet from the overwhelmingly brighter light from the host star. This chapter discusses these measurements and their interpretation. During the transit, the heated face of the planet is directed away from the observer, and as the planet proceeds around its orbit, a gradually increasing fraction of the heated face becomes visible. Consequently, we expect a small smooth modulation in the total amount of light from the system, with a maximum just before and after the secondary eclipse. This smooth modulation is the **phase variation**, which is the second subject of this chapter.

6.1 Expectations using simple approximations

This section follows the treatment of the paper 'Hot nights on extrasolar planets: mid-infrared phase variations of hot Jupiters' (2007, *Monthly Notices of the*

Royal Astronomical Society, **379**(2), 641–6) by N.B. Cowan, E. Agol and D. Charbonneau.

6.1.1 Reflected starlight

We stated in Section 5.1 that a fraction A of the incident starlight is reflected from the planet. This light is reflected from the day side of the planet as shown in Figure 5.1. The fraction of the starlight that is reflected is a function of wavelength, and this is expressed within the **wavelength-dependent geometric albedo**, p_λ. The amount of reflected light that is directed towards the observer varies with the **phase angle**, α, which is defined in Figure 6.2. Consequently, at wavelength λ, the ratio of the flux at the observer reflected from the exoplanet, $f_{\mathrm{P},\lambda}$, to the flux observed directly from the star, $f_{*,\lambda}$, is given by

$$\frac{f_{\mathrm{P},\lambda}(\alpha)}{f_{*,\lambda}} = \left(\frac{R_\mathrm{P}}{a}\right)^2 p_\lambda \, \Phi_\lambda(\alpha), \tag{6.1}$$

where a is the distance of the planet from the star, and $\Phi_\lambda(\alpha)$ is the **phase function** that gives the flux at phase angle α relative to that at superior conjunction (or at **opposition** for Solar System objects). By definition, $\Phi_\lambda(0) = 1$, and we will take Equation 6.1 as effectively defining the wavelength-dependent geometric albedo. The relationship between p_λ and A, the **Bond albedo**, involves integrating over wavelength and over the phase function to obtain the total flux reflected by the planet. The Bond albedo, A, is used to deduce the energy balance in the planet, while the wavelength-dependent geometric albedo, p_λ, is used to deduce the reflected spectrum observed at any instant. Equation 6.1 cannot be applied to Solar System planets in the same way as to exoplanets because it implicitly assumes that the distance from the observer to the star is the same as the distance from the observer to the planet; this is a good approximation for exoplanets, but not for Solar System objects.

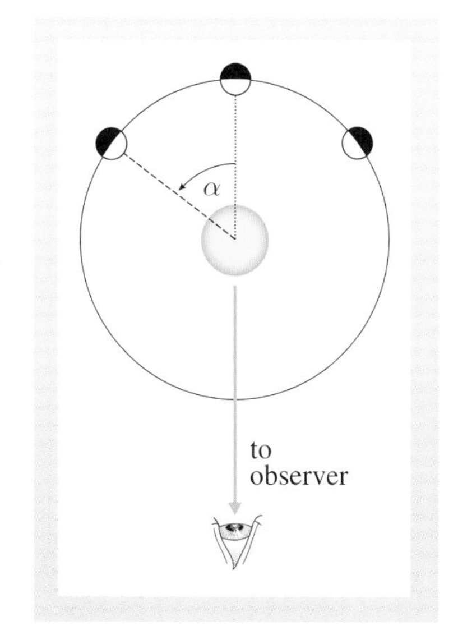

Figure 6.2 The phase angle, α, is zero at superior conjunction of the planet when it is furthest from the observer, and is the angle between the continuation of the line from the observer to the star, and the line from the star to the planet. α increases in the direction of the orbital motion. Note: the zero-point of α does *not* coincide with the zero-point of the orbital phase.

Φ_λ is an empirically determined curve showing how the reflected flux varies with phase angle α. This should not be confused with ϕ, the orbital phase.

- What is the value of α at superior conjunction or opposition?
- Referring to Figure 6.2, at superior conjunction $\alpha = 0$; this is also the value of α for Solar System objects at opposition.
- What is the value of $\Phi_\lambda(0)$?
- Since the phase function gives the flux relative to that at superior conjunction, $\Phi_\lambda(0) = 1$ by definition.

If no secondary eclipse occurred, the maximum in the reflected light spectrum would occur at superior conjunction, so the maximum value for the right-hand side of Equation 6.1 is given by $\Phi_\lambda(0) = 1$. The minimum value in the reflected light spectrum for a transiting system is effectively zero, when we view the unilluminated face of the exoplanet around inferior conjunction of the exoplanet. Consequently, it is clear that the ratio $f_{\mathrm{P},\lambda}(\alpha)/f_{*,\lambda}$ varies between 0 and ε_λ, where

$$\varepsilon_\lambda = p_\lambda \left(\frac{R_\mathrm{P}}{a}\right)^2 \tag{6.2}$$

is the amplitude of the reflected light spectrum.

6.1.2 Day side and night side equilibrium temperatures

As we saw in Section 5.1, Equation 5.5 implies that the temperatures of transiting hot Jupiters are $\sim 1000\,\mathrm{K}$. The Wien displacement law (maximizing flux per unit frequency interval) tells us that the emission from a black body at this temperature will peak at

$$\begin{aligned}\lambda_{\max} &= 5.1\times 10^{-3}\left(\frac{\mathrm{K}}{T}\right)\,\mathrm{m}\\ &\sim 5\times 10^{-6}\,\mathrm{m}\sim 5\,\mu\mathrm{m},\end{aligned}\tag{6.3}$$

so we expect the thermal emission from hot Jupiter planets to peak in the infrared, at wavelengths of several microns.

Worked Example 6.1

Approximating both the star and the planet as black body emitters, show that the equilibrium temperatures of the planet's day and night sides are given by

$$T_{\mathrm{day}}^4 = (1-P)(1-A)\,\frac{R_*^2}{2a^2}\,T_{\mathrm{eff}}^4 \tag{6.4}$$

and

$$T_{\mathrm{night}}^4 = P(1-A)\,\frac{R_*^2}{2a^2}\,T_{\mathrm{eff}}^4, \tag{6.5}$$

where P is the fraction of the absorbed energy that is transported to the night side of the planet, T_{eff} is the **effective temperature** of the star, and the other symbols have been defined previously. You may neglect the radiation powered by the Kelvin–Helmholtz contraction of the planet.

Solution

Obviously we need to follow the approach of Section 5.1, but relaxing the assumption that the energy is so efficiently transported to the night side that the temperature is the same on both hemispheres. In Section 5.1 the star was defined in terms of its luminosity, L_*, while the expressions that we seek to demonstrate are cast in terms of the stellar effective temperature, T_{eff}, and radius, R_*. Our first step, therefore, is to express the luminosity in terms of effective temperature and radius:

$$L_* = 4\pi R_*^2\sigma\,T_{\mathrm{eff}}^4. \tag{6.6}$$

We can obtain

$$\begin{aligned}\dot{E}_{\mathrm{abs}} &= (1-A)\pi R_{\mathrm{P}}^2 F_*\\ &= (1-A)\pi R_{\mathrm{P}}^2\frac{L_*}{4\pi a^2}\end{aligned}\tag{6.7}$$

by using Equations 5.3 and 5.1. Substituting in from Equation 6.6 we obtain

$$\begin{aligned}\dot{E}_{\mathrm{abs}} &= (1-A)\pi R_{\mathrm{P}}^2\frac{4\pi R_*^2\sigma\,T_{\mathrm{eff}}^4}{4\pi a^2}\\ &= (1-A)\pi R_{\mathrm{P}}^2\left(\frac{R_*}{a}\right)^2\sigma\,T_{\mathrm{eff}}^4.\end{aligned}\tag{6.8}$$

In equilibrium the energy absorbed must equal the energy emitted. The emitted energy is unequally divided between the day side and the night side hemispheres, but we can say

$$\begin{aligned}\dot{E}_{\rm abs} &= \dot{E}_{\rm emit,day} + \dot{E}_{\rm emit,night} = 2\pi R_{\rm P}^2 \sigma T_{\rm day}^4 + 2\pi R_{\rm P}^2 \sigma T_{\rm night}^4 \\ &= (1-P)\dot{E}_{\rm abs} + P\dot{E}_{\rm abs},\end{aligned} \tag{6.9}$$

where we have treated each hemisphere of the planet as a black body with surface area $2\pi R_{\rm P}^2$, and used the fraction of the energy transported to the night side, P, to express each hemisphere's rate of emission in terms of the energy absorbed from the star, $\dot{E}_{\rm abs}$. Thus we have

$$\begin{aligned}T_{\rm day}^4 &= \frac{(1-P)\dot{E}_{\rm abs}}{2\pi R_{\rm P}^2 \sigma} = \frac{(1-P)}{2\pi R_{\rm P}^2 \sigma}(1-A)\pi R_{\rm P}^2 \left(\frac{R_*}{a}\right)^2 \sigma T_{\rm eff}^4 \\ &= (1-P)(1-A)\,\frac{R_*^2}{2a^2}\,T_{\rm eff}^4, \qquad \text{(Eqn 6.4)}\end{aligned}$$

which is the first expression that we seek. Similarly,

$$\begin{aligned}T_{\rm night}^4 &= \frac{P\dot{E}_{\rm abs}}{2\pi R_{\rm P}^2 \sigma} \\ &= P(1-A)\,\frac{R_*^2}{2a^2}\,T_{\rm eff}^4. \qquad \text{(Eqn 6.5)}\end{aligned}$$

- What value of P would make $T_{\rm day} = T_{\rm night}$?
- $P = \frac{1}{2}$, so that $\dot{E}_{\rm emit,day} = \dot{E}_{\rm emit,night}$.

6.1.3 Ratio of star and planet fluxes at secondary eclipse

Near secondary eclipse, the planet presents its day side hemisphere to us. The equilibrium temperature of this hemisphere is $T_{\rm day}$, given in Equation 6.4. If we maintain our approximation that the planet radiates as a black body, then the flux at any wavelength, λ, emitted per unit surface area is simply the black body flux, $B_\lambda(T)$. On the other hand, stellar spectra are known to depart noticeably from black bodies and a better approximation is to use a **brightness temperature**, $T_{\rm bright}$, which is defined such that the flux at any wavelength, λ, emitted per unit surface area is $B_\lambda(T_{\rm bright})$. The relationship between the effective temperature of the star, $T_{\rm eff}$, and the brightness temperature depends on the spectral type of the star and the wavelength being considered, and is generally a factor of order unity. If we know the spectral type of the star, then in principle we know $T_{\rm bright}$ as a function of wavelength, λ. With these approximations (and assuming for simplicity that $i = 90^\circ$, so the night side hemisphere is completely hidden), the flux ratio between the star and the planet at superior conjunction is

$$\frac{f_{{\rm day},\lambda}}{f_{*,\lambda}} = p_\lambda \left(\frac{R_{\rm P}}{a}\right)^2 + \frac{B_\lambda(T_{\rm day})}{B_\lambda(T_{\rm bright})}\left(\frac{R_{\rm P}}{R_*}\right)^2, \tag{6.10}$$

where the first term is the reflected light contribution, and the second term is the thermal emission contribution. Equation 6.10 allows us to calculate the expected depth of secondary eclipse at any wavelength. For readers fatigued by equations, Figure 6.3 illustrates the comparison between $f_{\mathrm{day},\lambda}$ and f_{*}, λ.

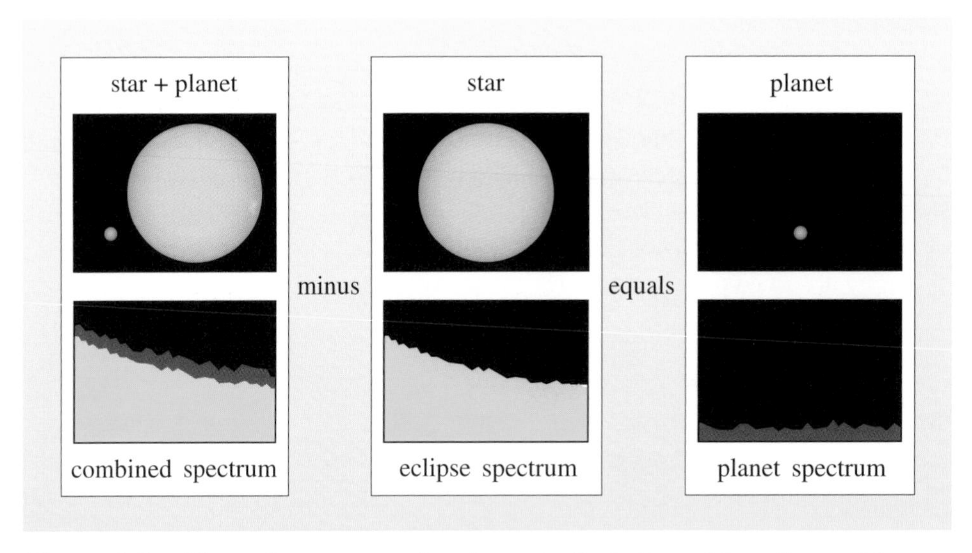

Figure 6.3 The stellar spectrum is schematically indicated in yellow, with the small amount of additional flux from the planet in red. For the purposes of illustration, the fraction of the total flux due to the planet has been hugely exaggerated.

As we saw in Figure 1.8 for Solar System planets, the reflected light from the planet peaks at shorter wavelengths than the thermal emission from the planet. This is in accordance with the second law of thermodynamics, which could be crudely stated as 'heat travels from hot to cold': insolation from the star cannot heat the day side of the planet to a temperature exceeding that of the star. Generally, for the Solar System planets the contribution of the reflected light to the contrast ratio (Equation 6.10) can be neglected at long wavelengths, while the thermal emission can be neglected at short wavelengths. If we generalize this statement to exoplanet analogues of the Solar System, the precise division between 'long' and 'short' wavelengths in this statement depends on the spectrum of the star: for cooler stars the reflection spectrum will peak further into the infrared. This clean separation of the thermal and reflected light does *not* hold for hot Jupiters, however. Their equilibrium temperatures are much higher than those of the Solar System planets, and the two components overlap substantially in wavelength.

6.1.4 Peak to trough amplitude of the orbital light curve

As the planet proceeds around its orbit, the visible fraction of the day side and night side hemispheres changes continuously. If the orbital inclination is 90°, i.e. exactly edge-on, then the modulation in the observed thermal flux will be maximized, with the visible disc of the planet being entirely composed of the cool night side hemisphere during the transit. The planet will appear as though it has thermal luminosity

$$\begin{aligned} L_{\text{P,night}} &= 4\pi R_{\text{P}}^2 \int B_\lambda(T_{\text{night}})\,\text{d}\lambda \\ &= 4\pi R_{\text{P}}^2 \sigma\, T_{\text{night}}^4, \end{aligned} \tag{6.11}$$

where we have used the well-known result for the integral of the Planck black body function, B_λ, over the entire spectrum. A similar result, with T_{night} replaced by T_{day}, would apply for the apparent thermal luminosity of the planet at superior conjunction of the planet, but at this phase the planet is obscured by the star during the secondary eclipse.

The actual luminosity of the planet is, of course,

$$L_{\text{P}} = \frac{L_{\text{P,night}}}{2} + \frac{L_{\text{P,day}}}{2}.$$

- What is the value of b for the case where the orbital inclination is exactly edge-on?
- The impact parameter is $b = 0$ when $i = 90°$.

At any wavelength, λ, the peak to trough difference in thermal flux from a planet in an edge-on orbit is therefore

$$\begin{aligned} \Delta f_{\text{P},\lambda} &= \frac{4\pi R_{\text{P}}^2}{4\pi d^2} \left[B_\lambda(T_{\text{day}}) - B_\lambda(T_{\text{night}})\right] \\ &= \left(\frac{R_{\text{P}}}{d}\right)^2 \left[B_\lambda(T_{\text{day}}) - B_\lambda(T_{\text{night}})\right]. \end{aligned} \tag{6.12}$$

If the orbit is not exactly edge-on, this flux difference will be diminished, as some of each hemisphere will be visible at all orbital phases; for an orbit in the plane of the sky, with $i = 0°$, there will be no modulation at all in the flux from the planet.

- What is the value of the orbital phase at the observed maximum of the thermal emission light curve?
- This occurs when the planet is farthest from us, at orbital phase $\phi = 0.5$. (Or just before and after this phase, since at exactly phase 0.5 the secondary eclipse occurs, so the thermal emission is not visible then.)
- What is the value of the phase angle, α, at the observed maximum of the thermal emission light curve?
- At orbital phase $\phi = 0.5$, $\alpha = 0°$ (or 0 radians), so the observed maximum of the thermal emission light curve occurs at (or just before and after) $\alpha = 0°$.

6.1.5 Contrast units and flux units

In Subsection 6.1.3 we used contrast units to give the expected depth of the secondary eclipse in Equation 6.10. This is the most natural unit to use, as the secondary eclipse will appear as a fractional loss in the observed light from the star–planet system. Conversely, we gave the amplitude of the thermal emission orbital light curve in flux units in Equation 6.12. We could easily convert this into contrast units by dividing through by the total flux of the star–planet system, which to a good approximation is simply the light from the star. In assessing whether a secondary eclipse or a modulation in the thermal flux from a planet is detectable, both contrast units and flux units are useful. An instrument must be capable of measuring at a precision sufficient to detect flux difference that constitutes the feature. Furthermore, it must be capable of making this measurement while simultaneously being exposed to the overwhelming light from the star: the instrument must have sufficient **dynamic range** to measure the tiny fractional change that constitutes the feature in contrast units.

Worked Example 6.2

(a) Show that the fractional depth of secondary eclipse in a bandpass centred on wavelength λ_c is

$$\frac{\Delta F_{SE}}{F} \approx \frac{f_{day,\lambda_c}}{f_{*,\lambda_c}} \approx p_{\lambda_c}\left(\frac{R_P}{a}\right)^2 + \frac{B_{\lambda_c}(T_{day})}{B_{\lambda_c}(T_{bright})}\left(\frac{R_P}{R_*}\right)^2. \qquad (6.13)$$

(b) Further, show that in bandpasses for which $\lambda_c \gg hc/(k\,T_{day})$ and in which the reflection component may be neglected,

$$\frac{\Delta F_{SE}}{F} \approx \frac{T_{day}}{T_{bright}}\left(\frac{R_P}{R_*}\right)^2. \qquad (6.14)$$

(c) Show that if the stellar spectrum is approximated as a black body, Equation 6.14 can be expressed as

$$\frac{\Delta F_{SE}}{F} \approx \left[\frac{(1-P)(1-A)}{2a^2}\right]^{1/4} \frac{R_P^2}{R_*^{3/2}}. \qquad (6.15)$$

Solution

(a) We have our expression for the flux from the planet in contrast units at superior conjunction:

$$\frac{f_{day,\lambda}}{f_{*,\lambda}} = p_\lambda\left(\frac{R_P}{a}\right)^2 + \frac{B_\lambda(T_{day})}{B_\lambda(T_{bright})}\left(\frac{R_P}{R_*}\right)^2. \qquad \text{(Eqn 6.10)}$$

The flux, ΔF_{SE}, that will be missing during the secondary eclipse is given by

$$\Delta F_{SE} = \int_{\lambda_l}^{\lambda_u} f_{day,\lambda}\, Q_\lambda\, d\lambda,$$

where λ_u, λ_l and Q_λ respectively give the limits of the bandpass and the weighting of the contribution of the flux at a particular wavelength to the integrated average flux. The total flux from the system is similarly

$$F = \int_{\lambda_l}^{\lambda_u} (f_{day,\lambda} + f_{*,\lambda})\, Q_\lambda\, d\lambda.$$

If we assume that we can take the fluxes at the central wavelength, λ_c, to be proportional to the integrated flux over the bandpass, and note that the flux from the star is much greater than the flux from the planet, then $F \propto f_{*,\lambda_c}$ and $\Delta F_{SE} \propto f_{day,\lambda_c}$. Thus

$$\frac{\Delta F_{SE}}{F} \approx \frac{f_{day,\lambda_c}}{f_{*,\lambda_c}} \approx p_{\lambda_c}\left(\frac{R_P}{a}\right)^2 + \frac{B_{\lambda_c}(T_{day})}{B_{\lambda_c}(T_{bright})}\left(\frac{R_P}{R_*}\right)^2, \qquad \text{(Eqn 6.13)}$$

where we have substituted from Equation 6.10.

(b) Equation 6.13 expresses the fluxes in terms of the **Planck function**

$$B_\lambda(T) = \frac{2hc^2}{\lambda^5} \frac{1}{\exp(hc/\lambda kT) - 1} \, \mathrm{W\,m^{-2}\,m^{-1}\,sr^{-1}}. \qquad (6.16)$$

This is a complicated expression, but in the regime $\lambda \gg hc/(k\,T_{\text{day}})$ it can be simplified significantly. For $x \ll 1$, $\exp(x)$ can be approximated by the first two terms in the Taylor expansion:

Note that sr^{-1} means 'per steradian'.

$$\exp(x) \approx 1 + x + \cdots .$$

If $\lambda \gg hc/kT$, then $hc/\lambda kT \ll 1$ and the exponent in the denominator of the Planck function can be approximated as

$$\exp\left(\frac{hc}{\lambda kT}\right) \approx 1 + \frac{hc}{\lambda kT}.$$

Using this approximation,

We explicitly include units in the expressions for $B_\lambda(T)$ to alert us to the 'per steradian' factor.

$$\begin{aligned} B_\lambda(T) &= \frac{2hc^2}{\lambda^5} \frac{1}{hc/\lambda kT} \, \mathrm{W\,m^{-2}\,m^{-1}\,sr^{-1}} \\ &= \frac{2ckT}{\lambda^4}, \end{aligned} \qquad (6.17)$$

which is known as the **Rayleigh–Jeans law**.

Since $T_{\text{day}} < T_{\text{eff}}$ (because heat flows from hot to cold), it is also almost certain that $T_{\text{day}} < T_{\text{bright}}$ as the deviations of the stellar spectrum from a black body are unlikely to be large compared to the difference between the stellar and planetary temperatures. Thus if $\lambda_{\text{c}} \gg hc/(k\,T_{\text{day}})$, then $\lambda_{\text{c}} \gg hc/(k\,T_{\text{bright}})$. This means that both the day side and the stellar black body flux will be well approximated by the Rayleigh–Jeans law. We are also told that the reflection component may be neglected, so

$$\begin{aligned} \frac{\Delta F_{\text{SE}}}{F} &\approx \frac{2ck\,T_{\text{day}}}{\lambda_{\text{c}}^4} \frac{\lambda_{\text{c}}^4}{2ck\,T_{\text{bright}}} \left(\frac{R_{\text{P}}}{R_*}\right)^2 \\ &\approx \frac{T_{\text{day}}}{T_{\text{bright}}} \left(\frac{R_{\text{P}}}{R_*}\right)^2, \end{aligned} \qquad \text{(Eqn 6.14)}$$

as required.

(c) If the star as well as the planet is approximated as having a black body spectrum, then $T_{\text{bright}} = T_{\text{eff}}$ for all wavelengths. In this case, we can use Equation 6.4 to express T_{day} in terms of T_{eff}:

$$T_{\text{day}}^4 = (1 - P)(1 - A)\,\frac{R_*^2}{2a^2}\,T_{\text{eff}}^4, \qquad \text{(Eqn 6.4)}$$

so

$$T_{\text{day}} = (1 - P)^{1/4}(1 - A)^{1/4} \left(\frac{R_*^2}{2a^2}\right)^{1/4} T_{\text{eff}}.$$

Substituting into Equation 6.14, therefore,

$$\frac{\Delta F_{\mathrm{SE}}}{F} \approx \frac{T_{\mathrm{day}}}{T_{\mathrm{eff}}}\left(\frac{R_{\mathrm{P}}}{R_*}\right)^2 \approx (1-P)^{1/4}(1-A)^{1/4}\left(\frac{R_*^2}{2a^2}\right)^{1/4}\left(\frac{R_{\mathrm{P}}}{R_*}\right)^2$$

$$\approx \left[\frac{(1-P)(1-A)}{2a^2}\right]^{1/4} R_*^{1/2}\left(\frac{R_{\mathrm{P}}}{R_*}\right)^2$$

$$\approx \left[\frac{(1-P)(1-A)}{2a^2}\right]^{1/4} \frac{R_{\mathrm{P}}^2}{R_*^{3/2}}, \qquad \text{(Eqn 6.15)}$$

as required.

Worked Example 6.2 gives us some simple analytic expressions for the expected secondary eclipse depth. We will apply some of these to real exoplanet systems later in the chapter.

6.1.6 Caveats on these expectations

In the above we have neglected the intrinsic thermal emission from the planet, which is powered by the Kelvin–Helmholtz contraction. The effective temperature of an isolated giant planet is ~ 100 K, so this is a justifiable approximation as far as the day side of a typical transiting hot Jupiter is concerned. Whether or not the intrinsic emission becomes important in the calculation of the night side hemisphere depends on the fraction, P, of heat that is transported to the night side.

We note that all the giant planets in our own Solar System have spectra that deviate wildly from the simple spectra implied by our assumptions here. The reflected light spectra of planets are governed by Rayleigh scattering and absorption by molecules and hazes. Non-equilibrium photo-chemically produced molecules can depress the reflected flux, and a variety of scattering processes contribute to determining the albedo: the physical and chemical processes at work are much more complex than our treatment has acknowledged! Nonetheless, the observations are near the limit of the capabilities of the current observatories, so it is important to minimize the complexity of the models used to interpret the results.

- Why must we minimize the complexity of models in this case?

- Because our assessment of whether a model fits is dependent on the number of free parameters, as mathematically expressed in Equation 3.37. A complex model with as many adjustable parameters as there are independently observed quantities can fit the data irrespective of whether the underlying physics has been accurately described by the model.

The work that we have done here in Section 6.1 will allow us to fully appreciate some of the most remarkable observations made so far this century. In exercises within this chapter we apply the equations derived above to real exoplanet systems, making predictions of observable quantities and using observations to make quantitative deductions about real exoplanets. In Sections 6.2 and 6.3 we will examine some of the observations. It is extremely exciting to be able to isolate and study the light from planets around other stars. The telescopes being used were not designed to do this, and few would have predicted in the early 1990s that this would be accomplished. These observational results can be used to constrain the values of the Bond albedo, A, the wavelength-dependent geometrical albedo, p_λ, and the amount of heat redistribution to the night side, P. That we can begin to *measure* these detailed quantities for planets orbiting around other stars is something that the author finds amazing.

Exercise 6.1 Estimate the expected secondary eclipse depth for HD 189733 b at 24 μm, using black body approximations for the emission of both the star and the planet. You may assume that no heat is redistributed to the night side of the planet and that the Bond albedo is $A = 0.05$. Selected parameters of HD 189733 b and its host star are given in Table 6.1. Give your answer in terms of the fractional depth, $\Delta F_{\rm SE}/F$. You should discuss any further approximations that you make.

Exercise 6.2 (a) Show that the peak to trough amplitude of the phase function in a bandpass centred on $\lambda_{\rm c}$, when expressed in contrast units, is

$$\frac{F_{\rm day} - F_{\rm night}}{F_*} \approx \frac{B_{\lambda_{\rm c}}(T_{\rm day}) - B_{\lambda_{\rm c}}(T_{\rm night})}{B_{\lambda_{\rm c}}(T_{\rm bright})} \left(\frac{R_{\rm P}}{R_*}\right)^2, \qquad (6.18)$$

making clear and justifying any approximations that you make.

(b) Discuss whether it is possible to deduce from observations the temperature difference between the day side and night side of a transiting hot Jupiter exoplanet.

(c) Discuss whether it is possible to deduce from observations the temperature difference between the day side and night side of a hot Jupiter exoplanet whose orbit is oriented so that it does not transit its host star. ■

Table 6.1 Selected parameters of HD 189733 b and its host star for use in Exercises 6.1 and 6.3.

Quantity	Value	Units
a	0.030 99	AU
$R_{\rm P}$	1.14	$R_{\rm J}$
R_*	0.788	$R_\odot$
$T_{\rm eff}$	5000	K
$T_{\rm bright}(16\,\mu{\rm m})$	4315	K

6.2 Secondary eclipses

6.2.1 The first detections

As we have quantified in Section 6.1, the detection of light from an exoplanet is challenging. Astrophysics has been fortunate that the discovery of the transiting hot Jupiter planets was made in time for NASA's Spitzer Space Telescope to observe them before it exhausted its supply of cryogens. Spitzer is an infrared telescope that was launched in 2003. Its original mission was primarily to observe objects heavily obscured by interstellar dust, but arguably its greatest contribution to astrophysics has been its stunning observations of exoplanets. Figure 6.4 shows the first detections: in each panel a model using the expected time of secondary eclipse, i.e. assuming that it occurs exactly half an orbit after the transit, has been plotted. With the aid of the model, the dip that occurs in each case at the secondary eclipse is clearly discernable, though without the prior knowledge of the expected timing, most people would be uncertain about the exact depth and timing of the secondary eclipse features in these light curves. Statistically these two secondary eclipses are each detected with a high degree of confidence, so the data in Figure 6.4 were rightly heralded in 2005 as the first detection of exoplanet light. Nonetheless, the announcement in 2006 of an absolutely clear secondary eclipse detection in HD 189733 b was most welcome.

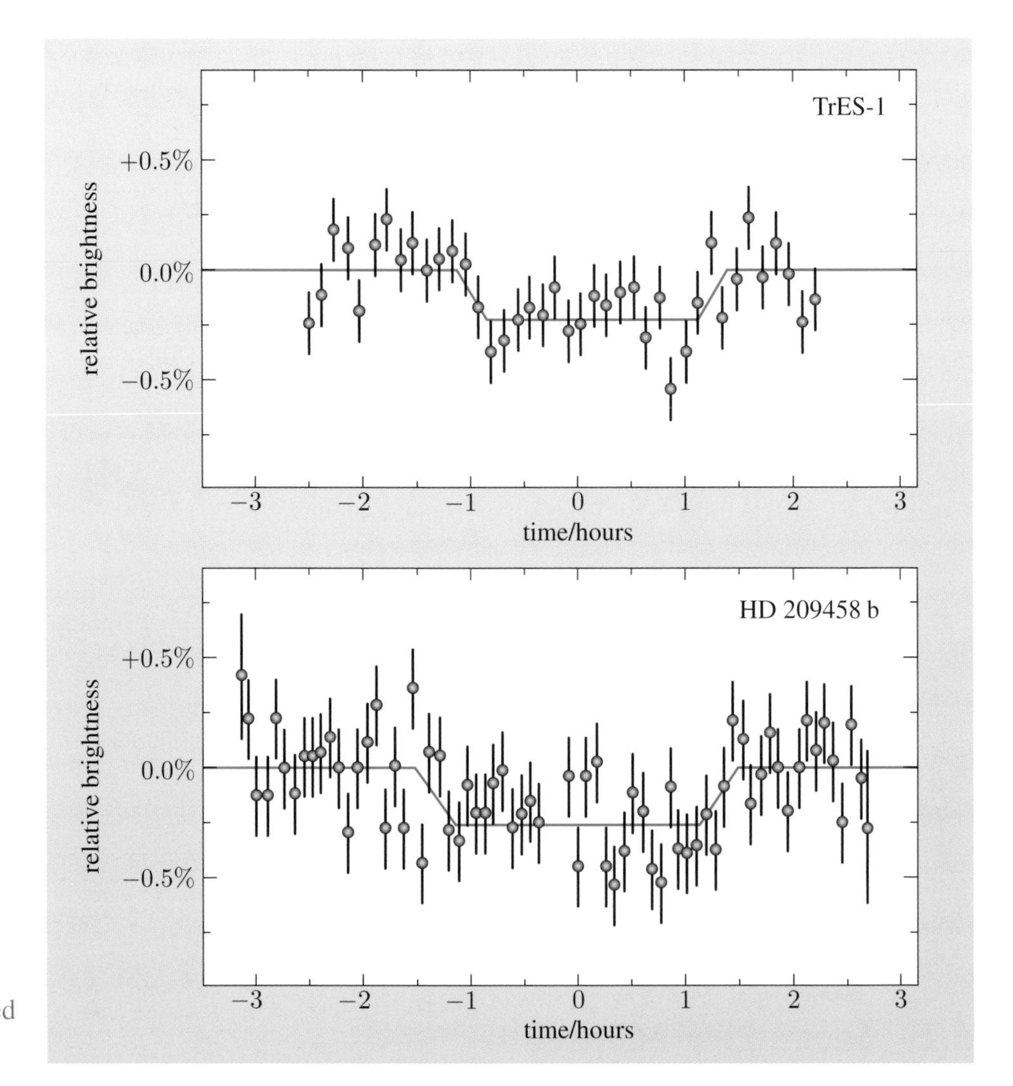

Figure 6.4 The first detections of the emission from exoplanets. Each panel shows the light curve of a secondary eclipse of an exoplanet. The solid lines indicate the fitted secondary eclipse model. (a) TrES-1 b observed at a wavelength of $8\,\mu$m. (b) HD 209458 b observed at $24\,\mu$m.

Exercise 6.3 Figure 6.5 shows the secondary eclipse of HD 189733 b in the $16\,\mu$m band of Spitzer.

(a) From Figure 6.5, estimate the depth of the secondary eclipse of HD 189733 b at $16\,\mu$m in contrast units.

(b) Show that if the reflection component is neglected and the planet is treated as a black body emitter in an orbit with $i = 90°$, using the full Planck function for both the star and the planet's emission leads to the expressions

$$\frac{\exp(hc/(\lambda_c k\, T_{\text{bright}})) - 1}{\exp(hc/(\lambda_c k\, T_{\text{day}})) - 1} = \frac{\Delta F_{\text{SE}}}{F} \left(\frac{R_*}{R_{\text{P}}}\right)^2 \tag{6.19}$$

and

$$T_{\text{day}} = \frac{hc}{\lambda_c k} \left[\log_e \left[\left(\exp\left(\frac{hc}{\lambda_c k\, T_{\text{bright}}}\right) - 1 \right) \frac{F}{\Delta F_{\text{SE}}} \left(\frac{R_{\text{P}}}{R_*}\right)^2 + 1 \right] \right]^{-1}, \tag{6.20}$$

which allow the day side temperature of the planet to be calculated from the secondary eclipse depth and the brightness temperature of the star in the observed bandpass.

(c) Ignoring the reflection component of the light from the day side of the planet, and using the parameters in Table 6.1, use your answer from part (a) and one of the expressions from part (b) to calculate the temperature of the day side hemisphere of HD 189733 b.

(d) At the wavelength in question, $16\,\mu$m, is the Rayleigh–Jeans law an acceptable approximation for either the star HD 189733 or the planet HD 189733 b? Discuss your answer quantitatively and include an assessment of the empirical uncertainties involved.

(e) If the wavelength of the observation were shorter, would the Rayleigh–Jeans law be a better or a worse approximation?

(f) If the brightness temperature of the star at $16\,\mu$m were higher, would the Rayleigh–Jeans law be a better or a worse approximation at this wavelength? ■

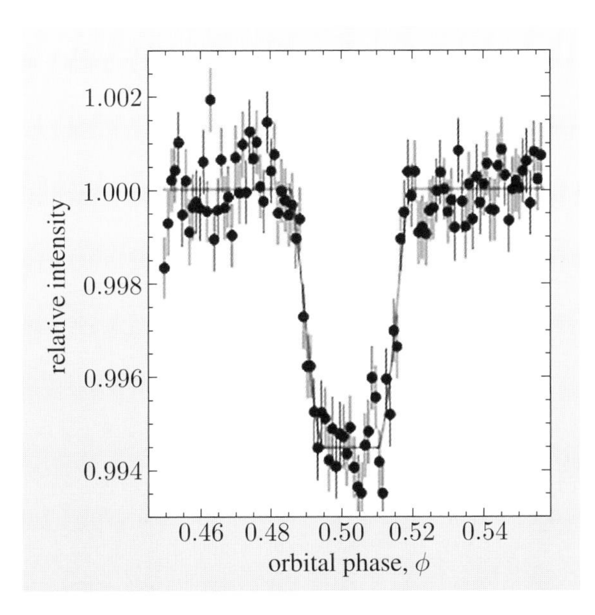

Figure 6.5 The secondary eclipse of HD 189733 b at $16\,\mu$m, announced in 2006.

Phase curve or orbital light curve?

In Section 6.1 we introduced the concept of the phase function. This describes how the detected light from the planet varies with the phase angle, α. Of course, the phase angle changes with time because the exoplanet orbits around the host star, so we could equivalently describe the variation in the detected light as an **orbital light curve**. The two ways of describing exactly the same phenomenon arose because of the heritage from the two fields of planetary science and astrophysics. Planetary scientists studying Solar System objects use the phase angle; this makes sense, because Solar System observations depend on the locations of both the *instrument* and the object being studied. The reflected light from a Solar System asteroid depends on the angle subtended at the Sun by the asteroid and the instrument, *not* simply on the orbital phase of the asteroid. On the other hand, astrophysicists studying binary star systems use the orbital phase. This provides a complete characterization because distances within the Solar System are small compared with interstellar distances; the phase angle for binary stars and exoplanets depends only on their orbital phase because all our observations are effectively made from a fixed point. The author feels that it makes most sense to adopt the orbital phase, rather than the phase angle, as the standard independent variable in studies of exoplanet emission. This view is, admittedly, biased by a heritage in astrophysics. Both are used by exoplanet researchers, so you need to know the meanings of both.

6.2.2 The Kepler mission

In 2009 NASA launched a new satellite, Kepler, which is designed to search for transiting Earth-sized planets. As we saw in Subsection 1.4.1, the Earth's transit of the Sun would have a fractional depth of 8×10^{-5}, so this requires very precise photometry. Kepler has to date (September 2009) produced only preliminary results, but these are impressive. Figure 6.6 shows the Kepler observations of the previously known giant transiting planet host, HAT-P-7.

Stop Press! By January 2010 Kepler had announced the discovery of 5 planets.

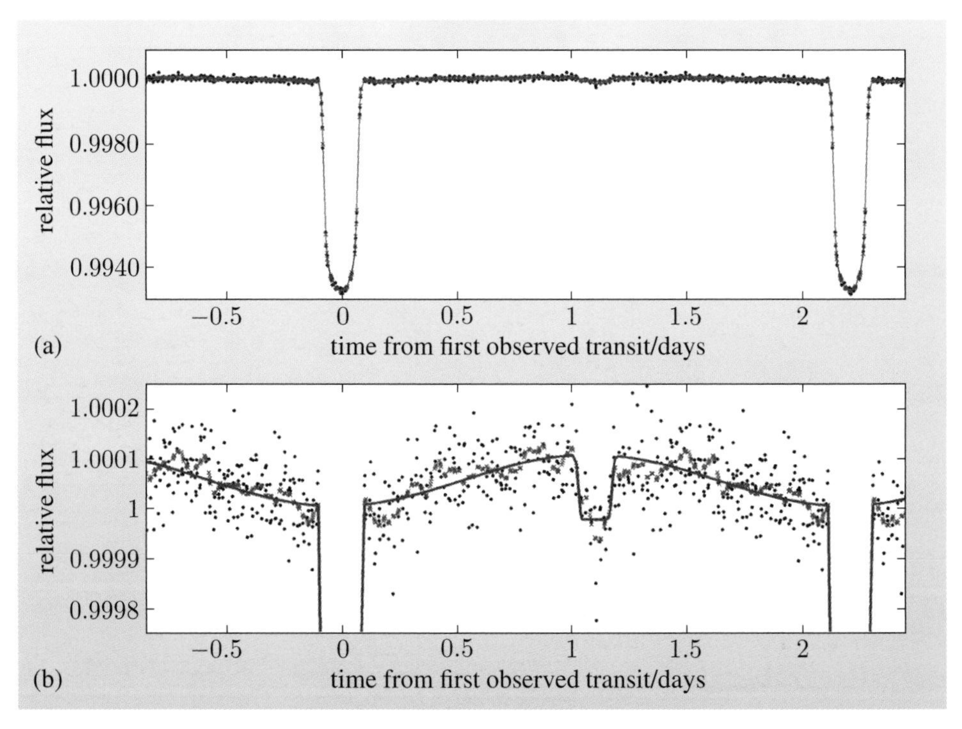

Figure 6.6 Optical photometry of HAT-P-7 from 10 days of observations by the Kepler satellite. The bandpass of these data is 420–890 nm, i.e. optical light. (a) The folded light curve, revealing the transit. (b) The orbital light curve rescaled to emphasize the secondary eclipse, which has a fractional depth of $(1.30 \pm 0.11) \times 10^{-4}$.

In the upper panel we see the 0.67% deep transit of HAT-P-7 b measured with exquisite precision. The lower panel shows the same data but with a magnified vertical scale showing the flux variations in the out-of-transit data. This reveals the modulation due to the orbital light curve of the planet, and the secondary eclipse. The fractional depth of the secondary eclipse is $(1.30 \pm 0.11) \times 10^{-4}$, i.e. less than twice the transit depth that would be produced by the Earth. These data show that Kepler can clearly detect the secondary eclipses of giant exoplanets, and suggest that it should be able fulfil its primary mission of detecting Earth-like planets, assuming that they are common enough to be found around the stars that Kepler will observe.

6.2.3 The emergent spectrum: exoplanets with and without hot stratospheres

From a secondary eclipse measurement in a single photometric band, we can use the known radius, R_P, to determine the brightness temperature of the planet at that band. We can interpret this within the context of the simple model that we outlined in Section 6.1. By combining measurements in several different photometric bands, we can begin to build up a picture of the spectral energy distribution of the light from the planet. We will refer to these as **emergent** spectra. Figure 6.7 shows a collection of Spitzer secondary eclipse measurements that have been used to do this.

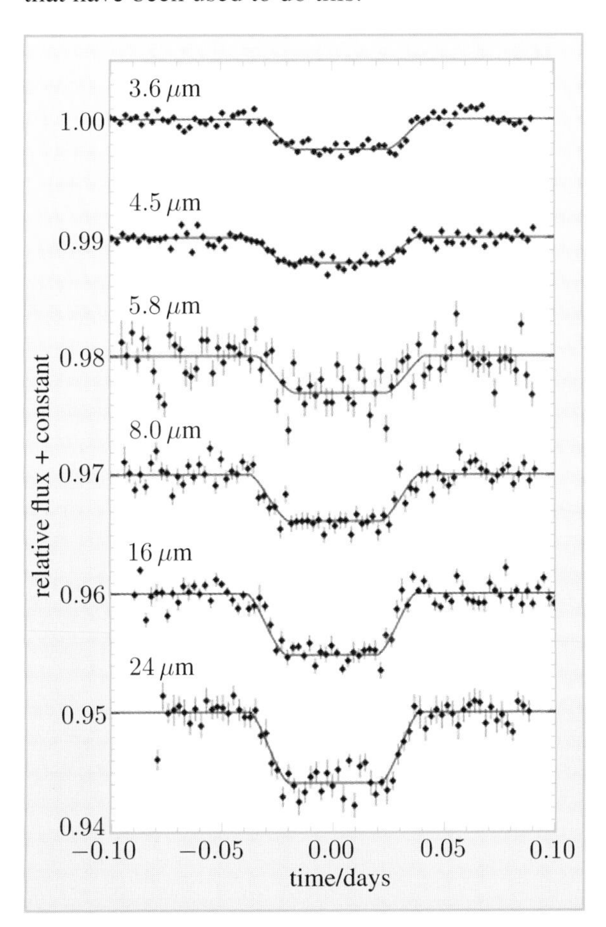

Figure 6.7 Spitzer observations of the secondary eclipse of HD 189733 b in six distinct infrared bands. Each light curve is labelled with the central wavelength of the band, and they have all been rebinned to 3.5-minute time resolution, normalized and offset. As expected, the fractional depth of the secondary eclipse increases as wavelength increases.

Data like these have been collected for several planets, and a montage of the resulting emergent spectra, composed primarily of the planets' thermal emission, is shown in Figure 6.8. Along with the planets' emergent spectra, Figure 6.8 also plots various model predictions from the original research papers. The details of all these models are beyond the scope of this book, and we will only summarize the key points here. Readers interested in a thorough discussion should refer to the original journal papers.

- Why is the fractional depth of secondary eclipse generally expected to increase with wavelength?

- Because planets are cooler than stars, so in general their emission becomes relatively stronger at longer wavelengths.

- HD 189733 is one of the least massive host stars among known transiting exoplanets (January 2010). How does the depth of transit depend on the host star mass (assuming that all host stars are on the main sequence, and that R_P is the same in all cases)?

- On the main sequence, stellar radius increases with stellar mass. A low-mass host star will therefore be smaller than a high-mass host star, and a planet of a given size will produce a deeper transit of a lower-mass host star.

- How does the depth of secondary eclipse depend on the host star mass (assuming that all host stars are on the main sequence, and that R_P and T_{eq} are the same in all cases)?

- Lower-mass main sequence stars are both smaller and dimmer than higher-mass main sequence stars. For a given R_P and T_{eq}, therefore, the planet will produce a larger fractional contribution to the total star plus planet emission. Consequently, for a planet of a given size and temperature, the secondary eclipse will be deeper for lower-mass host stars.

Figure 6.8a shows the emergent spectrum of HD 189733 b. This planet is less irradiated than the majority of the known hot Jupiters, and the measured emergent spectrum between 2 μm and 24 μm is pretty consistent with the emission predicted for a metal-rich planetary atmosphere. Figure 6.8b,c show the emergent spectra for HD 209458 b and XO-1 b, and in each case the dashed coloured lines show the predictions for the planet emitting as a perfect black body at two different temperatures. In both HD 209458 b and XO-1 b, some of the measured contrast ratios are consistent with the lower-temperature black body, while others are consistent with the higher-temperature black body. As we have already noted, the emission from the planet is governed by physics far more complex than our treatment in Section 6.1. Whether or not any particular photon escapes from the planet's atmosphere depends on whether it is absorbed or scattered before reaching interplanetary space. The probabilities for absorption and scattering are wavelength-dependent, so the actual emergent spectrum is not likely to be an simple black body of a single temperature. The models plotted as solid coloured lines in Figure 6.8b,c,d,e contain a prominent bump at 4 μm that is due to a low opacity in this part of the spectrum, so it is relatively easy for these photons to escape.

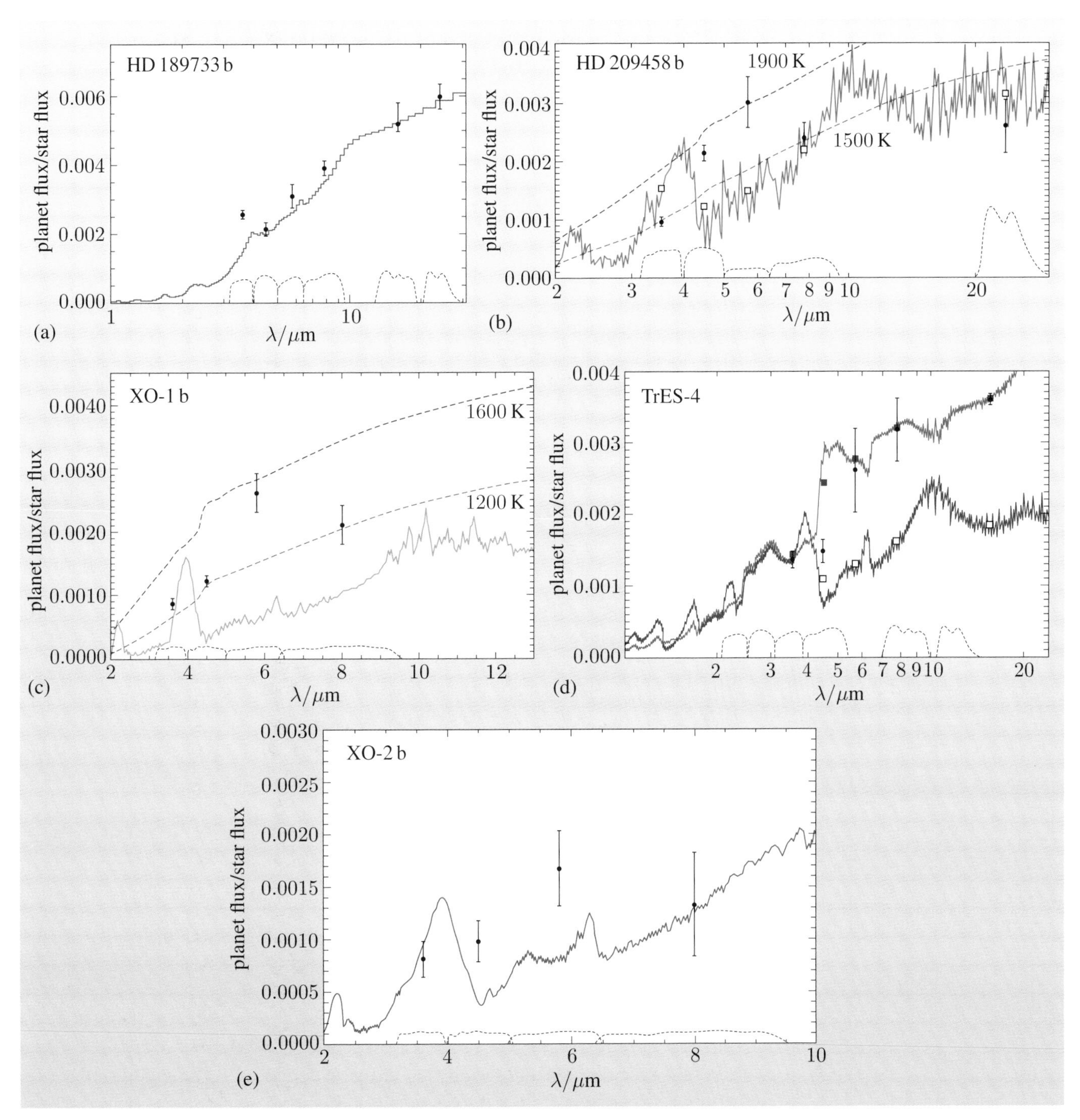

Figure 6.8 Exoplanet emission derived from secondary eclipses. Filled circles with error bars show emission in contrast units against wavelength. The dashed black lines indicate the relevant Spitzer photometric bands. (a) HD 189733 b. A model with a metal-rich planet atmosphere (green line) fits the data well. (b) HD 209458 b. The dashed lines correspond to the planet emitting as a black body at 1500 K or 1900 K. The red line is a predicted emission spectrum from one of the earliest models. (c) XO-1 b. As panel (b) but with temperatures of 1200 K and 1600 K. (d) TrES-4 b. The purple line is a simple model with $P = 0.1$. The blue line is also a model with $P = 0.1$, in this case with an additional optical absorber at high altitudes, which produces a temperature inversion. (e) XO-2 b. The green line is the predicted emission spectrum of the planet without a high-altitude opacity source and with $P = 0.3$. Note that (a), (b) and (d) are shown on logarithmic scales, while (c) and (e) are linear. Squares in some panels show the predictions of the spectral models.

- Why does having a low-mass host star cause HD 189733 b to be less irradiated than other hot Jupiters with similar semi-major axes?
- Because main sequence stars have luminosities that increase strongly with mass.

The level in the atmosphere from which photons escape is called the **photosphere**, and at different wavelengths the photosphere can occur at different depths within the planet's atmosphere. The simplest interpretation of Figure 6.8c is that the temperature at the photospheric depth of the $5.8\,\mu$m emission is higher than that of the photospheric depth in the other bands. Similarly for Figure 6.8b, except in this case the $4.5\,\mu$m emission also appears to come from a layer at higher temperatures. The model plotted in blue in Figure 6.8d includes an opacity source high in the atmosphere that causes a **temperature inversion** as illustrated in Figure 6.9. As in the Earth's **stratosphere**, the temperature in this layer is hotter higher up because the higher layers are heated due to absorption of the radiation from the host star. This absorption of energy is directly due to the opacity of the high-altitude absorber.

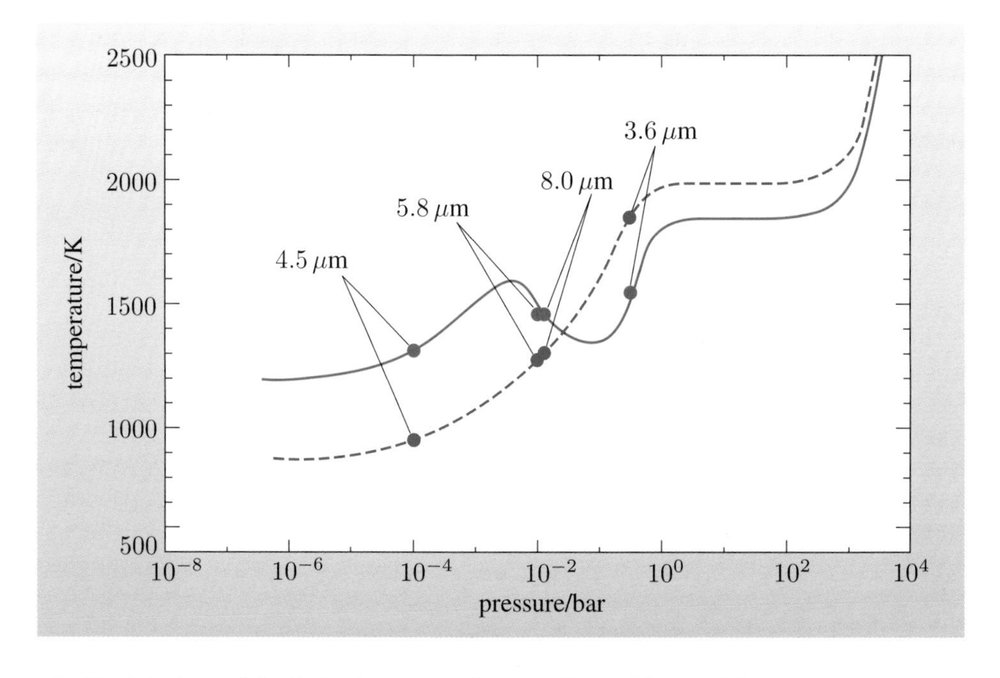

Figure 6.9 Temperature versus pressure profiles in an atmospheric model for XO-2 b. The dashed line corresponds to no high-altitude opacity source, and does not have a temperature inversion. The solid line corresponds to a model with a high-altitude opacity source, which induces a temperature inversion at pressures around 10^{-2} bar. The Earth's atmosphere has such a temperature inversion: the stratosphere. The $5.8\,\mu$m and $8\,\mu$m emissions in this model for XO-2 b come from the layers of the atmosphere at pressures around 10^{-2} bar, consequently the presence of the inversion increases the brightness temperature at these wavelengths.

- Explain how altitude in the atmosphere varies in Figure 6.9.
- Hydrostatic equilibrium means that pressure decreases with height, so the altitude decreases from left to right across Figure 6.9.

The data in Figure 6.8a suggest that HD 189733 b does not have a temperature inversion in its atmosphere, while the planets shown in the remaining panels, HD 209458 b, XO-1 b, XO-2 b and TrES-4 b, all have emergent spectra that seem consistent with a temperature inversion caused by a high-altitude absorber. What might cause the high-altitude opacity? There is a strong clue present in stellar atmospheres of temperatures comparable to these hot Jupiter planets. Among M dwarf stars, molecules TiO and VO are present in the gas phase and produce prominent molecular absorption bands in the red and infrared parts of the spectrum. The L class brown dwarfs are cooler, and in their atmospheres TiO and

VO condense, so these molecular absorption features are absent. It is thought that the same chemistry might operate in hot Jupiter atmospheres: in planets such as HD 209458 b, XO-1 b, XO-2 b, WASP-1 b and TrES-4 b, the atmosphere is hot enough for TiO and VO to remain in the gas phase, creating a hot, strongly absorbing stratosphere. Conversely, in HD 189733 b the atmosphere is cooler and TiO and VO condense out, leaving the atmosphere without the corresponding temperature inversion.

The hotter giant planets may also have opaque silicate and iron clouds in their upper atmospheres, and these may reflect the incident starlight with an optical albedo of $A \sim 0.6$. Conversely, the cooler transiting giant planets, with $900\,\mathrm{K} < T_{\mathrm{eq}} < 1500\,\mathrm{K}$, may have much less reflective atmospheres.

6.2.4 Orbital eccentricity

The timing of the secondary eclipse can reveal if the orbit is eccentric. Figure 6.10 shows the secondary eclipse of the Neptune-mass transiting planet GJ 436 b. The secondary eclipse occurs at phase 0.587 ± 0.005, indicating a significantly eccentric orbit. Combining this constraint with the radial velocity measurements implies that $e = 0.150 \pm 0.012$. The radial velocity measurements themselves implied a significant eccentricity, but the secondary eclipse phase allows a much more precise determination of e.

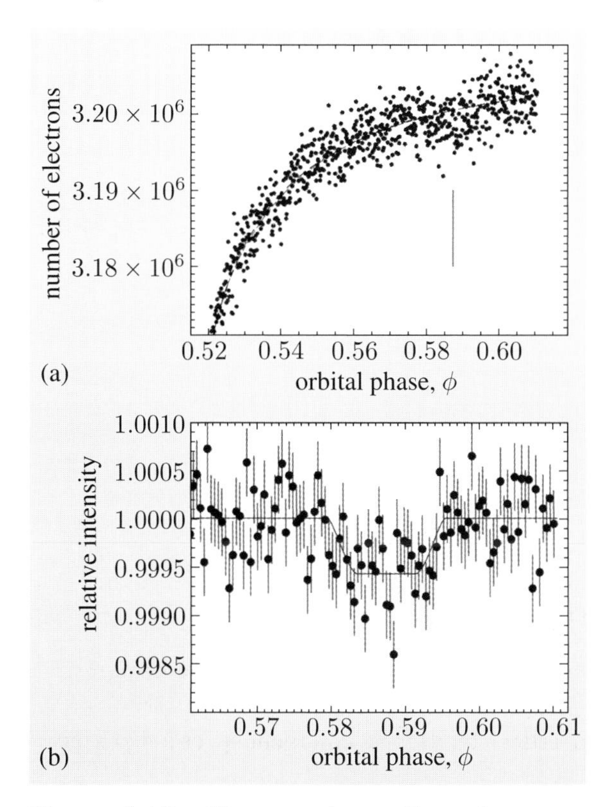

Figure 6.10 The secondary eclipse photometry of GJ 436 b. (a) The photometric signal from the detector; the vertical line marks the time of mid-eclipse. There is a very clear upwards trend in the data, and this is an artefact of the Spitzer detector. This 'ramp' is removed by fitting a curve to the data and dividing by this to make the out-of-eclipse light curve flat. (b) The data after the 'ramp' have been removed; the orbital phase scale has been expanded relative to (a).

Tidal locking

The Earth's Moon always presents the same hemisphere to an Earth-bound observer. This is because the Moon's spin rate is exactly synchronized with its orbit, a phenomenon known as **tidal locking**. Tidal locking is generally expected for objects in close orbits. Obviously, the gravitational forces are strong for objects orbiting close to each other, and the magnitude of the gravitational force will vary significantly across a body if the size of the orbit is not very much greater than the size of the body. This variation of the gravitational force due to the Moon is what causes the ocean tides on the Earth, and the effect of these tides is to dissipate energy and transfer angular momentum. Tidal bulges are also raised by gravity in solid bodies, and the associated dissipation is responsible for synchronizing the Moon's orbit and spin periods. Many of the hot Jupiter exoplanets are expected to be tidally locked, and thus to always present the same hemisphere to their host stars.

6.3 Phase curves

We have already seen one beautiful example of a phase curve in Figure 6.6. These are optical data, so the light from the planet is dominated by reflected starlight. To see the orbital modulation in the thermal emission from the planet, infrared phase curves are required, and in practice this means using Spitzer. There is a caveat in the interpretation of the phase curves from Spitzer: as Figure 6.10 shows, the Spitzer detectors have a slow sensitivity variation with comparable timescale and larger amplitude than the orbital variation in emission from the planet. These variations are, fortunately, believed to be reproducible, so they can be removed in the data reduction process.

6.3.1 The $24\,\mu$m phase curve of υ And b, a non-transiting hot Jupiter

Figure 6.11 shows Spitzer $24\,\mu$m photometry of υ And. This star is host to three known planets (January 2010), the innermost being υ And b, which has an orbital period of 4.617 days. No transits have been detected for this system. Once the data in Figure 6.11 were reduced, they revealed a clear modulation of the total $24\,\mu$m flux, consistent with a dependence on the orbital phase of υ And b. There are five measurements in this figure, each of which has been plotted twice to make the dependence on orbital phase apparent. The solid line shows a model fit that assumes that each element of the planet promptly and instantaneously re-radiates the absorbed stellar flux. In such a model, the **sub-stellar point**, i.e. the position on the planet where the centre of the star's disc appears exactly overhead, must have the highest brightness temperature. It is also possible that the stellar radiation penetrates deep within the planet's atmosphere and is redistributed around the planet before it re-emerges as thermal emission. The data in Figure 6.11 show a possible slight offset in phase between the observations and the model, and the best-fit allowing the position of maximum brightness temperature to vary freely is shown as a dotted line.

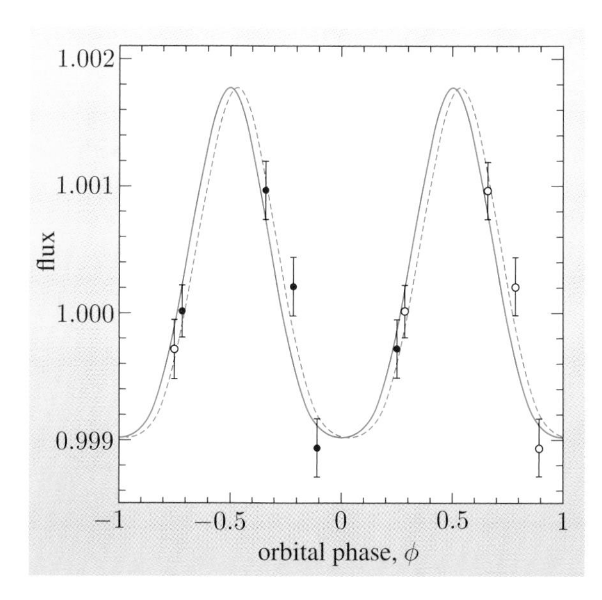

Figure 6.11 The orbital light curve of the planet υ And b at 24 μm. The vertical axis shows total 24 μm flux normalized to a mean value of unity, while the horizontal axis is the orbital phase of the innermost planet, υ And b.

- Why is the sub-stellar point hottest in the instantaneous re-radiation approximation?

❍ The spherical shape of the planet means that the insolation per unit surface area is highest at the sub-stellar point, where the line to the star's centre is perpendicular to the surface. Everywhere else the planet's surface is inclined to this line. The sub-stellar point is also the closest to the star.

Figure 6.11 is clear evidence for a difference in the 24 μm brightness temperature between the two hemispheres of υ And b. There is a 0.2% difference in the flux measurements, with the lowest flux being observed closest to inferior conjunction of the planet, when our view will largely be of the planet's night side. As this system does not transit, some part of both the day side and the night side hemispheres will be visible at all times: the modulation that we see is due to the relative fractions of each changing smoothly as the planet moves around its orbit.

The measurements in Figure 6.11 provide a lower limit on the true difference between the day side and night side fluxes. This limit is already enough to conclude that very little transport of energy to the night side occurs in υ And b: i.e. the value of P in this case is low. Figure 6.12 illustrates this conclusion. The simple model that we outlined in Section 6.1 is used to calculate the day side to night side flux difference *in contrast units*, i.e. expressed as a fraction of the total 24 μm flux from the system. The observed flux difference will be a function of the (unknown) orbital inclination: as the inclination decreases, at any phase the visible disc becomes more equally composed of the two hemispheres. Consequently, if the orbital inclination, i, is low, the observed flux difference implies a larger difference in the brightness of the two hemispheres than would be required if i were high. Figure 6.12 shows the flux difference between the two hemispheres implied by the observed flux difference as a function of the (unknown) orbital inclination. The vertical dashed line shows the value of the contrast corresponding to $P = 0$. The unshaded region of the graph indicates the range of parameters that are consistent with the observations (according to the simple model). It is clear that the maximum possible contrast, with $P = 0$, is consistent with the observations.

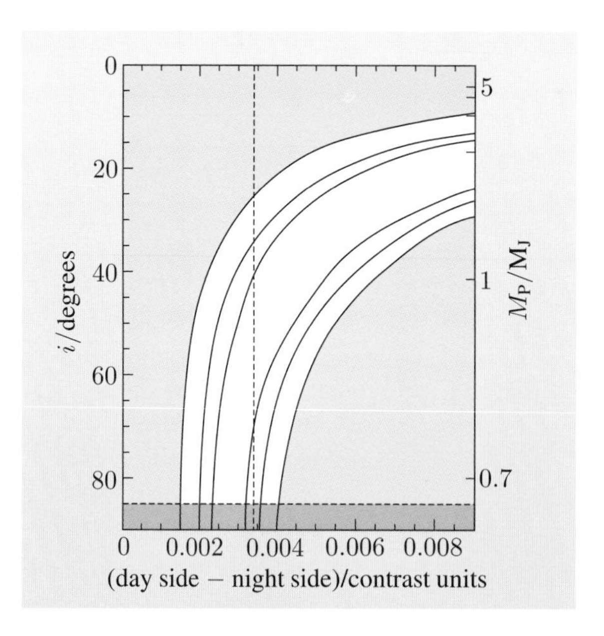

Figure 6.12 Constraints on the 24 μm flux difference between the day side and night side of υ And b. The horizontal axis shows the flux difference in contrast units implied by the light curve in Figure 6.11 as a function of the (unknown) orbital inclination. The additional vertical axis gives the planet mass implied by the radial velocity data for each value of the orbital inclination. Very high inclinations corresponding to the dark shaded area at the bottom of the graph are excluded because υ And b does not transit its host star. The vertical dashed line shows the maximum flux difference expected, i.e. it corresponds to $P = 0$. The curves give 1σ, 2σ and 3σ confidence contours, with the light shaded region excluded by the simple model at the 3σ level.

6.3.2 The 8 μm phase curve and a map of HD 189733 b

We have already seen examples of the secondary eclipse of HD 189733 b. This planet transits a relatively nearby star of high apparent brightness, but relatively low luminosity compared to the majority of known transit host stars. These two factors and the semi-major axis of 0.03 AU conspire to make this system one of the most favourable for detections of the thermal infrared emission from the planet itself. Consequently, this chapter is rather rich with examples of observations of this particular planet! Figure 6.13 shows the exquisite 8 μm phase curve of HD 189733 b. Even in Figure 6.13a, which is scaled to show the transit, the gradual increase in the flux from the system as the planet proceeds around its orbit is visible.

- What causes the gradual increase in flux between the transit egress and the secondary eclipse ingress, shown in Figure 6.13?

- ❍ During transit we are viewing the night side hemisphere of the planet. As the planet proceeds towards secondary eclipse, the day side becomes gradually more visible, until just before secondary eclipse our view is almost entirely of the day side hemisphere.

Figure 6.13b has an expanded vertical axis, so most of the transit ingress and egress, and the entire transit floor, are outside the range shown. The data are of sufficiently high signal-to-noise ratio that the secondary eclipse is very clearly measured. This allows the 8 μm flux from the star to be determined exactly, and the light curve has been normalized so that stellar flux (rather than the total flux from the system) is exactly unity.

- Using Figure 6.13b, estimate the 8 μm flux from the day side of HD 189733 b in contrast units.

- ❍ The flux level just before and after secondary eclipse is 1.0033, so there is an additional flux of $(1.0033 - 1)F_*$ from the day side hemisphere of the planet. In contrast units, the flux of the day side of HD 189733 b is 0.0033.

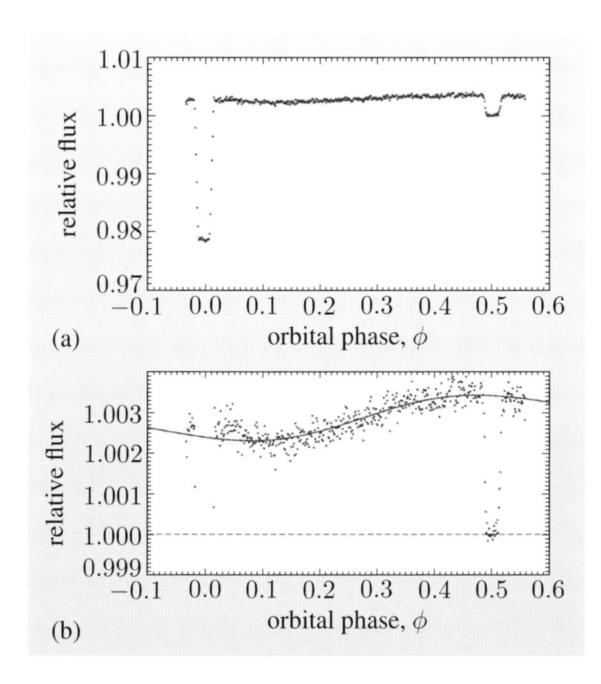

Figure 6.13 A partial $8\,\mu$m orbital light curve of HD 189733 b. (a) Scaled to show the transit. (b) The same data scaled to show detail of the secondary eclipse and the variation with orbital phase.

If the night side of HD 189733 b were completely dark, then just before and after transit we would expect that the flux would be almost entirely the stellar flux. The orbital inclination of HD 189733 b is $i = 86°$, so there would be only a tiny crescent of the day side visible to us during the transit. This tiny crescent would produce a tiny fraction of the flux of the nearly full day side visible just before and after secondary eclipse. Consequently, if the night side of HD 189733 b were completely dark, the modulation in Figure 6.13b would be expected to have significantly higher amplitude, with the total flux around phase $\phi = 0.0$ being only a tiny fraction of its value of 0.0033 around phase $\phi = 0.5$.

● Using Figure 6.13b, estimate the $8\,\mu$m flux from the night side of HD 189733 b in contrast units.

❍ The flux level just before and after transit is 1.0025, so there is an additional flux of $(1.0025 - 1)F_*$ from the night side hemisphere and a tiny crescent of the day side hemisphere. In contrast units, the flux of the night side of HD 189733 b is about 0.0025 (ignoring the contribution of the crescent of the day side hemisphere).

● What can you conclude about the contrast of the day side and night side hemispheres of HD 189733 b at $8\,\mu$m?

❍ The difference in flux between the two hemispheres is less than one-third of the day side hemisphere's flux.

The data in Figure 6.13 can be quantitatively interpreted within the framework of Section 6.1. The secondary eclipse depth implies that $T_{\text{day}} = 1205.1 \pm 9.3$ K if the planet is assumed to emit as a black body as it was in Section 6.1. The dimmest hemisphere presented to us in the course of the orbit has $T_{\text{dimmest}} = 973 \pm 33$ K. There are two reasons why we have used T_{dimmest} rather than T_{night}:

- At $\phi = 0.5$ the visible crescent of the day side will increase the temperature that we deduce above that of the uncontaminated night side.

The values for these temperatures are quoted from the original research paper. We note that there are much bigger fractional uncertainties in quantities like $T_{\text{bright}}(8\,\mu\text{m})$ for the star, which means that the true uncertainty in these temperatures is larger.

- The minimum flux from the planet does not occur at $\phi = 0.0$ as the simplest model would predict; the minimum flux in Figure 6.13 occurs at around phase $\phi = 0.12$, after the transit has occurred.

Since the minimum in the light curve in Figure 6.13 does not occur at phase $\phi = 0.0$, we know that the dimmest face of the planet is not that which we see during transit. The light curve clearly gives us information on the brightness distribution across the 8 μm photosphere of the planet. The easiest way to communicate distributions of quantities across a surface is with a map, and the data in Figure 6.13 were used to produce the first map of the emission from an exoplanet. We show this map in Figure 6.14a.

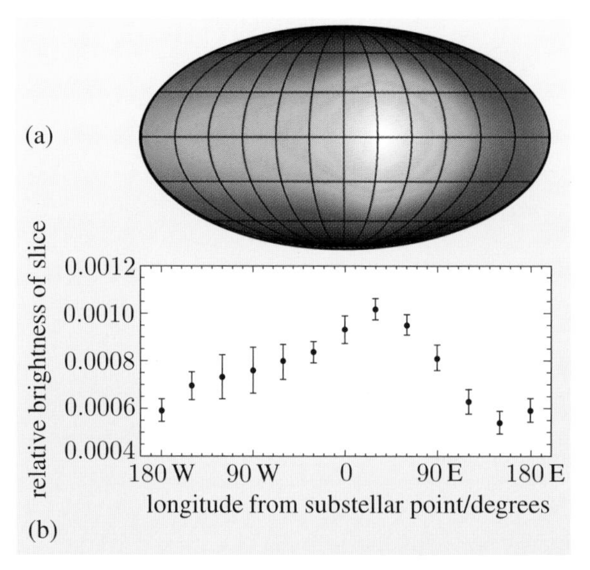

Figure 6.14 (a) The first ever map of an exoplanet. The longitudinal variation of the 8 μm brightness of HD 189733 b is shown, with the brightest regions shown in the lightest tones. The sub-stellar point is in the centre of the map, and an arbitrary variation of brightness with latitude has been imposed. The brightest longitude is offset from the sub-stellar point. (b) The same data presented as a graph of brightness in each longitudinal slice versus longitude from the sub-stellar point. The maximum and minimum 8 μm brightness both occur on the same hemisphere.

The map has been produced by assuming that the planet is tidally locked, i.e. the same face of the planet always faces the star, and the (incorrect but simplifying) assumption that the orbital inclination is exactly 90° has been adopted. With these assumptions one can take the light curve in Figure 6.13b and use it to infer how the brightness of the planet varies with longitude; Figure 6.14b shows the result as a graph. This is an example of an **indirect imaging technique**, and like most applications of these in astrophysics, the graph in Figure 6.14b is not unique: it is the smoothest longitudinal variation that is consistent with the light curve in Figure 6.13b. There is no information on the variation with latitude, and the map in Figure 6.14a was generated from the data shown in Figure 6.14b by arbitrarily assuming a sinusoidal modulation of the brightness with latitude.

To explain the relatively low contrast between the day side and night side of HD 189733 b, **advection** of heat to the night side is required. Atmospheric winds can carry heat around the planet, and may also explain the unexpected proximity of the locations of the minimum and maximum temperatures in Figure 6.14a; both occur within the same hemisphere.

Summary of Chapter 6

1. The amplitude of the reflected light spectrum is

$$\varepsilon_\lambda = p_\lambda \left(\frac{R_P}{a}\right)^2, \qquad \text{(Eqn 6.2)}$$

where p_λ is the wavelength-dependent geometric albedo.

2. The thermal emission from hot Jupiter planets peaks at wavelengths of several microns. Approximating both planet and star as black body emitters, the equilibrium temperatures of the planet's day side and night side are given by

$$T_{\text{day}}^4 = (1-P)(1-A)\,\frac{R_*^2}{2a^2}\,T_{\text{eff}}^4 \qquad \text{(Eqn 6.4)}$$

and

$$T_{\text{night}}^4 = P(1-A)\,\frac{R_*^2}{2a^2}\,T_{\text{eff}}^4, \qquad \text{(Eqn 6.5)}$$

where A is the Bond albedo and P is the fraction of the absorbed energy that is transported to the night side.

3. Stellar spectra depart from black bodies, so the flux from a star at wavelength λ is most concisely defined by specifying a wavelength-dependent brightness temperature, T_{bright}. The flux ratio between the star and the planet at superior conjunction is then

$$\frac{f_{\text{day},\lambda}}{f_{*,\lambda}} = p_\lambda \left(\frac{R_P}{a}\right)^2 + \frac{B_\lambda(T_{\text{day}})}{B_\lambda(T_{\text{bright}})}\left(\frac{R_P}{R_*}\right)^2, \qquad \text{(Eqn 6.10)}$$

where the first term is the reflected light contribution, and the second term is the thermal emission contribution. In some cases the thermal emission contribution can be neglected at short wavelengths and the reflected light contribution can be neglected at long wavelengths. For hot Jupiter exoplanets, however, the day side is hot enough that the two components overlap.

4. If the reflection component is neglected and the planet is treated as a black body emitter, the day side temperature of the planet can be derived from the depth of the secondary eclipse:

$$T_{\text{day}} = \frac{hc}{\lambda_c k}\left[\log_e\left[\left(\exp\left(\frac{hc}{\lambda_c k\,T_{\text{bright}}}\right) - 1\right)\frac{F}{\Delta F_{\text{SE}}}\left(\frac{R_P}{R_*}\right)^2 + 1\right]\right]^{-1}. \qquad \text{(Eqn 6.20)}$$

5. The Rayleigh–Jeans law holds for $\exp(hc/\lambda kT) \ll 1$, and is a simplified version of the Planck function. In the Rayleigh–Jeans regime for both the star and the planet,

$$\frac{\Delta F_{\text{SE}}}{F} \approx \frac{T_{\text{day}}}{T_{\text{bright}}}\left(\frac{R_P}{R_*}\right)^2. \qquad \text{(Eqn 6.14)}$$

6. The Kepler satellite detects the secondary eclipses of hot Jupiter transiting exoplanets in the optical. The depth of secondary eclipse in contrast units is generally expected to increase with wavelength, though the albedo and the atmospheric opacity may have complex dependence on wavelength, causing exceptions to this. Secondary eclipses allow exoplanets' emergent spectral energy distributions to be deduced. The infrared spectral energy distributions suggest that some exoplanets, including HD 209458 b, XO-1 b, XO-2 b, WASP-1 b and TrES-4 b, have temperature inversions in their atmospheres, while others, including HD 189733 b, do not. The hot stratospheres are attributed to absorption of stellar radiation by a high-altitude opacity source. This is tentatively identified as gaseous TiO and VO in the atmospheres of the more irradiated planets.
7. In the context of exoplanets, the 'phase curve' and the 'orbital light curve' are two different ways of referring to the same phenomenon. The 24 μm phase curve of υ And b reveals substantial differences in the brightness of the day side and night side hemispheres, suggesting that P is low for this planet. υ And b does not transit, so this can only be quantified as a function of the (unknown) orbital inclination, i. The 8 μm phase curve of HD 189733 b, on the other hand, has little contrast between the day side and night side hemispheres, suggesting in this case that atmospheric winds transport energy around the planet, i.e. P is closer to 0.5 than to 0.0.
8. Phase curves can be used to generate maps of the brightness of the photosphere. This is an example of an indirect imaging technique. Generally, the smoothest and/or simplest map that is consistent with the data is adopted.
9. Tidal locking is expected to occur for bodies in close orbits.

Chapter 7 Transit timing variations and orbital dynamics

Introduction

In Chapter 1, we introduced the idea of planetary orbits, Kepler's third law, and expressions for the reflex radial velocity of stars being orbited by planets. In this chapter we focus on the dynamics of the orbits of exoplanets. We begin with a reminder of how orbits are described and consider how the resulting equations of motion may be solved. This leads on to an examination of orbital eccentricities and a consideration of the effect that an orbit's eccentricity can have on the detectability of transits. We next give a brief summary of other aspects of the orbital dynamics of known exoplanets, including systems with multiple planets and systems containing multiple stars.

In previous chapters of this book you have seen how the observation and measurement of transits can lead to the detection and characterization of exoplanets. However, despite improvements in technology over the last few years, and the consequent improvement in signal-to-noise ratio of the measurements obtained, the technique is still mainly restricted to studying the transits of relatively large planets in close orbits around their parent star. Only two transiting 'super-Earth' mass planets, CoRoT-7 b and GJ 1214 b, have been detected (February 2010). However, there is a way in which transits of giant exoplanets can be used to infer the presence and properties of other planets or moons in the system, including (possibly) small, terrestrial-sized bodies. The orbits of planetary systems are governed by gravity, and generally speaking the dominant influence on the orbit of a planet will be the mass of the star that lies at the centre of the planetary system. However, each of the planets or other bodies in a system will also exert their own gravitational influence on all the rest of the planets or other bodies, and as a result will perturb their orbits to some extent. This perturbation may change the period or orientation of the orbit of a transiting planet, and so give rise to **transit timing variations (TTVs)**, which may be measured. Hence the gravitational perturbation can provide the clue to the detection of further planets or other bodies, in addition to the planet detected by transit.

The second half of this chapter therefore shows how the orbital dynamics of exoplanets may be revealed by variations in the timing of exoplanet transits. We describe how transit timing variations may be used to search for other planets in the system, including possible terrestrial planets in the habitable zone. In particular we examine the role of resonances and how these enhance the detectability of transit timing variations. We also describe how terrestrial-planet-sized exomoons and exotrojans could be discovered using similar techniques. We conclude the chapter with a discussion of the implications for the formation and evolution of planetary systems, including the migration of planets.

7.1 Newton's law of gravity

We begin the discussion of planetary orbital dynamics with a review of the underlying physical law that describes planetary motion. Newton's law of gravity is able to predict the behaviour of planetary orbits to a high degree of precision. The different behaviour predicted by Einstein's theory of gravity, general relativity, becomes apparent only when considering situations such as the orbital dynamics of massive compact objects in relatively small orbits, for example binary pulsars. For the purposes of this book, we may ignore general relativity. The effects that we are considering are adequately described by Newton's law of gravity, at the level of precision with which we may measure them.

General relativity causes a small contribution ($< 2\%$) to the rate of precession of Mercury's orbit. Exoplanet measurements have insufficient precision to detect effects like this.

7.1.1 Two bodies

In the simplest case, a system of two bodies, Newton's law of gravity gives the gravitational force exerted on a body of mass M_1 by a body of mass M_2 as

$$\boldsymbol{F}_1 = \frac{GM_1M_2}{d_{12}^2}\widehat{\boldsymbol{d}}_{12},$$

where d_{12} is the separation between the two bodies, and $\widehat{\boldsymbol{d}}_{12}$ is the unit vector that points from the body of mass M_1 to the body of mass M_2, namely $\widehat{\boldsymbol{d}}_{12} = \boldsymbol{d}_{12}/d_{12}$. If we write the position vectors of the two bodies as $\boldsymbol{d}_1$ and $\boldsymbol{d}_2$, referred to an arbitrary origin, then $\boldsymbol{d}_{12} = \boldsymbol{d}_2 - \boldsymbol{d}_1$ and its magnitude is $d_{12} = |\boldsymbol{d}_2 - \boldsymbol{d}_1|$. Hence we can write

$$\boldsymbol{F}_1 = GM_1M_2\frac{\boldsymbol{d}_2 - \boldsymbol{d}_1}{|\boldsymbol{d}_2 - \boldsymbol{d}_1|^3}. \tag{7.1}$$

- What is the force exerted on the body of mass M_2 by the body of mass M_1?

- The force has the same magnitude but acts in the opposite direction, i.e.

$$\boldsymbol{F}_2 = GM_1M_2\frac{\boldsymbol{d}_1 - \boldsymbol{d}_2}{|\boldsymbol{d}_1 - \boldsymbol{d}_2|^3} = -\boldsymbol{F}_1.$$

Now, by Newton's second law of motion, the acceleration of a body resulting from a force acting on that body is equal to the force divided by the mass of the body. So, using the 'double dot' notation to signify the second derivative of a position with respect to time, i.e. an acceleration, we may write Equation 7.1 as

$$\ddot{\boldsymbol{d}}_1 = GM_2\frac{\boldsymbol{d}_2 - \boldsymbol{d}_1}{|\boldsymbol{d}_2 - \boldsymbol{d}_1|^3}. \tag{7.2}$$

- What is the acceleration of the body of mass M_2 in this situation?

- The acceleration is simply

$$\ddot{\boldsymbol{d}}_2 = GM_1\frac{\boldsymbol{d}_1 - \boldsymbol{d}_2}{|\boldsymbol{d}_1 - \boldsymbol{d}_2|^3}.$$

An equation such as Equation 7.2 may be solved as a function of time, and the resulting orbits of each body may be derived. As we said in Chapter 1, the solutions of the equation for Newton's law of gravity are orbits that are conic sections (circles, ellipses, parabolas or hyperbolas).

7.1.2 Orbits

For a two-body system, such as a planet orbiting a star, the differential equation implied by Newton's law of gravity is exactly soluble. Although the mathematics is relatively straightforward, it is a lengthy derivation. We therefore simply state the result that the orbit of one body (e.g. a planet) around another (e.g. a star) may be described by

$$r = \frac{a(1-e^2)}{1+e\cos\theta}, \tag{7.3}$$

where r indicates, for example, the distance from the star to the orbiting planet, and the angle θ is referred to as the true anomaly, as noted in Chapter 1.

As you also saw Chapter 1, when observing the motion of a planet in such an orbit, the projection of the distance r onto the line of sight depends on two further angles: i, the orbital inclination, and ω_{OP}, which describes the orientation of the elliptical orbit and is sometimes referred to as the **longitude of pericentre** (see Figures 1.12 and 1.13). The resulting distance is

$$z = r\sin(\theta + \omega_{\mathrm{OP}})\sin i. \tag{7.4}$$

While the orbital dynamics of a two-body system can be calculated exactly, this is not generally possible in the case of three massive bodies.

7.1.3 Three bodies

In the case of three massive bodies, each one is subject to the gravitational attraction of the other two. No simple solution exists for the resulting set of equations. If one of the three bodies is significantly less massive than the other two, approximate solutions may be found to this so-called **restricted three-body problem**. Such analyses are often used to investigate the dynamics of a system such as a star plus a planet plus a satellite, where the mass of the satellite is taken to be negligible when compared with the mass of the star or planet.

Exercise 7.1 How does the magnitude of the force of gravity acting on the Earth due to the Sun compare with the magnitude of the force of gravity acting on the Earth due to Jupiter? Calculate the ratio of the magnitudes of the two forces when the Earth is at its closest to Jupiter. For simplicity, you may assume that the orbits of the Earth and Jupiter are circular, with $a_{\oplus} = 1\,\mathrm{AU}$ and $a_{\mathrm{J}} = 5\,\mathrm{AU}$, respectively, and that $M_{\oplus} \sim 3\times10^{-6}\,M_{\odot}$ and $M_{\mathrm{J}} \sim 10^{-3}\,M_{\odot}$. ■

In the case of the Solar System, although the gravitational force acting on a planet due to the pull of Jupiter is small, it is not negligible when compared with the gravitational pull of the Sun, as Exercise 7.1 indicates. Clearly then, a planet such as Jupiter will influence the motion of a planet like the Earth to some extent.

It is straightforward enough to write down the equations of motion of each object in a three-body system. Using subscripts 0, 1 and 2 to represent the three objects, we have

$$\boldsymbol{F}_0 = GM_0M_1\frac{\boldsymbol{d}_1 - \boldsymbol{d}_0}{|\boldsymbol{d}_1 - \boldsymbol{d}_0|^3} + GM_0M_2\frac{\boldsymbol{d}_2 - \boldsymbol{d}_0}{|\boldsymbol{d}_2 - \boldsymbol{d}_0|^3}, \tag{7.5}$$

$$\boldsymbol{F}_1 = GM_1M_0\frac{\boldsymbol{d}_0 - \boldsymbol{d}_1}{|\boldsymbol{d}_0 - \boldsymbol{d}_1|^3} + GM_1M_2\frac{\boldsymbol{d}_2 - \boldsymbol{d}_1}{|\boldsymbol{d}_2 - \boldsymbol{d}_1|^3}, \tag{7.6}$$

$$\boldsymbol{F}_2 = GM_2M_0\frac{\boldsymbol{d}_0 - \boldsymbol{d}_2}{|\boldsymbol{d}_0 - \boldsymbol{d}_2|^3} + GM_2M_1\frac{\boldsymbol{d}_1 - \boldsymbol{d}_2}{|\boldsymbol{d}_1 - \boldsymbol{d}_2|^3}. \tag{7.7}$$

- What is the sum of the three forces above?

- $\boldsymbol{F}_0 + \boldsymbol{F}_1 + \boldsymbol{F}_2 = \boldsymbol{0}$. This tells us that the mutual gravitational attraction of the three bodies produces no net forces on the system as a whole.

Since the force on each object is equal to the mass of the object multiplied by its acceleration, we may write these force equations in terms of accelerations. For convenience we express each acceleration using dot notation as $\ddot{\boldsymbol{d}}$ (i.e. the second derivative of its position with respect to time), and obtain

$$\ddot{\boldsymbol{d}}_0 = GM_1\frac{\boldsymbol{d}_1 - \boldsymbol{d}_0}{|\boldsymbol{d}_1 - \boldsymbol{d}_0|^3} + GM_2\frac{\boldsymbol{d}_2 - \boldsymbol{d}_0}{|\boldsymbol{d}_2 - \boldsymbol{d}_0|^3}, \tag{7.8}$$

$$\ddot{\boldsymbol{d}}_1 = GM_0\frac{\boldsymbol{d}_0 - \boldsymbol{d}_1}{|\boldsymbol{d}_0 - \boldsymbol{d}_1|^3} + GM_2\frac{\boldsymbol{d}_2 - \boldsymbol{d}_1}{|\boldsymbol{d}_2 - \boldsymbol{d}_1|^3}, \tag{7.9}$$

$$\ddot{\boldsymbol{d}}_2 = GM_0\frac{\boldsymbol{d}_0 - \boldsymbol{d}_2}{|\boldsymbol{d}_0 - \boldsymbol{d}_2|^3} + GM_1\frac{\boldsymbol{d}_1 - \boldsymbol{d}_2}{|\boldsymbol{d}_1 - \boldsymbol{d}_2|^3}. \tag{7.10}$$

From here on, we will assume that the three bodies in question are a star of mass $M_0 = M_*$ at position $\boldsymbol{d}_0 = \boldsymbol{d}_*$, and two planets of mass M_1 and M_2. The total mass of the system is therefore $M_{\text{total}} = M_* + M_1 + M_2$, and in general we will assume that the masses of the planets are much less than the mass of the star. We will also assume that the planet of mass M_1 is on a smaller (inner) orbit than the planet of mass M_2.

Jacobian coordinates are named for the mathematician Carl Gustav Jacob Jacobi (1804–1851).

In analyzing the motion of three bodies that mutually interact via gravity, it is often useful to use a different coordinate system, referred to as a **Jacobian** coordinate system. In this scheme, the coordinates describe the motion of the centre of mass of the system, the motion of the inner planet with respect to the star (usually referred to as the 'inner binary') and the motion of the outer planet with respect to the barycentre of the inner binary (usually referred to as the 'outer binary'). The three coordinates are then

$$\boldsymbol{r}_0 = \boldsymbol{0}, \tag{7.11}$$

$$\boldsymbol{r}_1 = \boldsymbol{d}_1 - \boldsymbol{d}_*, \tag{7.12}$$

$$\boldsymbol{r}_2 = \boldsymbol{d}_2 - \frac{M_*\boldsymbol{d}_* + M_1\boldsymbol{d}_1}{M_* + M_1}. \tag{7.13}$$

The equations of motion of the inner binary and the outer binary may then be derived by combining Equations 7.8–7.13. The procedure is straightforward, but requires lengthy algebraic manipulation, so we simply state the results:

$$\ddot{\boldsymbol{r}}_1 = -\frac{GM_*}{1-\mu_1}\frac{\boldsymbol{r}_1}{r_1^3} - GM_{\text{total}}\,\mu_2\frac{\boldsymbol{r}_1 - \boldsymbol{r}_{21}}{|\boldsymbol{r}_1 - \boldsymbol{r}_{21}|^3} - GM_{\text{total}}\,\mu_2\frac{\boldsymbol{r}_{21}}{r_{21}^3}, \tag{7.14}$$

$$\ddot{\boldsymbol{r}}_2 = -\frac{GM_*}{1-\mu_2}\frac{\boldsymbol{r}_{21}}{r_{21}^3} - GM_{\text{total}}\,\mu_1\frac{\boldsymbol{r}_{21} - \boldsymbol{r}_1}{|\boldsymbol{r}_{21} - \boldsymbol{r}_1|^3}, \tag{7.15}$$

where we have written the reduced masses of each component as $\mu_1 = M_1/M_{\text{total}} \approx M_1/M_*$ and $\mu_2 = M_2/M_{\text{total}} \approx M_2/M_*$, and $\boldsymbol{r}_{21} = \mu_1\boldsymbol{r}_1 + \boldsymbol{r}_2$.

As you might imagine, similar approaches may be employed to write down the equations of motion in a system consisting of four, five or more bodies. However, the algebra is even more involved and the resulting equations even more complex than those above.

7.1.4 Solving the equations of motion

Broadly speaking, there are two approaches to solving Equations 7.14 and 7.15, and the analogous equations for systems of more than three bodies.

The first approach is simply to feed the equations into a sophisticated computer program and then step through the motion of the bodies using small increments of time. These computations are known as ***N*-body simulations**, and are an example of **numerical modelling**. At each time step, the position and velocity of each body in the system are calculated, and the force that each imparts to every other body is calculated, thus allowing the positions and velocities to be calculated at the next time step. To achieve a high degree of accuracy, very small time steps must be used, in particular when bodies undergo close encounters with each other. During close encounters, the gravitational force changes rapidly with time, so great care must be taken to ensure that accuracy is not lost by using too large a time step. Unfortunately, the smaller the time step and the larger the number of bodies concerned, the larger the amount of computation time that is required.

It is also important that these simulations correctly treat objects that are expelled from the system as a result of gravitational encounters with other bodies, as well as objects that collide with other bodies. Often such computer programs deal with two classes of objects: massive bodies and test particles. Massive bodies feel the influence of gravity from other objects in the model, and also exert gravitational influence themselves. Test particles are treated as 'massless' objects and only react to the gravitational force of other objects; they do not exert any gravitational influence of their own. Stars and planets are generally regarded as massive bodies, but asteroids and material that constitutes discs or rings are usually treated as massless test particles. Treating certain objects as massless test particles can greatly simplify and speed up the running of a simulation.

The most sophisticated orbital modelling programs, running on some of the fastest modern computers, typically run at something like 50 million times 'real time' speed. This may sound impressive, but it means that a million years of 'planetary system time' will still take a week to run on the computer, and a million years is still a relatively short interval in the lifetime of a planetary system, which may evolve over billions of years. Furthermore, we never know the parameters of a given planetary system precisely. Consequently, many different simulations may need to be performed, each having a slightly different set of initial conditions for the planetary masses, orbital radii, eccentricities, inclinations and orientations. For this reason, the numerical approach is not always practical.

An alternative approach to numerical modelling is to try to solve the equations of motion analytically, using an approximation that renders the original equations more tractable. One such approach has already been referred to: the restricted three-body problem considers one of the three bodies to be essentially massless (it is a test particle), and then one only has to consider forces exerted by the other two (massive) objects. With this approximation, some useful results may be derived analytically, i.e. purely by manipulating the equations obtained. Such analytical approaches are considered in the later sections of this chapter, which look at how a second planet can influence the motion of a transiting planet.

7.2 The shapes of orbits

The shape of an exoplanetary orbit is described by a single parameter, the eccentricity, e. In this section we first look at what mechanisms might change the eccentricity of a planetary orbit over time, then consider how the eccentricity might affect the detectability of transits.

Tidal forces and tidal dissipation

In 1850 Edouard Roche calculated the radius at which a moon would be disrupted by **tidal forces** due to the gravity of the planet it orbits. The tidal force arises because different parts of the orbiting body experience different gravitational forces. The part of the moon closest to the planet experiences the strongest gravitational force towards it. We will refer to the difference between the gravitational force due to the planet on the near side and the far side of the moon as the differential gravitational force, or the tidal force.

The tidal force will tend to elongate the orbiting moon, and if it is strong enough it will ultimately disrupt the moon. We can obtain an estimate of the minimum distance at which a moon can orbit a planet without being disrupted by equating the tidal force to the force of self-gravity. We could do this by considering the tidal force due to the planet as the difference in magnitude (the directions are identical) between the planet's gravitational force acting on identical test particles, of mass $m_{\rm test}$, at the nearest and farthest points on the moon's surface:

$$\begin{aligned}|\text{tidal force}| &= \frac{GM_{\rm P}\,m_{\rm test}}{(a_{\rm M}-R_{\rm M})^2} - \frac{GM_{\rm P}\,m_{\rm test}}{(a_{\rm M}+R_{\rm M})^2}\\ &= GM_{\rm P}\,m_{\rm test}\left(\frac{1}{(a_{\rm M}-R_{\rm M})^2} - \frac{1}{(a_{\rm M}+R_{\rm M})^2}\right).\end{aligned}$$

It is clear that the tidal force becomes bigger as the size of the moon, $R_{\rm M}$, increases relative to the orbital separation, $a_{\rm M}$. We can compare this tidal force to the self-gravitational force binding the test particles to the moon:

$$F_{\rm grav} = \frac{GM_{\rm M}\,m_{\rm test}}{R_{\rm M}^2}.$$

If we do this, we discover that there is a limiting distance $d_{\rm R}$, within which an orbiting moon will be disrupted because the tidal force exceeds the self-gravitational force. In fact, the disruption would occur slightly further out than the distance produced by the argument above, because catastrophic oscillations would destroy the body. The **Roche limit**, within which an orbiting moon would be disrupted, is given by

$$d_R = R_M \left(\frac{2M_P}{M_M}\right)^{1/3}. \tag{7.16}$$

The mathematical theory of astronomical tides was developed in the late nineteenth century by the English astronomer and mathematician Sir George Darwin (1845–1912), son of the naturalist Charles Darwin. Planets, as well as moons, are subject to tidal forces. In particular, the hot Jupiter exoplanets are subject to strong tidal forces due to their proximity to their host stars, and are expected to be elongated so that they are spheroidal, or rugby ball shaped. As the planet rotates, and orbits around the star, so the deformed shape will rotate too, generating *tidal waves* on the surface of the planet. However, the intrinsic physical structure of the planet (whether gaseous, icy or rocky) will tend to resist the motion. This resistance may be expressed in terms of a **quality factor**, Q, which encapsulates the response of the planet to the tide-raising gravitational potential. In Subsection 4.4.5 we encountered Q and labelled it the 'tidal dissipation parameter', as it entered there as a parameter governing the tidal heating of a gas giant planet, hence affecting the predicted radius. One definition of the quality factor is the reciprocal of the lag angle associated with the tidal waves on the planet.

Tidal forces will be felt on the star too. So the star, as well as the planet, will have an associated quality factor. As a result of the redistribution of the mass of the planet, caused by the tide, there will be an additional gravitational potential due to the deformed planet itself. The ratio of the additional potential produced by the redistribution of the planet's mass (as a result of the tide) to the deforming gravitational potential that caused the mass redistribution in the first place, is known as the **dynamical Love number**, k_d. Both the star and the planet will have their own Love numbers. The ratios of the quality factor and the dynamical Love number therefore depend on the physical characteristics of the star and the planet, and describe their overall response to tides.

The shortest-period exoplanets, for example transiting planets with orbital periods of less than a day (WASP-18 b, WASP-19 b and CoRoT-7 b), are subject to large tidal forces. These planets will raise tidal bulges on the stellar surface. If the orbital period is shorter than the stellar rotation period, which is likely since the orbital period is so short, this redistribution of mass causes an additional torque that drains angular momentum from the planetary orbit. If the tidal quality factors have the values that we expect from studies of binary stars and giant planets in the Solar System, this drain in angular momentum should cause a measurable change in the orbital periods of the shortest-period hot Jupiters on a timescale of a decade or so. The interpretation of these period changes is complicated, however, by the **Applegate mechanism** in which the stellar analogue of the solar activity cycle changes the angular momentum distribution within a star; this also affects the orbital period.

7.2.1 Circularization of orbits

Exoplanetary orbits are altered via tidal friction. Tides are familiar to us here on Earth through the response of Earth's oceans to the gravitational pull of the Sun and Moon. However, similar effects will arise between any orbiting bodies as long as they can be tidally deformed. In the case of an exoplanet orbiting close to its parent star, the result of these tidal interactions can be to alter both the size of the orbit (i.e. its semi-major axis, a) and the shape of the orbit (i.e. its eccentricity, e). In general, both the semi-major axis and the eccentricity of the orbit will decrease with time. The process of reducing the eccentricity of an orbit is referred to as **circularization** and when the eccentricity has reduced such that the orbit becomes circular, it is said to have **circularized**.

The **circularization timescale** for an orbit is defined by

$$\tau_{\text{circ}} = \frac{e}{|\dot{e}|}, \tag{7.17}$$

where $|\dot{e}|$ is the magnitude of the rate of change of the eccentricity with time. A detailed mathematical treatment of tides, originating from work by George Darwin in the nineteenth century (see the box above), gives rise to the following equation for a simple system consisting of a single planet in orbit around a star:

$$\tau_{\text{circ}} = \frac{3\omega^{-1}a^5}{18\widehat{s} + 7\widehat{p}}, \tag{7.18}$$

where ω is the mean orbital angular speed of the planet, a is the semi-major axis of the orbit, and $\widehat{s}$ and $\widehat{p}$ are two parameters that depend on the physical characteristics of the star and the planet, respectively, and describe their response to tides.

Since $\tau_{\text{circ}} \propto \omega^{-1}a^5$ and $\omega^{-1} = P/2\pi$, Kepler's third law clearly gives $\tau_{\text{circ}} \propto a^{3/2} \times a^5$, or $\tau_{\text{circ}} \propto a^{13/2}$. Therefore the circularization timescale has a very strong dependence on the size of the orbit.

- If the orbit of one exoplanet has a semi-major axis that is ten times smaller than that of another exoplanet, by what factor will its circularization timescale be smaller (all other parameters being equal)?
- Since $10^{13/2} \sim 3 \times 10^6$, the circularization timescale will be about 3 million times shorter.

Circularization timescale

Since $\tau_{\text{circ}} \propto a^{13/2}$, small orbits will circularize very quickly.

The $\widehat{s}$ and $\widehat{p}$ parameters in Equation 7.18 are given by

$$\widehat{s} = \frac{9}{4}\frac{k_{\text{d}*}}{Q_*}\frac{M_{\text{P}}}{M_*}R_*^5, \tag{7.19}$$

$$\widehat{p} = \frac{9}{2}\frac{k_{\text{dP}}}{Q_{\text{P}}}\frac{M_*}{M_{\text{P}}}R_{\text{P}}^5, \tag{7.20}$$

where the numerical factors $Q_*/k_{\text{d}*}$ and $Q_{\text{P}}/k_{\text{dP}}$ for the star and the planet combine the dynamical Love number, k_{d}, and a tidal quality factor, Q. These factors are determined analytically as $Q_*/k_{\text{d}*}$ = typically a few $\times\ 10^7$ to a few $\times\ 10^9$ for a star, and $Q_{\text{P}}/k_{\text{dP}}$ = a few $\times\ 10^4$ to a few $\times\ 10^5$ for a planet.

- To calculate the circularization timescale due to planetary tides alone, we can set $\widehat{s}=0$ in Equation 7.18. What is the equation for the circularization timescale $\tau_{\text{circ,P}}$ in this case?

- If $\widehat{s}=0$, then Equation 7.18 reduces to

$$\tau_{\text{circ,P}} = \frac{3\omega^{-1}a^5}{7\widehat{p}}.$$

- To calculate the circularization timescale due to stellar tides alone, we can set $\widehat{p}=0$ in Equation 7.18. What is the equation for the circularization timescale $\tau_{\text{circ},*}$ in this case?

- If $\widehat{p}=0$, then Equation 7.18 reduces to

$$\tau_{\text{circ},*} = \frac{3\omega^{-1}a^5}{18\widehat{s}}.$$

- What is the ratio of these two timescales?

- The ratio of these two timescales is

$$\frac{\tau_{\text{circ,P}}}{\tau_{\text{circ},*}} = \frac{18\widehat{s}}{7\widehat{p}}, \qquad (7.21)$$

which is independent of the size of the orbit under consideration.

Worked Example 7.1

(a) Derive an expression for the ratio $\widehat{s}/\widehat{p}$.

(b) Consider an exoplanetary system consisting of a star similar to the Sun with $M_* = 1\,\text{M}_\odot$ and $R_* = 1\,\text{R}_\odot$, and a Jupiter-like planet of mass $1\,\text{M}_\text{J}$ and radius $1\,\text{R}_\text{J}$. If $Q_*/k_{\text{d}*}$ and Q_P/k_{dP} take typical values of 10^8 and 10^5, respectively, explain whether tidal evolution of the eccentricity is mainly determined by the planetary or stellar tide in this case.

(c) What is the circularization timescale of the orbit, assuming an orbital period of 4 days?

Solution

(a) From Equations 7.19 and 7.20,

$$\begin{aligned}\frac{\widehat{s}}{\widehat{p}} &= \frac{9}{4}\frac{k_{\text{d}*}}{Q_*}\frac{M_\text{P}}{M_*}R_*^5 \times \frac{2}{9}\frac{Q_\text{P}}{k_{\text{dP}}}\frac{M_\text{P}}{M_*}R_\text{P}^{-5} \\ &= \frac{1}{2}\frac{k_{\text{d}*}}{Q_*}\frac{Q_\text{P}}{k_{\text{dP}}}\left(\frac{M_\text{P}}{M_*}\right)^2\left(\frac{R_*}{R_\text{P}}\right)^5.\end{aligned}$$

(b) We can assume that $M_* = 1.99\times10^{30}\,\text{kg}$, $M_\text{P} = 1.90\times10^{27}\,\text{kg}$, $R_* = 6.96\times10^8\,\text{m}$ and $R_\text{P} = 7.15\times10^7\,\text{m}$. So the equation above becomes

$$\begin{aligned}\frac{\widehat{s}}{\widehat{p}} &= \frac{1}{2}\frac{k_{\text{d}*}}{Q_*}\frac{Q_\text{P}}{k_{\text{dP}}}\left(\frac{1.90\times10^{27}}{1.99\times10^{30}}\right)^2\left(\frac{6.96\times10^8}{7.15\times10^7}\right)^5 \\ &= 0.040\times\frac{k_{\text{d}*}}{Q_*}\frac{Q_\text{P}}{k_{\text{dP}}}.\end{aligned}$$

Since Q_*/k_{d*} and Q_P/k_{dP} for the star and the planet are 10^8 and 10^5, respectively, it is clear that $\widehat{s}/\widehat{p} \ll 1$. Therefore $\tau_{circ,P} \ll \tau_{circ,*}$ and the tidal evolution of the eccentricity is mainly determined by the planetary tides.

(c) The parameters $\widehat{s}$ and $\widehat{p}$ are given by Equations 7.19 and 7.20 as

$$\begin{aligned}\widehat{s} &= \frac{9}{4}\frac{k_{d*}}{Q_*}\frac{M_P}{M_*}R_*^5 \\ &= \frac{9}{4} \times 10^{-8} \times \left(\frac{1.90 \times 10^{27}}{1.99 \times 10^{30}}\right) \times \left(6.96 \times 10^8\,\mathrm{m}\right)^5 \\ &= 3.509 \times 10^{33}\,\mathrm{m}^5\end{aligned}$$

and

$$\begin{aligned}\widehat{p} &= \frac{9}{2}\frac{k_{dP}}{Q_P}\frac{M_*}{M_P}R_P^5 \\ &= \frac{9}{2} \times 10^{-5} \times \left(\frac{1.99 \times 10^{30}}{1.90 \times 10^{27}}\right) \times \left(7.15 \times 10^7\,\mathrm{m}\right)^5 \\ &= 8.807 \times 10^{37}\,\mathrm{m}^5.\end{aligned}$$

So, as noted in part (b), the planetary tide is the dominant contribution to the circularization of the orbit.

To calculate the circularization timescale of the orbit, we need the mean orbital angular speed, ω, and the semi-major axis, a, of the orbit. The mean orbital angular speed is just

$$\omega = \frac{2\pi}{P} = \frac{2\pi}{4 \times 24 \times 3600}\,\mathrm{rad\,s^{-1}} = 1.818 \times 10^{-5}\,\mathrm{rad\,s^{-1}},$$

while from Kepler's third law (Equation 1.1),

$$\begin{aligned}a &= \left(\frac{GM_*P^2}{4\pi^2}\right)^{1/3} \\ &= \left(\frac{(6.673 \times 10^{-11}\,\mathrm{N\,m^2\,kg^{-2}}) \times (1.99 \times 10^{30}\,\mathrm{kg}) \times (4 \times 24 \times 3600\,\mathrm{s})^2}{4\pi^2}\right)^{1/3} \\ &= 7.379 \times 10^9\,\mathrm{m} = 0.0493\,\mathrm{AU}.\end{aligned}$$

Now, the circularization timescale of the orbit is given by Equation 7.18 as

$$\begin{aligned}\tau_{circ} &= \frac{3\omega^{-1}a^5}{18\widehat{s} + 7\widehat{p}} \\ &\sim \frac{3\omega^{-1}a^5}{7\widehat{p}} \\ &= \frac{3 \times (1.818 \times 10^{-5}\,\mathrm{s^{-1}})^{-1} \times (7.379 \times 10^9\,\mathrm{m})^5}{7 \times 8.807 \times 10^{37}\,\mathrm{m}^5} \\ &= 5.85 \times 10^{15}\,\mathrm{s}.\end{aligned}$$

The circularization timescale is therefore of order 200 million years. Compared to the lifetime of a star or a planet, exoplanets in close orbits around their star will circularize quickly.

For most situations, the orbital circularization timescale is determined only by the planetary tide, and the stellar tide is insignificant. In this case, Equation 7.18 reduces to

$$\tau_{\mathrm{circ}} = \frac{3\omega^{-1}a^5}{7\widehat{p}}.$$

Then substituting for $\widehat{p}$ from Equation 7.20 and putting $\omega^{-1} = P/2\pi = (a^3/GM_*)^{1/2}$, where we have used Kepler's third law, we obtain the relationship

$$\tau_{\mathrm{circ}} = \frac{2}{21}\frac{Q_{\mathrm{P}}}{k_{\mathrm{dP}}}\left(\frac{a^3}{GM_*}\right)^{1/2}\frac{M_{\mathrm{P}}}{M_*}\left(\frac{a}{R_{\mathrm{P}}}\right)^5. \qquad (7.22)$$

Exercise 7.2 Using Equation 7.22, what is the circularization timescale for the transiting exoplanet WASP-17 b, which has a period of 3.74 days and an eccentricity of 0.13? (You may assume that $Q_{\mathrm{P}}/k_{\mathrm{dP}} = 10^5$, $a = 0.051\,\mathrm{AU} = 7.63 \times 10^9\,\mathrm{m}$, $M_{\mathrm{P}} = 0.49\,\mathrm{M_J} = 9.31 \times 10^{26}\,\mathrm{kg}$, $M_* = 1.2\,\mathrm{M_\odot} = 2.39 \times 10^{30}\,\mathrm{kg}$, $R_{\mathrm{P}} = 1.74\,\mathrm{R_J} = 1.24 \times 10^8\,\mathrm{m}$.) ■

As indicated above, a planet such as WASP-17 b should have circularized on a relatively short timescale and it is therefore surprising to find it, and other short period exoplanets, with such an eccentric orbit. As noted earlier, these planets may be relatively young, or they may be subject to dynamical interactions with other planets, which we consider in the next subsection.

7.2.2 Increasing the eccentricity of orbits

The mean eccentricity of the known exoplanets increases with orbital semi-major axis and period (see Figure 7.1), as does that of binary star systems. The vast majority of very short period exoplanets are indeed in circular orbits, as expected due to their short circularization timescales. For most of the exoplanet population, therefore, no special mechanism need be invoked to explain their eccentricities. However, around half the known exoplanets do have significant orbital eccentricities ($e > 0.1$), and around 10% of these are in short-period ($P < 10$ days) or close ($a < 0.1$ AU) orbits. As noted earlier, this includes a handful of transiting exoplanets. Since these short-period, eccentric orbit planets should circularize on a short timescale, we can conclude that either they are simply young planets, or some mechanism must have acted to increase their eccentricity.

Various suggestions have been made to explain this, including resonant interactions between planets, close encounters between planets, and interactions with a companion star in the system. The first two processes would be expected to produce higher eccentricities for lower-mass planets, but there are currently too few examples in total to meaningfully examine their distribution in mass (February 2010). If interactions with a companion star are responsible, then it would imply that some exoplanetary systems have undetected distant stellar companions. Currently there is therefore no satisfactory explanation for the existence of short-period, eccentric orbit planets, but the picture should become clearer as more examples are discovered.

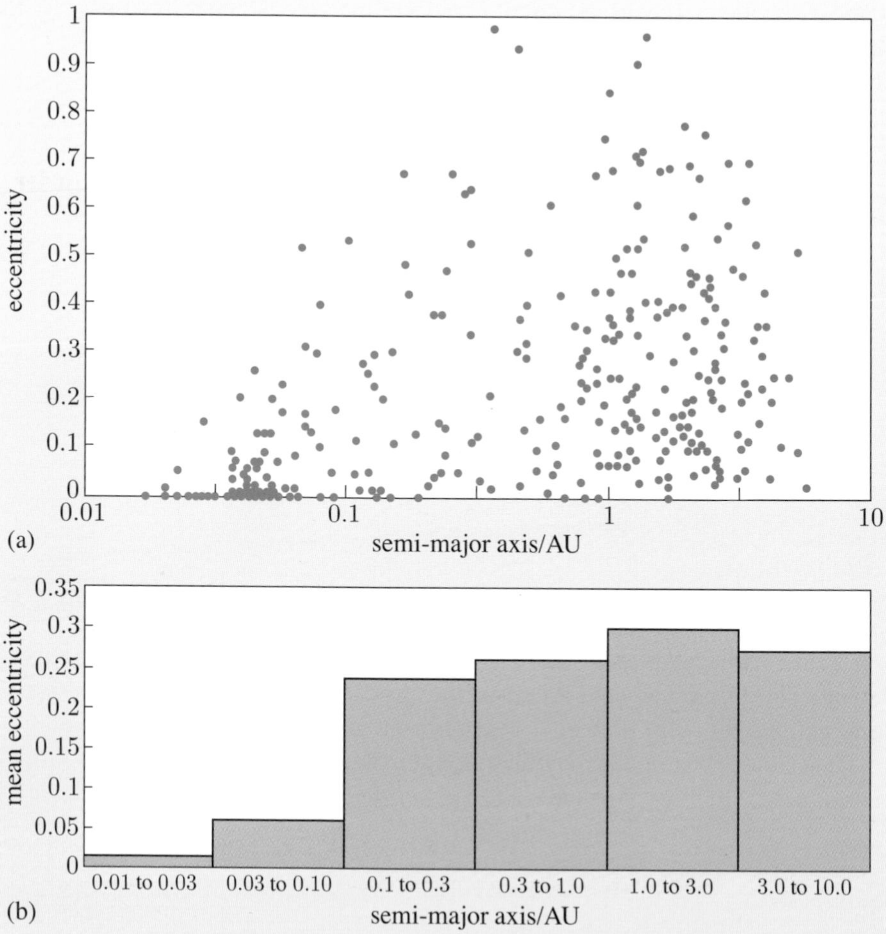

Figure 7.1 The distribution of orbital eccentricities of exoplanets as a function of their semi-major axis. The upper panel shows all the data with a logarithmic spacing on the horizontal axis, while the lower panel shows the mean orbital eccentricity in a series of logarithmically spaced bins. Although the mean eccentricity increases with orbital semi-major axis, there are some planets in close orbits with significant eccentricities.

Orbital resonances

A **mean motion orbital resonance** is said to exist when the orbital periods of two bodies in a planetary system exhibit a simple integer ratio between them. Such resonances can either stabilize the orbit, rendering it immune to change, or destabilize the orbit, causing it to change rapidly.

Stabilization of an orbit happens when the two bodies never approach very close to each other. An example of this in the Solar System is the orbit of Pluto (and the other objects in similar orbits known as 'plutinos') which is in a 2 : 3 mean motion orbital resonance with Neptune: for every 2 orbits of Pluto, Neptune completes 3 orbits around the Sun, but the two bodies are never close together. Similarly, in the main asteroid belt, mostly between the orbits of Mars and Jupiter, asteroids with mean motion orbital resonances with Jupiter of 3 : 2 and 4 : 3 are abundant.

- Given that Jupiter has an orbital period of about 12 years, what are the orbital periods of asteroids that exist in 3 : 2 and 4 : 3 mean motion orbital resonances with Jupiter?

- An asteroid in a 3 : 2 mean motion orbital resonance with Jupiter will complete 3 orbits of the Sun for every 2 orbits of Jupiter. Hence its orbital period is about 2×12 years$/3 = 8$ years. Similarly, an asteroid in a $4 : 3$ mean motion orbital resonance with Jupiter will complete 4 orbits of the Sun for every 3 orbits of Jupiter. Hence its orbital period is about 3×12 years$/4 = 9$ years.

A resonance of the form $j + 1 : j$, as these are, is referred to as a **first-order resonance**. There are also asteroids in a $1 : 1$ resonance with the orbit of Jupiter — these are the **Trojan** asteroids, which lead and trail Jupiter, roughly $60°$ ahead of and behind the planet in its orbit.

The individual asteroids in these locations are named after characters in Homer's *Iliad*, which describes events surrounding the legendary Battle of Troy.

Such mean motion orbital resonances are also seen in exoplanets. The exoplanets GJ 876 b and GJ 876 c are in a $1 : 2$ mean motion orbital resonance with periods of about 60 days and 30 days, respectively.

As noted above, mean motion orbital resonances can also destabilize an orbit. Taking an example again from the main asteroid belt in the Solar System, at orbital period resonances of $3 : 1$, $5 : 2$, $7 : 3$ and $2 : 1$ with the orbital period of Jupiter, we find a lack of asteroids. These are known as the Kirkwood gaps and correspond to regions of the asteroid belt where interactions with Jupiter have cleared asteroids away. Similarly, in the rings of Saturn there is a gap between the rings known as the Cassini division, which corresponds to a $2 : 1$ resonance with the orbital period of Saturn's moon Mimas.

A different sort of resonance is that known as a **secular resonance**. Eccentric orbits may precess, or change their orientation with time. Typically, the longitude of pericentre of the orbit may precess or the longitude of the ascending node of the orbit may precess. A secular resonance refers to the situation when this precession is synchronized between the orbits of two bodies.

A final sort of resonance is termed a **Kozai resonance**, which occurs when the eccentricity and inclination of an orbit oscillate with time. Typically, the eccentricity will increase while the inclination decreases, and vice versa.

7.2.3 The effect of eccentricity on the detectability of transits

As discussed in Subsection 1.4.2, the transit probability depends on the separation between the star and the planet. For a circular orbit, the transit probability is simply $p = (R_* + R_{\rm P})/a$. Planets in circular orbits with small orbital separations will therefore be more likely to give rise to transits than planets in orbits with larger separations. If the planetary orbit is eccentric, the separation between the star and the planet will clearly vary around the orbit, and so will the speed at which the planet moves relative to the star. Hence, if a transiting exoplanet has an eccentric orbit, then the transit probability will depend on the orientation and shape of the orbit. In fact, for an eccentric orbit, the transit probability is simply

$$p = \frac{R_* + R_{\rm P}}{a(1 - e^2)}. \qquad (7.23)$$

- For an exoplanetary orbit with a significant eccentricity, how does the transit probability compare with that of a system with a circular orbit with the same semi-major axis?

- $p \propto (1 - e^2)^{-1}$, so if $e > 0$, the transit probability will be increased.

However, orbital eccentricity affects the *detectability* of transits in two opposing ways. To illustrate this, consider a system in which inferior conjunction of the planet occurs when it is close to the pericentre in an eccentric orbit. In this case, the chance of a transit occurring will be *higher* than in the case of a circular orbit with the same semi-major axis (because the star and the planet are *closer*), but the chance of spotting the transit will be *lower* than in the case of an equivalent circular orbit (because the planet will be moving *faster*, so the transit duration will be *shorter*). If, instead, inferior conjunction occurs near to apocentre, then the chance of a transit occurring will be lower, but the duration of the transit will be longer.

According to simulations by Christopher Burke (2008, *Astrophysical Journal*, **679**, 1566–73), for a distribution of eccentricities that matches that of the known exoplanets with $P > 10$ days, the probability for a transit to occur is around 25% higher than for the equivalent circular orbit, but the typical duration of those transits is about 88% of the duration of the transits for the equivalent circular orbits. Clearly, these two effects nearly cancel each other out, but the net effect is calculated as an overall 4% increase in the detectability of transiting exoplanets, compared to a situation in which all exoplanets have circular orbits. However, in the case when intrinsic stellar variability or correlated measurements are the dominant noise source, it turns out that the detectability of transits does not depend on the duration of the transit. In this case the transit yield just depends on the first factor above and is therefore around 25% higher than what would be expected if all exoplanets had circular orbits.

7.3 Orbital dynamics of known exoplanetary systems

In this section we briefly survey the orbital dynamics of the known exoplanetary systems. We begin by looking at which exoplanetary systems could harbour terrestrial planets within the habitable zone of their star. We then consider the exoplanetary systems where mean motion resonances exist between multiple planets in a given system, and finally move on to look at the exoplanets that might exist in systems containing multiple stars.

The habitable zone

The **habitable zone** of a star is conveniently defined as the range of distances from that star within which liquid water could exist on the surface of a terrestrial planet. Liquid water is believed to be essential for life, because it acts as a solvent in biochemical reactions. For the Solar System, the habitable zone extends from just outside the orbit of Venus, to just inside the orbit of Mars, as shown in Figure 7.2. The Earth lies within the habitable zone and is neither too hot nor too cold for liquid water to exist on its surface.

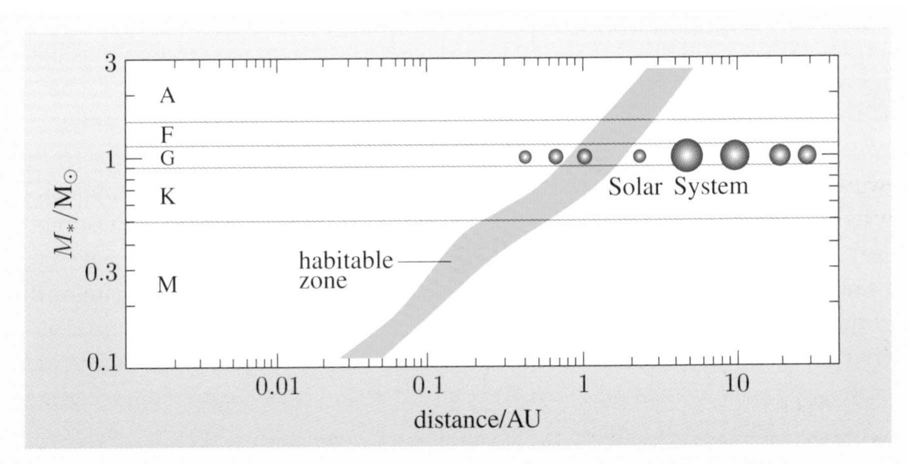

Figure 7.2 The location of the habitable zone depends primarily on the luminosity of the star. For stars on the main sequence, habitable zones have the approximate locations and extents shown here.

Although the location of the habitable zone depends primarily on the luminosity of the star, in practice, its precise location will also depend on properties of the planet, such as the extent of cloud cover and the composition and density of its atmosphere. Nonetheless, an appreciation of the extent of the habitable zone and its dependence on the stellar spectral type may be gained from a simple calculation of the equilibrium temperature as a function of distance from the star, as illustrated by the following exercise.

Exercise 7.3 Estimate the approximate width of the habitable zone in stars of different spectral type by considering (a) an A type star with $L_* = 10\,L_\odot$ and (b) a K type star with $L_* = 0.1\,L_\odot$. In each case calculate the distance from the star (in AU) at which the equilibrium temperature of a planet would be 273 K and 373 K, corresponding to the freezing temperature and boiling temperature of water, respectively. Assume that the albedo of the planet is $A = 0$, which corresponds to the limiting case of a perfect absorber. ■

As indicated by the solution to Exercise 7.3, stars on the main sequence of lower mass than the Sun will be less luminous than the Sun, so their habitable zones will be closer to the star and narrower than the Sun's habitable zone. Conversely, stars of higher mass than the Sun will be more luminous than the Sun, so their habitable zones will be further away from the star and wider than the Sun's habitable zone.

7.3.1 Exoplanets in habitable zones

Among the currently known exoplanets, very few lie in the habitable zone of their star. A good example of one that does is GJ 581 d. This planet has mass about $7.1\,M_\oplus$ and orbits its host star at a distance of about 0.22 AU with a period of 66.8 days. The host star, which is only 6.3 parsecs from the Earth, is an M3 dwarf, which means that it has a mass of about $0.3\,M_\odot$, an effective surface temperature of about 3200 K and a luminosity of only $0.013\,L_\odot$. The habitable zone of this star extends from roughly 0.1 AU to 0.3 AU, depending on the criteria used to define it and assumptions about the nature of the planets residing there. GJ 581 d is therefore well within the habitable zone and, assuming that it has a similar composition to the Earth, is a planet with a radius of only about twice that of the Earth. Given that GJ 581 d orbits a relatively low mass star, the protoplanetary disc from which the planet formed would itself have been of relatively low mass. As a result there would probably have been a relatively small mass present in the form of metals from which to form the planetary system. The discovery team therefore suggested that GJ 581 d is probably too massive to have a completely rocky composition. They further speculated that it may be an icy planet that has migrated closer to the star, and could even be covered by a deep-water ocean.

Because of selection effects, many of the known exoplanets orbit very close to their parent star, interior to the star's habitable zone. However, it is interesting to ask the question: *Could (terrestrial?) planets be present in the habitable zones of such systems or will the known giant planets have ejected Earth-mass planets, or prevented their formation?* Barrie Jones, Nick Sleep and David Underwood considered this question in their paper 'Which exoplanetary systems could harbour habitable planets?' (2006, *International Journal of Astrobiology*, **5**(3), 251–9).

Jones et al. define a critical distance from a giant planet within which an Earth-mass planet would be ejected from its orbit. This critical distance is related to the concept of the **Hill sphere**, defined as the volume of space around a planet where the planet, rather than the star, dominates the gravitational attraction. The **Hill radius** is simply the radius of the Hill sphere and is given by

$$R_{\rm H} = a\left(\frac{M_{\rm P}}{3M_*}\right)^{1/3}. \tag{7.24}$$

We discuss this formula later (in Subsection 7.6.2).

Using numerical simulations, Jones et al. establish that the critical distance is about $3R_{\rm H}$ either side of the giant planet's orbital path. The numerical factor varies: the critical distance is shorter interior to the planet's orbit, because the gravitational influence of the central star is stronger there (cf. Figure 7.3). The critical distance also depends on the eccentricity, but for simplicity we will consider only the simple case of circular orbits.

The known giant planets may lie interior to the habitable zone, exterior to it, or within it. There are three cases to consider.

- If neither critical distance from the giant planet lies within the habitable zone, then the entire habitable zone is available for a planet to exist in a stable orbit.
- If the critical distance from a giant planet lies within the habitable zone, then only part of the habitable zone is available to harbour a stable orbit.
- If the entire habitable zone lies in the region between a giant planet and its critical distance, then nowhere in the habitable zone can house a stable orbit.

These three cases give rise to the six possible configurations in Figure 7.3. Configurations 1 and 2 correspond to the first case above; configurations 3, 4 and 5 correspond to the second case; and configuration 6 corresponds to the third case.

Jones et al. initially assume that the presence of a giant planet interior to the habitable zone will not necessarily have prevented the formation of a terrestrial planet within the habitable zone, even though the giant planet must have migrated through that region to reach its present position (see the final section of this chapter). They establish that about 60% of the known exoplanetary systems could have a terrestrial planet somewhere in the habitable zone on a stable orbit. This falls to about 50% under the requirement that these orbits have remained in the habitable zone for over a billion years, as the stars have evolved. However, if the migration of a giant planet through the habitable zone to its present position prevents the formation or survival of a terrestrial planet in that region, then only about 7% of the known systems could have terrestrial planets there now on stable orbits. This 7% comprises the systems where the giant planet currently lies exterior to the habitable zone.

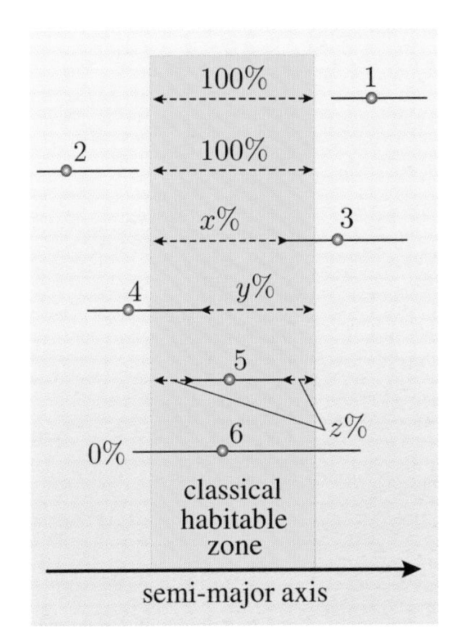

Figure 7.3 The critical distance from a giant planet within which an Earth-mass planet would be ejected from its orbit is indicated with solid lines. The habitable zone is the shaded band. There are six possible configurations shown, five of which permit the existence of Earth-mass planets on stable orbits within the habitable zone. The locations of these are indicated with dashed lines. (From Jones et al., Figure 3.)

7.3.2 Multiple exoplanet systems

About 10% of the known exoplanetary systems harbour more than one planet (2009). In most of these, two planets have been found, but there are nine systems with three known planets, two systems with four known planets (GJ 581 and HD 160691), and one system with five known planets (55 Cnc). Of the known multi-planet systems containing transiting exoplanets, only the inner planet produces transits, and the outer planet has been found by radial velocity measurements, not transit timing variations, in each case.

In about one-third of the multiple planet systems, two or more of the planets appear to be in mean motion resonant orbits. These systems, with the resonant orbits highlighted in bold, are shown in Table 7.1. Resonant orbits are reasonably common among exoplanets in general, and there is every reason to expect that they will be equally common among the systems with transiting exoplanets.

Table 7.1 Multiple planet systems exhibiting resonant orbits.

System	Orbital period/days					Resonance
	b	c	d	e	f	
47 UMa	**1083**	**2190**				2 : 1
55 Cnc	**14.65**	**44.34**	5218	2.817	260	3 : 1
GJ 876	**60.94**	**30.1**	1.938			1 : 2
HD 102272	**127.6**	**520**				4 : 1
HD 108874	**395.4**	**1605.8**				4 : 1
HD 128311	**448.6**	**919**				2 : 1
HD 43564	**226.9**	**342.9**				3 : 2
HD 60532	**201.3**	**604**				3 : 1
HD 73526	**188.3**	**377.8**				2 : 1
HD 82943	**441.2**	**219**				1 : 2

7.3.3 Exoplanets in multiple star systems

A large proportion of the stars seen in the night sky are not in fact single, isolated stars. Many of them, although appearing to the eye (or even through a telescope) as a single point of light, actually comprise two (or more) stars orbiting around each other. In most cases, the binary (or multiple) nature cannot be resolved purely by imaging, and often spectroscopy is needed to determine that more than one star is present. A binary, or multiple, star system is one in which the two, or more, stars are gravitationally bound together, undergoing stable orbital motion.

Detailed surveys of stars in the local neighbourhood have established that stellar multiplicity is more common among higher-mass stars. The fraction of stellar systems containing single stars is about 25% for high-mass (O/B/A type) stars, rising to around 50% for stars similar to the Sun (F/G type), and reaching around 75% for low-mass (M type) red dwarf stars.

The most common multiple star systems are simple binary stars, with the two stars in orbit around their common centre of mass. Exoplanets can exist in stable orbits in binary star systems; Raghavan et al. (2006, *Astrophysical Journal*, **646**, 523–42) reported that at least 23% of known exoplanetary systems are in fact binary star systems. In principle, stable orbits for exoplanets can exist around either one of the components of a binary star, or around the binary star system itself — these are known as **circumbinary orbits**. All the confirmed exoplanets known in binary star systems orbit around one of the two stars, with the second star much more distant. However, there has been a suggestion that the subtle changes seen in the eclipse times of the eclipsing binary star HW Virginis *may* be indications of two circumbinary planets in orbits of 9.1 and 15.8 years. If this proves to be true, it may indicate that circumbinary planets are common around close binary stars.

When considering more than two stars, the only stable configurations are those referred to as **hierarchical multiple** star systems. These can be illustrated by diagrams such as those shown in Figure 7.4, which are known as 'mobile diagrams' due to their similarity with hanging mobiles. Hierarchical systems have only two branches descending from each node. A simple binary star is illustrated by the schematic diagram of Figure 7.4a.

A hierarchical triple star system consists of a close binary star orbiting a more distant third star, and is illustrated schematically by Figure 7.4b. The nearest star system to the Sun, α Centauri, is actually a triple star: α Cen A and α Cen B form a close binary, and they themselves are in orbit with a more distant third star, proxima Centauri (currently the nearest star to the Sun). A binary star with a planet orbiting one of the two stars is also a hierarchical triple system: the planet plus star constitute a close binary, with the second star orbiting at a larger distance.

Five known exoplanets are in triple star systems (January 2009). In principle these exoplanets could exist in stable orbits around any one of the three stars, around the close binary, or around the entire triple system itself. In practice, each of the known exoplanets in triple star systems is in orbit around the distant, single star of the hierarchical triple. In effect, this forms a hierarchical quadruple system consisting of a close binary star and a binary formed by the third star and the planet. The two 'binaries' then orbit around their common centre of mass in an architecture shown schematically by Figure 7.4c.

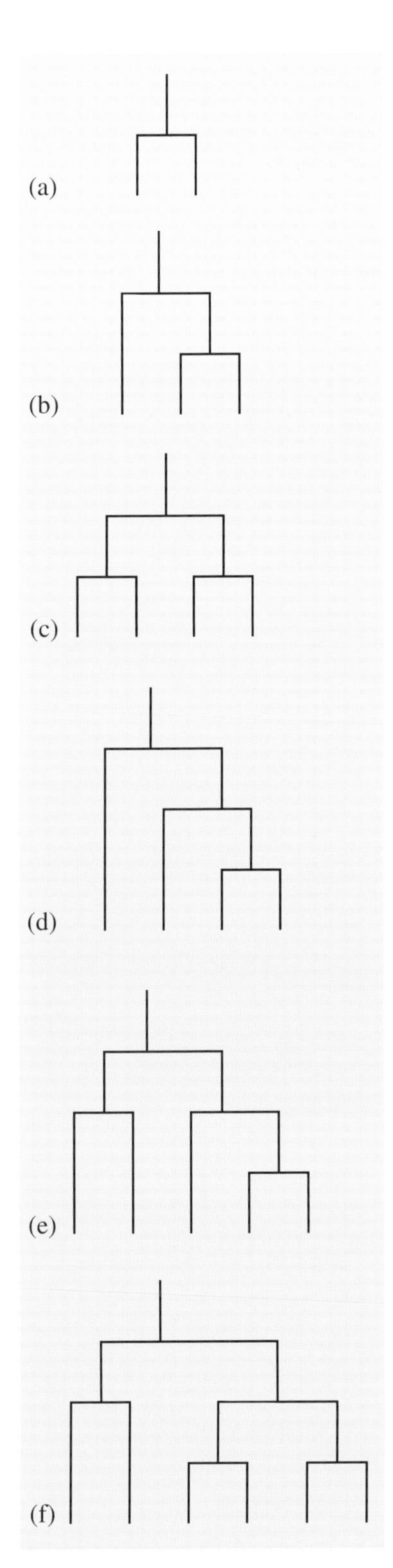

Figure 7.4 'Mobile diagrams' illustrating various multiple star hierarchies. (a) A binary star. (b) A triple system (which can be thought of as either three stars or a binary star plus a planet). (c) A quadruple system consisting of a binary-binary. (d) A quadruple system consisting of a triple plus a single body. (e) A pentuple system consisting of a binary-binary with one of the binaries orbited by a fifth body. (f) A sextuple system consisting of a binary-binary in orbit with another binary.

- If a triple star system had a planet orbiting the close binary stars (instead of the distant single star), what would the mobile diagram look like?

❍ It would resemble the diagram in Figure 7.4d. The planet orbiting the close binary star forms a hierarchical triple, and this triple system is then in orbit with a distant single star.

A hierarchical quadruple star system could have either of the arrangements shown in Figure 7.4c,d. An example of the former is the star HD 98800, which consists of two binary stars orbiting around each other, as shown in Figure 7.5. A circumbinary disc of material is observed around one of the binary stars, comprising two distinct belts. It is speculated that there may be a planet located in the gap between the two belts, which has cleared out a space in the disc. If such a planet were present, the mobile diagram of the system would resemble that in Figure 7.4e.

Figure 7.5 An artist's impression of the HD 98800 system. It consists of two binary stars orbiting each other; one of the binaries has a circumbinary disc, within which it is speculated that a planet may exist.

A final example of a multiple hierarchical stellar system is Castor, or α Gem. Through a telescope, Castor appears as a visual binary star, but when observed spectroscopically, each component of the visual binary is revealed as a spectroscopic binary star. If this were the limit of the hierarchy, this would be a quadruple system similar to HD 98800 described above. However, this quadruple system is orbited by a further star system that is fainter and more distant, which is itself a close binary star. Hence Castor is actually a sextuple system, as illustrated schematically in Figure 7.4f.

Exercise 7.4 What are the possible orbits for an exoplanet in the Castor system? Sketch a mobile diagram for each type of system architecture. ■

7.4 Transit timing variations due to another planet

Here and in Section 7.5 we follow one of the first papers to investigate transit timing variations (TTVs): 'On detecting terrestrial planets with timing of giant planet transits' by Eric Agol, Jason Steffen, Re'em Sari and Will Clarkson (2005,

Monthly Notices of the Royal Astronomical Society, **359**, 567–79). We begin by looking at how the transit timings of a planet are affected by the presence of a second planet on an interior orbit, i.e. lying between the transiting planet and the star. For simplicity, we assume that the orbits of both planets are aligned in the same plane and that we are viewing these orbits exactly edge-on.

7.4.1 Interior planets on circular orbits

In the simplest situation both planets are in circular orbits. We will label the inner planet with subscript 1 and the outer planet (whose transits we measure) with subscript 2. We can treat the system as consisting of an 'inner binary' comprising the star and inner planet, which orbit around their common barycentre, and an 'outer binary' comprising the inner binary and the outer planet, which orbit around their common barycentre. The transits of the outer planet across the face of the star will occur approximately when it lines up with the barycentre of the inner binary. However, the motion of the inner binary displaces the star from perfect alignment with the outer planet; sometimes the transit will occur early, and sometimes it will occur late, as shown in Figure 7.6.

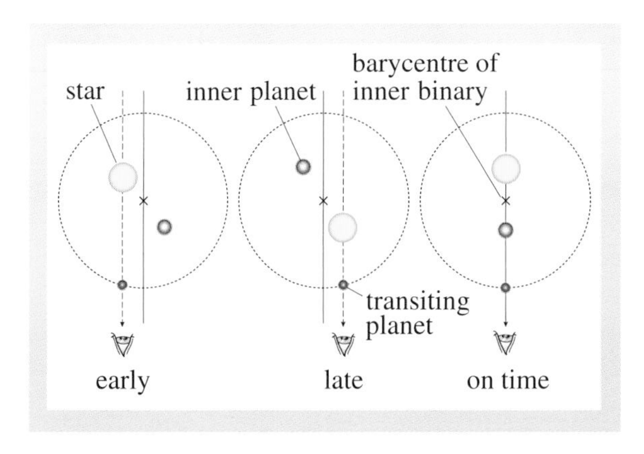

Figure 7.6 The presence of a planet on an inner orbit will alter the transit timings of an outer transiting planet, making some transits occur earlier and others occur later than the otherwise predicted times. Not drawn to scale.

For the inner binary, composed of the star and inner planet (mass M_1), we can write

$$a_* M_* = a_1 M_1,$$

where a_* is the orbital radius of the star around the barycentre of the inner binary, and a_1 is the orbital radius of the inner planet. The maximum TTV will occur when the inner planet is at quadrature. In this situation, the inner planet, star and outer planet form a right-angled triangle, with the star at the right angle. The maximum displacement in the plane of the sky of the star from the barycentre of the inner binary is then

$$x_{*,\mathrm{max}} = \frac{a_1 M_1}{M_*} \approx a_1 \mu_1,$$

where μ_1 is the reduced mass of the inner planet. Since we are assuming the planets to be in circular orbits, the displacement of the star as a function of time is simply

$$x_*(t) \approx a_1 \mu_1 \sin[2\pi(t - T_\mathrm{c})/P_1], \qquad (7.25)$$

where P_1 is the orbital period of the inner planet and T_c is the predicted time of the centre of the transit of the outer planet, in the absence of the inner (perturbing) planet. The offset in the time of the transit of the outer planet is just this distance divided by the relative speeds of the outer planet and star, i.e. $\delta t_2 = x_*/(v_2 - v_*)$. We can assume that the speed of the star is much less than the speed of the outer planet, so $v_* \ll v_2$ and therefore $\delta t_2 \approx x_*/v_2$. Since the orbits are circular, $v_2 = 2\pi a_2/P_2$, where a_2 and P_2 are the radius and period of the orbit of the outer binary, respectively. So we have

$$\delta t_2 \approx \frac{P_2 a_1 \mu_1 \sin[2\pi(t - T_c)/P_1]}{2\pi a_2}. \tag{7.26}$$

Exercise 7.5 Consider a system with a star of mass $M_* = \mathrm{M}_\odot$ and two planets of masses similar to those of the Earth and Jupiter, i.e. $M_1 \sim 3 \times 10^{-6}\,\mathrm{M}_\odot$ and $M_2 \sim 10^{-3}\,\mathrm{M}_\odot$. Both planets lie in circular orbits in the same plane, and the outer planet transits the star every 30 days, while the inner planet has an orbital period of 7 days. The radii of the two orbits are 0.19 AU and 0.072 AU.

(a) What is the maximum offset in the transit time of the outer planet?

(b) If the inner planet had a mass 10 times that of the Earth, how would the maximum offset of the transit time of the outer planet differ? ■

As Exercise 7.5 shows, in general a terrestrial planet that is interior to a transiting giant planet will cause TTVs of at most a few seconds in a thirty-day orbit. Such a measurement would be extremely challenging to make, but in principle may be achievable with future generations of space-based transit detection telescopes.

The approximate time at which each transit occurs is just some integer, n, times the orbital period of the outer planet, i.e. $t = nP_2$, so the argument of the sine function in Equation 7.26 is approximately $2\pi(nP_2 - T_c)/P_1$. For circular orbits, if the periods of the two planets are in a simple ratio so that $P_2 = jP_1$, where j is an integer, then the TTVs disappear, because the value of the sine function is the same for each transit. Consequently, δt_2 is the same for each transit. It is only by detecting a variation in δt_2 that the presence of the inner planet may be deduced.

In practice, the best way to detect the effect may be to observe many transits and calculate the standard deviation of the observed transit timing deviations over many orbits. This may be calculated as the square root of the mean of the squares of all the transit timing deviations observed, namely

$$\sigma_2 = \left\langle (\delta t_2)^2 \right\rangle^{1/2} \approx \frac{P_2 a_1 \mu_1}{2^{3/2}\pi a_2}, \tag{7.27}$$

where the $\langle\ \rangle$ notation indicates a mean value.

7.4.2 Interior planets on eccentric orbits

If the orbits of the planets are eccentric, rather than circular, the transit timing offsets can be slightly larger. The TTVs may be calculated by solving the equations of motion for each Jacobian coordinate, Equations 7.14 and 7.15. When the orbits are coplanar, as we are assuming here, eight parameters specify the positions: the eccentricity of each orbit (e_1, e_2); the semi-major axis of each orbit

(a_1, a_2); the longitude of pericentre of each orbit ($\omega_{\text{OP1}}, \omega_{\text{OP2}}$); and the true anomaly at a given time for each planet (θ_1, θ_2).

As in the case of circular orbits, the offset in the time of a given transit of the outer planet is $\delta t_2 \approx x_*/v_2$, where in this case the position of the star with respect to the barycentre of the inner binary is

$$x_* \approx \mu_1 a_1 \cos[\theta_1(t) + \omega_{\text{OP1}}],$$

and the speed of the outer planet perpendicular to the line of sight at the time of each transit is

$$v_2 \approx \frac{2\pi a_2(1 - e_2 \sin \omega_{\text{OP2}})}{P_2\sqrt{1 - e_2^2}}.$$

So, combining these two equations, the transit timing offset for an outer planet in an eccentric orbit due to the presence of an inner planet is

$$\delta t_2 \approx \frac{P_2 \mu_1 a_1 \cos[\theta_1(t) + \omega_{\text{OP1}}]\sqrt{1 - e_2^2}}{2\pi a_2(1 - e_2 \sin \omega_{\text{OP2}})}. \tag{7.28}$$

Exercise 7.6 Consider the same system parameters as in Exercise 7.5(a), but now with an orbital eccentricity for the outer planet of (a) $e_2 = 0.1$ and (b) $e_2 = 0.5$. What is the maximum transit timing offset in each case? ■

As in the case of circular orbits, the TTV due to an interior planet simply scales with the mass of the perturbing planet. Nonetheless, even with a transiting planet in an eccentric orbit, a terrestrial planet interior to a transiting giant planet can cause TTVs of only a few seconds at most.

7.4.3 Exterior planets

We now consider the situation where we have an inner, transiting planet (labelled 1) in a relatively short circular orbit, and another planet exterior to this (labelled 2) in a much larger, eccentric orbit. The presence of the outer planet causes a slight increase in the period of the orbit of the inner (transiting) planet, by an amount that depends on the instantaneous distance from the outer planet to the star:

$$\delta P_1 \approx \mu_2 \left(\frac{a_1}{r_2}\right)^3 P_1,$$

where r_2 is the distance from the outer planet to the star, and μ_2 is the reduced mass of the outer planet, given by $M_2/M_{\text{total}} \approx M_2/M_*$, with M_2 and M_* being the masses of the outer planet and the star, respectively. If the outer planet were in a circular orbit, this slight increase in period would be constant with time, and so undetectable. However, if the outer planet is in an eccentric orbit, its distance from the star varies with time and it causes a periodic change in the orbital period P_1 of the inner binary, with a period of P_2.

The maximum TTV for the inner planet is approximately

$$\delta t_1 \approx \mu_2 e_2 \left(\frac{a_1}{a_2}\right)^3 P_2. \tag{7.29}$$

Exercise 7.7 What is the maximum TTV in the orbit of an inner transiting planet in the following system? Stellar mass $M_* = \mathrm{M_\odot}$, inner (transiting) planet mass $M_1 = 10^{-3}\,\mathrm{M_\odot}$, outer (perturbing) planet mass $M_2 = 0.5 \times 10^{-3}\,\mathrm{M_\odot}$; semi-major axis and period of inner planet $a_1 = 0.05\,\mathrm{AU}$, $P_1 = 4$ days; semi-major axis and period of outer planet $a_2 = 0.42\,\mathrm{AU}$, $P_2 = 100$ days; eccentricity of outer planet $e_2 = 0.5$. ■

As Exercise 7.7 shows, the typical TTVs introduced by an exterior planet are of the order of a few seconds.

7.5 Transit timing variations for planets in resonant orbits

In the previous section, we assumed that the perturbation provided by the gravitational pull of one planet on another will change the transit timing of that planet, but will not significantly change its orbital parameters. This will not be the case if the two planets are in orbital resonance with each other.

To investigate how TTVs would arise in planets that are in a mean motion orbital resonance, we begin by considering a $j+1:j$ (first-order) resonance, with the planetary orbits initially having zero eccentricity, and one of the planets significantly less massive than the other ($M_1 \ll M_2$). The transiting planet is on an interior orbit to the perturbing planet, so $P_2 < P_1$. The two planets will undergo conjunctions (i.e. line up with each other) every j orbits of the outer planet and will have the same longitude each time a conjunction occurs. This will increase the eccentricity of the less massive planet and so change its period and semi-major axis. The change in period causes the longitude of the conjunction to change with time, until a shift of about 180° has accumulated relative to the initial position, at which time the eccentricity will decrease again. This is referred to as a **libration cycle**, and will cause the timing of the transit of the less massive planet to change with time.

Numerical simulations of this effect carried out by Agol et al. (2005, *Monthly Notices of the Royal Astronomical Society*, **359**, 567–79) show that a good fit to the data is provided by the expression

δt_2 here gives the maximum TTV.

$$\delta t_2 \approx \frac{P_2}{4.5j}\frac{M_1}{M_1+M_2}, \qquad (7.30)$$

where M_1 is the mass of the outer perturbing planet and M_2 is the mass of the inner transiting planet. Agol et al. show that Equation 7.30 agrees with their numerical modelling to better than 10% accuracy, even when $j = 2$. Agol et al. also calculate the libration period in this situation as

$$P_{\mathrm{lib}} \approx 0.5j^{-4/3}\mu_2^{-2/3}P_2, \qquad (7.31)$$

where μ_2 is the reduced mass of the transiting planet, given by $M_2/M_{\mathrm{total}} \approx M_2/M_*$.

Exercise 7.8 Consider two planets, of mass $M_2 = 10^{-3}\,\mathrm{M_\odot}$ (i.e. Jupiter-mass) and $M_1 = 3 \times 10^{-6}\,\mathrm{M_\odot}$ (i.e. Earth-mass), in orbit around a star of mass $M_* = \mathrm{M_\odot}$. Assume that both orbits are initially circular and coplanar. The orbital periods of the two planets are $P_2 = 12$ days and $P_1 = 16$ days, so that they are in a $4:3$ resonance (i.e. a first-order resonance with $j = 3$).

(a) What is the maximum TTV of the inner, more massive planet?

(b) What is the libration period for this TTV? ■

As this exercise indicates, the TTVs for planets in resonant orbits can be vastly larger than for simple non-resonant orbits, as calculated in earlier sections. A TTV of a few minutes in an orbit of a few days is measurable even with current technology. We therefore anticipate detections of terrestrial planets in resonant orbits with transiting giant planets via TTVs.

Figure 7.7 shows the TTVs that would arise according to Equations 7.30 and 7.31. The situations represented there are for the planet and star masses in Exercise 7.8, but for a range of values of j (i.e. for different mean motion resonances). Note that the simple model implied by Equations 7.30 and 7.31 is unlikely to be very accurate when $j < 2$.

● If the period of the (inner) more massive planet is $P_2 = 12$ days for each of the curves in Figure 7.7, what is the period of the (outer) less massive planet in each case?

❍ For $j = 1$, the resonance is $2:1$, so $P_1 = 12 \times 2/1 = 24$ days.
For $j = 2$, the resonance is $3:2$, so $P_1 = 12 \times 3/2 = 18$ days.
For $j = 3$, the resonance is $4:3$, so $P_1 = 12 \times 4/3 = 16$ days.
For $j = 4$, the resonance is $5:4$, so $P_1 = 12 \times 5/4 = 15$ days.

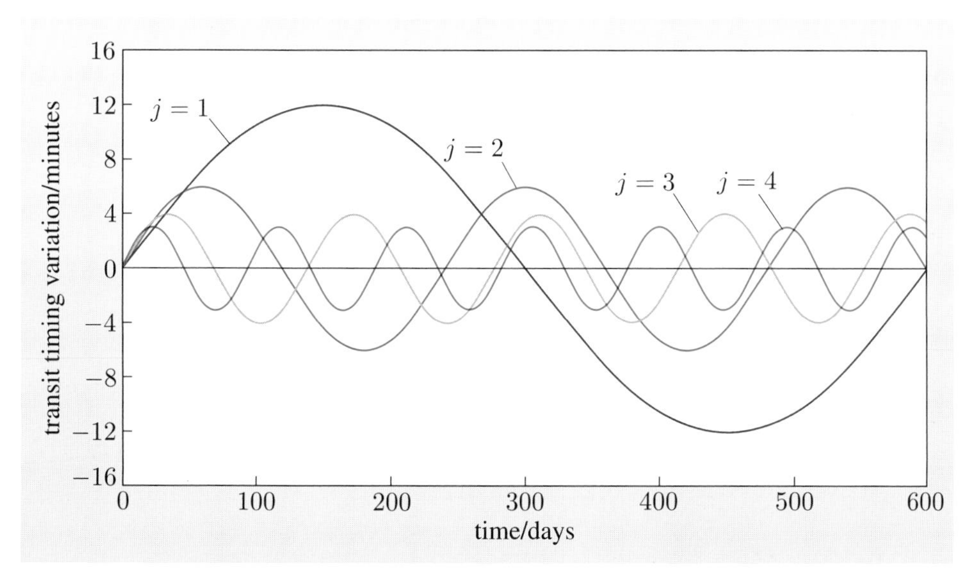

Figure 7.7 TTVs for planets in resonant orbits. The curves shown here are for the system outlined in Exercise 7.8, that is, two planets of mass $M_2 = 10^{-3}\,\mathrm{M_\odot}$ (i.e. Jupiter-mass) and $M_1 = 3 \times 10^{-6}\,\mathrm{M_\odot}$ (i.e. Earth-mass) in orbit around a star of mass $M_* = \mathrm{M_\odot}$. Both orbits are initially circular and coplanar. The orbital period of the (inner) transiting planet is $P_2 = 12$ days, and the curves correspond to first-order mean motion resonances with $j = 1, 2, 3, 4$.

In the above, we considered planets that were initially on circular orbits. It turns out that when the eccentricity of either orbit is significant, then resonances higher than those of first order become important. When the ratio of the semi-major axes of the two orbits becomes large, and the eccentricity of the outer orbit approaches 1, resonances of the form $1 : j$ become as important as the first-order resonances discussed previously.

7.6 Exomoons

The previous section showed that terrestrial planets in resonant orbits can be detected through the TTV effect that they have on transiting giant planets. In principle such terrestrial planets may be detected in the habitable zone of their star by this technique. There is another way of populating the habitable zone of a star with habitable bodies: terrestrial-sized moons of giant planets that lie in the habitable zone.

Exomoons will also cause perturbations in the transits of the planet around which they are orbiting. In the next subsection we describe how such perturbations arise and how, in principle, their measurement may lead to the detection of habitable terrestrial exomoons. In practice, however, it is very difficult to have an exomoon in a stable orbit around a transiting hot Jupiter, for the reasons that we outline in Subsection 7.6.2.

7.6.1 Transit timing variations and transit duration variations due to exomoons

In this subsection we follow the description given by David Kipping in his paper 'Transit timing effects due to an exomoon' (2009, *Monthly Notices of the Royal Astronomical Society*, **392**, 181).

If an exoplanet has an exomoon, the two objects will orbit around their common barycentre. The planet–moon barycentre will execute an orbit around the star; this orbit will obey Kepler's third law and the other equations appropriate to the orbits of moonless planets. Hence the planet will 'wobble' slightly as it orbits around the star. Consequently, there will be a time lag between the planet being at the mid-point of its transit (the time that we measure and call T_c) and the barycentre of the planet–moon system being at the mid-point of its transit (the time that we would otherwise predict). This, therefore, gives rise to a TTV that in principle is measurable.

Kipping presents a derivation for the magnitude of the TTV in the general case of non-circular orbits. For the simplest case of an exomoon in a circular orbit, the maximum amplitude of the TTV is

$$\delta t_{\mathrm{M}} \approx \frac{a_{\mathrm{M}} M_{\mathrm{M}}}{2^{1/2} M_{\mathrm{P}}} \frac{P}{2\pi a}, \tag{7.32}$$

where M_{M} and a_{M} are the mass of the moon and the semi-major axis of its orbit around the planet, while P and a are the period and semi-major axis of the planet–moon barycentre's orbit around the star. Recognizing that we may write $\mu_{\mathrm{M}} = M_{\mathrm{M}}/M_{\mathrm{P}}$ for the reduced mass of the moon with respect to that of the planet, Equation 7.32 has the same form as the amplitude of the TTV derived for the case of a perturbing planet, in Equation 7.26, apart from the extra factor of '$2^{1/2}$' on the bottom line here.

Hence the amplitude of this TTV is proportional to the product of the exomoon's mass (M_M) and its orbital radius (a_M). So, even if such a TTV could be measured, the result cannot be used to uniquely determine the mass of the exomoon. If we knew the period of the exomoon's orbit around its planet, we could use Kepler's third law to derive a_M and so solve for the mass of the exomoon that way. As we will see in Subsection 7.6.2, for the orbit to be stable, the orbital period of the exomoon must be less than about 10% of the orbital period of the exoplanet around its star.

There is, however, another transit timing effect produced by the exomoon. The velocity of the exoplanet around its star will be the vector sum of the orbital velocity of the planet–moon barycentre around the star and the wobble velocity caused by the exoplanet and exomoon orbiting around the planet–moon barycentre. The small perturbation to the overall velocity of the planet provided by the 'wobble' will probably be different at the time of each transit, unless the ratio of the exoplanet's orbital period to the exomoon's orbital period is a small integer. At some transits the wobble velocity will add to the planet's overall velocity, while at other transits it will subtract from it. The net result is that the observed *duration* of the transit of the exoplanet, T_{dur}, will vary with time. The recently-discovered transiting planet HAT-P-14 b, which has a grazing impact parameter, could potentially accommodate an exomoon in a stable orbit, causing transit duration variations of up to 45 s.

For circular orbits, the maximum variation in the transit duration due to this extra velocity component is

$$\delta T_M \approx \left(\frac{a}{a_M}\right)^{1/2} \frac{M_M}{[M_P(M_P + M_*)]^{1/2}} \frac{T_{dur}}{2^{1/2}}, \tag{7.33}$$

where T_{dur} is the unperturbed duration of the transit. (The derivation of Equation 7.33 in the general case of eccentric orbits is given by Kipping.)

Now, since the variation in the time of transit is proportional to $a_M M_M$ but the variation in the duration of the transit is proportional to $a_M^{-1/2} M_M$, the combination of these two effects allows both the mass and orbital radius of the exomoon to be determined independently. Using Kepler's third law, the orbital period of the exomoon can then also be determined.

Exercise 7.9 Consider a transiting exoplanet system with the following parameters. A Jupiter-mass ($M_P = 10^{-3}\,M_\odot$) planet transits a solar-mass star once every 4 days in a circular orbit with a radius of 0.05 AU. The transit duration is 3 hours.

(a) A TTV of 30 seconds and a transit duration variation of 15 seconds are observed. If these are assumed to be produced by an exomoon orbiting the exoplanet, what would be the mass and orbital radius of the exomoon causing this perturbation?

(b) What would be the orbital period of the exomoon around the exoplanet? ■

Exercise 7.9 shows that TTVs and transit duration variations of 10–30 seconds in the transit of a hot Jupiter would imply the presence of an exomoon with the mass of a terrestrial planet, orbiting extremely close to the planet. As we will show in the next subsection, many such orbits are likely to be unstable and it is unlikely that we would actually find an exomoon orbiting a hot Jupiter.

Might we be able to find terrestrial-planet-sized exomoons orbiting transiting planets in the habitable zone of another star? The following exercise considers this possibility.

Exercise 7.10 Consider a Jupiter-mass ($M_\text{P} = 10^{-3}\,\text{M}_\odot$) exoplanet orbiting a 0.5 solar-mass star once every 46 days in a circular orbit with a radius of 0.2 AU (i.e. within the star's habitable zone). If the planet was observed to transit its host star for 5 hours every orbit, what would be the maximum TTV and the maximum transit duration variation caused by a terrestrial mass ($M_\text{M} = 3 \times 10^{-6}\,\text{M}_\odot$) exomoon orbiting the planet at a distance of $a_\text{M} = 10^9$ m? ■

In practice it is likely to be very difficult to observe transits of a planet in a 46-day orbit. However, as Exercise 7.10 shows, if such a system is detected, then the presence of a terrestrial-sized exomoon orbiting that planet (in the star's habitable zone) could, in principle, be detected by virtue of the TTVs and transit duration variations that it induces.

7.6.2 The stability of the orbit of an exomoon

We noted earlier that even though TTVs from an exomoon orbiting a hot Jupiter might be measurable in principle, in practice such orbits are likely to be unstable. In this subsection we look at the maximum and minimum orbital radii at which stable orbits can exist for moons orbiting planets.

An important concept for this issue is that of the Hill sphere mentioned earlier. A moon can have a stable orbit around a planet if it is within the planet's Hill sphere, but cannot do so outside this. The approximate size of the Hill sphere may be derived as follows, where we indicate quantities referring to the planet by subscript P and quantities referring to the moon by subscript M.

Kepler's third law states that the orbital period and orbital radius of a planet around a star are related by

$$P^2 = \frac{4\pi^2 a^3}{G(M_* + M_\text{P})} \approx \frac{4\pi^2 a^3}{GM_*}.$$

Assuming that the orbits are circular, the orbital speed is just $v = 2\pi a/P$. So we have

$$v = \left(\frac{GM_*}{a}\right)^{1/2},$$

and the orbital angular speed of the planet is $\omega = v/a$, so

$$\omega = \left(\frac{GM_*}{a^3}\right)^{1/2}.$$

Similarly, the angular orbital speed of an exomoon around an exoplanet is

$$\omega_\text{M} = \left(\frac{GM_\text{P}}{a_\text{M}^3}\right)^{1/2}.$$

The Hill radius, R_{H}, is approximately the orbital radius of the exomoon, a_{M}, at which the orbital angular speed of the exomoon around an exoplanet equals the orbital angular speed of the exoplanet around the star, i.e.

$$\left(\frac{GM_*}{a^3}\right)^{1/2} = \left(\frac{GM_{\mathrm{P}}}{a_{\mathrm{M}}^3}\right)^{1/2} \approx \left(\frac{GM_{\mathrm{P}}}{R_{\mathrm{H}}^3}\right)^{1/2}.$$

So

$$R_{\mathrm{H}} \approx a\left(\frac{M_{\mathrm{P}}}{M_*}\right)^{1/3}.$$

A rigorous derivation of the Hill radius gives an extra numerical factor, and the usual formula used is that noted earlier:

$$R_{\mathrm{H}} = a\left(\frac{M_{\mathrm{P}}}{3M_*}\right)^{1/3}. \qquad \text{(Eqn 7.24)}$$

The Hill radius is the *maximum* distance at which a moon can orbit a planet and still remain in a stable orbit.

Exercise 7.11 What is the Hill radius for a Jupiter-mass planet ($M_{\mathrm{P}} = 10^{-3}\,\mathrm{M}_{\odot}$) in a 4-day orbit, of radius 0.05 AU, around a solar-mass star? ■

Exercise 7.11 shows that the maximum radius at which an exomoon might orbit a hot Jupiter is about 520 thousand kilometres. In the box on tidal forces and tidal dissipation (page 224) we introduced the Roche limit, which is the distance within which a moon will be pulled apart by the tidal forces of the planet:

$$d_{\mathrm{R}} = R_{\mathrm{M}}\left(\frac{2M_{\mathrm{P}}}{M_{\mathrm{M}}}\right)^{1/3}. \qquad \text{(Eqn 7.16)}$$

The Roche limit is the *minimum* distance at which a moon can orbit a planet.

Exercise 7.12 What is the Roche limit for a moon with the mass and radius of the Earth ($M_{\mathrm{M}} = 3 \times 10^{-6}\,\mathrm{M}_{\odot}$, $R_{\mathrm{M}} = 6.4 \times 10^{6}\,\mathrm{m}$) orbiting a Jupiter-mass planet ($M_{\mathrm{P}} = 10^{-3}\,\mathrm{M}_{\odot}$)? ■

Exercise 7.12 shows that the Roche limit for a terrestrial-sized moon orbiting a Jupiter-sized planet is about 56 thousand kilometres. Now consider: if the planet is a hot Jupiter (as described in the earlier exercise), is there any stable radius at which this moon might orbit the planet?

As we have seen, the Roche limit is 56 thousand kilometres, while the Hill radius (assuming that the planet is in a 4-day orbit around a solar-mass star) is about 520 thousand kilometres. There is therefore a stable zone between these two extremes, and the exomoon in Exercise 7.9 orbits its planet in this zone with a radius of about 400 thousand kilometres. However, we also note that the radius of such a hot Jupiter is likely to be at least the radius of Jupiter itself (possibly larger), which is about 70 thousand kilometres. So the Roche limit actually lies within the atmosphere of the planet in this case. Clearly, other forces acting on the exomoon (such as atmospheric drag) will be important if it orbits close to the exoplanet. Ultraviolet observations of exospheric gas surrounding HD 209458 b and WASP-12 b suggest that an exomoon orbiting at only a few planetary radii from a hot Jupiter will likely suffer orbital decay and rapidly spiral into the planet itself. In contrast, Phobos orbits Mars at a distance of only about 3 planetary radii, but Mars has very little atmosphere, so there is little drag.

7.7 Exotrojans

Trojans are bodies that orbit around a star on the same trajectory as another planet, but ahead (or behind) in the orbit by about 60°. These locations are known as the L4 and L5 **Lagrangian points** (or sometimes **Trojan points**). As shown in Figure 7.8, they lie at the corners of equilateral triangles, with the star and the planet marking the other two corners.

- Given that the distance from the Trojans to the planet and to the star are equal, what is the ratio of the gravitational force on them due to the star relative to that of the planet?

❍ The gravitational forces are in the same ratio as the mass of the star to the mass of the planet.

The above implies that the net gravitational force on the Trojans acts in a direction that passes through the centre of mass of the star–planet system (see Figure 7.8). Consequently, bodies at the Trojan positions remain in stable orbits.

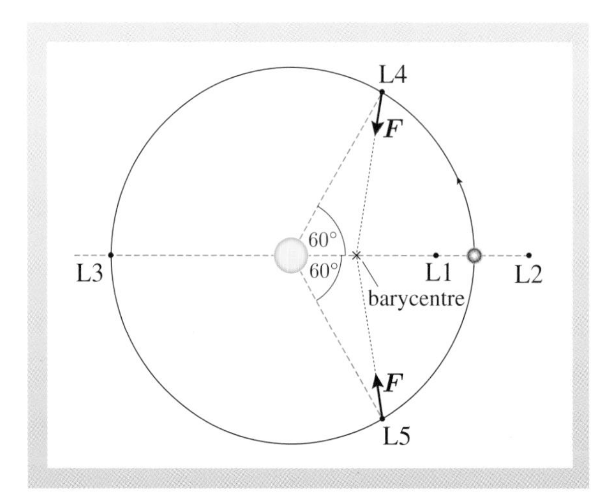

Figure 7.8 The L4 and L5 Lagrangian points in the orbit of a planet may harbour Trojans.

7.7.1 The orbits of exotrojans

Since Trojans exist in our own Solar System, in Jupiter's orbit and also in the orbits of Mars and Neptune, it seems likely that exotrojans may also exist, co-orbiting stars along with exoplanets.

- What is the mean motion orbital resonance of a Trojan with respect to the planet with which it co-orbits?

❍ It is a 1 : 1 mean motion orbital resonance — the Trojan and the planet will necessarily have the same orbital period as each other.

There are various ways to consider detecting exotrojans. One way would be to detect the transit signal of an exotrojan directly.

● If an exotrojan were to give rise to a transit signal of its own, at what point in the orbit (relative to the position of the planetary transit) might one look for the transit signal?

❍ Assuming that the exoplanet and exotrojans are in circular orbits, the transit due to the exotrojan would precede or lag the exoplanet transit by about one-sixth of the orbital period.

However, the orbits of exotrojans may not be at precisely the same orientation as the orbit of the exoplanet, so a transit may not be visible. Exotrojans undergo librations around the Lagrangian points, so their exact transit times can be difficult to predict. Finally, exotrojans may be rather small and so give rise to unmeasurable transit signals with current technology.

Another possibility for their detection is the effect that an exotrojan may have on the radial velocity signature. Even if an exotrojan does not itself transit the star, it may cause an offset between the zero-crossing time of the radial velocity curve and the time of mid-transit of the transiting planet. An exotrojan may also conceivably give rise to an anomalous deviation in the Rossiter–McLaughlin effect signal. None of these possibilities is as likely as detecting the effect of an exotrojan on the timing of the exoplanetary transit. We discuss this next.

7.7.2 Transit timing variations due to exotrojans

An exploration of the possible TTV effects caused by exotrojans is given by Eric Ford and Matthew Holman in 'Using transit timing observations to search for Trojans of transiting extrasolar planets' (2007, *Astrophysical Journal*, **664**, L51–4). We follow their formulation.

We first note that if the exoplanet and any exotrojans have the same orbital eccentricities, and if the exotrojans remain fixed precisely at the L4 and L5 points, then the transit timing signal of the exoplanet would be identical to the case with no exotrojans present, and the additional bodies would be undetectable. It is only when the exotrojans oscillate or librate around the L4 and L5 points that TTVs will arise as a result of small changes to the period of the transiting planet.

The TTV due to an exotrojan is

$$\delta t \approx \mu_{\mathrm{T}} P \frac{\Delta\psi(t_i)}{2\pi}, \tag{7.34}$$

where μ_{T} is the reduced mass of the Trojan, i.e. $M_{\mathrm{T}}/(M_{\mathrm{P}} + M_{\mathrm{T}})$, P is the orbital period of the exoplanet, and $\Delta\psi(t_i)$ is the angular displacement (in radians) of the exotrojan from the L4 or L5 point at the time of the ith transit. In the case of Trojans in Jupiter's orbit, typical libration amplitudes are of order $\sim 5°–25°$, so it is plausible to expect similar variations to occur in exoplanetary systems too.

Exercise 7.13 What is the likely maximum TTV due to an Earth-mass exotrojan librating around the L4 point of a Jupiter-mass exoplanet with an angular displacement amplitude of $25°$, if the exoplanet is in a 4-day orbit? ■

The typical libration timescale of the exotrojans is estimated as

$$P_{\text{lib}} \approx \left(\frac{4}{27\mu_{\text{P}}}\right)^{1/2} P, \tag{7.35}$$

where $\mu_{\text{P}} = M_{\text{P}}/(M_* + M_{\text{P}}) \approx M_{\text{P}}/M_*$. So for an exotrojan in orbit with a Jupiter-mass planet around a solar-mass star, the libration timescale is about 12 times the planet's orbital period. In the example in Exercise 7.13, TTV would show a periodic change with an amplitude of about 72 seconds over the course of about 12 transits, or roughly 50 days.

- Exotrojans might co-orbit with an exoplanet around a star. What other locations in exoplanetary systems might harbour Trojan objects?

- Conceivably, there could be Trojan planets co-orbiting with a low-mass star around a more massive star in a binary star system. There could also be exotrojan moons co-orbiting with exomoons around a giant exoplanet. In this case the giant planet, exomoon and exotrojans would form the three corners of the triangle.

7.8 The formation, evolution and migration of planets

As noted in Chapter 1, until 1995 we knew of only one planetary system around a main sequence star, and that was our own Solar System. It was therefore natural to assume that other planetary systems would resemble our own: i.e. having small, rocky planets close to the star and gas giant planets at larger orbital radii, possibly with one or more asteroid belts too. Since the discovery of the first exoplanet, our ideas about what a planetary system might look like have changed dramatically. As you have seen, many of the known exoplanetary systems have giant planets that are orbiting relatively close to their parent star: around two-thirds of the known Jupiter-mass planets have orbital semi-major axes less than 1 AU, and about one-third lie closer than 0.1 AU to their star.

The widely accepted **core accretion theory** for the formation of giant planets states that such planets form in a region of the protoplanetary disc that is cool enough (below about 150 K) for water, methane and ammonia to condense into solid ice grains and so add to the other solid material present there. This region is referred to as beyond the **snowline** and lies beyond about 4 AU from a solar-mass protostar. It is believed that planetary cores of mass up to about ten times that of the Earth form in this way, and these cores then accumulate a massive envelope of gaseous material. Clearly, the question then arises: if giant exoplanets formed at distances greater than a few AU from their stars, how do many of them come to lie significantly closer than 1 AU from their stars?

The answer is believed to be that giant planets migrate inwards from where they formed due to gravitational interactions with the remaining protoplanetary disc material. In the case of our own Solar System, it seems that Jupiter and Saturn underwent only a relatively small migration, perhaps because they formed quite late, only shortly before the remaining gas was dispersed. Many of the known exoplanets must have undergone significant migration through the region within the snowline where terrestrial planets might be expected. It is suggested that the formation and migration of giant planets occurred within the first 1–10 million years after the formation of the star, and that terrestrial planet formation occurred later, perhaps between 10 and 100 million years after the star formed.

- What is different about the formation process of terrestrial planets that lie interior to the orbit of a giant planet, compared with the formation process of those that lie exterior to the orbit of a giant planet?

- Terrestrial planets that lie interior to the orbit of a giant planet would form from a region of the protoplanetary disc that has been little affected by the migration of the giant planet. Terrestrial planets that lie exterior to the orbit of a giant planet would form from a region of the protoplanetary disc through which a giant planet had previously migrated.

As noted earlier, terrestrial planets in orbits exterior to hot Jupiters can have stable orbits, but the question remains: does the migration of the giant planet through the terrestrial planet zone disrupt the protoplanetary disc such that there is no material left to build terrestrial planets from? The obvious answer might seem to be that the migration of the giant planet would clear the inner regions of the system of any planet-forming material, leaving nothing behind from which to form terrestrial planets. However, some recent numerical simulations of this process (particularly those of Richard Nelson and Martin Fogg) suggest that this clearing out of the inner protoplanetary disc does not, in fact, happen. These simulations show that the migrating giant planet compresses the disc inwards, compacting it and causing objects in the disc to enter first-order mean motion resonances with the giant planet. Subsequent close encounters between many of these objects and the giant planet then expel them into an exterior orbit. The overall effect of the giant planet's migration is therefore not to destroy and disperse the inner disc, but to cause a new disc of solid material to build up exterior to the final orbit of the giant planet. The formation of terrestrial planets can then continue in this region, and indeed the planets formed here may be abundant in water as a result of the inward mixing of material that has been brought with the giant planet from beyond the snowline. Exterior terrestrial planets may possibly be quite common in systems containing hot Jupiters.

In addition to the expulsion of material into exterior orbits, some of the remaining compressed inner disc material remains interior to the migrating giant planet and can subsequently form terrestrial-mass or Neptune-mass objects very close to the star. Numerical simulations of the migration process show that about half of the disc material is scattered outwards where it can form cooler terrestrial planets, and about half is compressed inwards where it may form further hot planets.

So, even though many of the known exoplanetary systems contain hot Jupiters, such systems may well have further terrestrial-mass exoplanets lying in exterior orbits, possibly within the star's habitable zone. Furthermore, if these putative terrestrial exoplanets are in resonant orbits with a transiting hot Jupiter, it is entirely possible (perhaps even highly likely) that they may be discovered as a result of the periodic TTVs that they induce in the orbit of the hot Jupiter.

Summary of Chapter 7

1. The orbits of exoplanetary systems may be described using Newton's law of gravity. In the case of a two-body system, the orbits may be solved exactly:

$$r = \frac{a(1-e^2)}{1+e\cos\theta}. \qquad \text{(Eqn 7.3)}$$

   Where there are more than two massive bodies, approximate solutions may be found in some cases. Alternatively, numerical modelling techniques may be used to find the orbital solutions.

2. Tidal forces arise when the size of an orbiting body is not negligible compared to the orbital separation; tidal forces tend to elongate the orbiting body. The Roche limit

$$d_R = R_M \left(\frac{2M_P}{M_M}\right)^{1/3} \qquad \text{(Eqn 7.16)}$$

   is the minimum orbital separation for a moon of mass M_M and radius R_M orbiting around a planet of mass M_P. If the moon's orbit were smaller than d_R, the moon would be disrupted by tidal forces.

3. Tidal interactions cause dissipation. This will cause heating, and will tend to reduce the size of a planetary orbit and circularize it. For short-period systems, the circularization timescale is much less than the lifetime of the star. The circularization timescale is

$$\tau_{circ} = \frac{2}{21}\frac{Q_P}{k_{dP}}\left(\frac{a^3}{GM_*}\right)^{1/2}\frac{M_P}{M_*}\left(\frac{a}{R_P}\right)^5, \qquad \text{(Eqn 7.22)}$$

   where k_d is the dynamical Love number, and Q is a tidal quality factor. Q_P/k_{dP} characterizes the effect of tidal interactions and typically takes values from a few $\times\ 10^4$ to a few $\times\ 10^5$ for a planet. For the shortest-period transiting exoplanets, tidal interactions are predicted to cause changes in orbital period that would be measurable over a decade or so.

4. The mean eccentricity of exoplanetary orbits increases with semi-major axis (or period), but there are some eccentric orbit, short-period planets that should have circularized unless they have been perturbed by interactions with other planets or companion stars.

5. The transit probability of a planet increases with eccentricity as

$$p = \frac{R_* + R_P}{a(1-e^2)}. \qquad \text{(Eqn 7.23)}$$

   However, since planets move faster near pericentre, the typical duration of transits is reduced for eccentric orbits. If systematic errors dominate, the net effect is a selection effect in favour of finding eccentric transiting exoplanets.

6. The habitable zone of a star is the region within which liquid water might exist on the surface of a terrestrial exoplanet. GJ 581 d lies within the habitable zone. Many known exoplanetary systems could harbour additional planets on stable orbits within the habitable zone, as long as the migration of a giant planet through the habitable zone has not prevented the formation of a planet there.

7. Several of the known exoplanetary systems contain multiple exoplanets, and a few of these have exoplanets in resonant orbits with each other. Of the known multi-planet systems containing transiting exoplanets, only the inner planet produces transits, and the outer planet has been found by radial velocity measurements, not transit timing variations (2009).
8. The only stable systems are those that may be described as hierarchical multiple systems. These may be conveniently represented using mobile diagrams. Many exoplanets exist in binary star systems, and a few exist in triple star systems. In each case the known exoplanets orbit a single star within the system, constituting an inner binary that is itself in orbit with a more distant single star or close binary star.
9. A second planet in a system with a transiting exoplanet will perturb the orbit of the first planet and so give rise to transit timing variations (TTVs). These variations will in general be small (of order a few seconds) and are unlikely to be detected. The maximum offset in the transit time of an outer planet (2) due to planet 1 on an interior orbit is
$$\delta t_2 \approx \frac{P_2 a_1 \mu_1 \sin[2\pi(t - T_c)/P_1]}{2\pi a_2}, \qquad \text{(Eqn 7.26)}$$
where both orbits are assumed circular, and we are using μ to indicate the reduced mass. A perturbing planet (2) exterior to the orbit of a transiting exoplanet (1) will cause observable TTVs only if the outer planet's orbit is eccentric. In this case the maximum TTV will be
$$\delta t_1 \approx \mu_2 e_2 \left(\frac{a_1}{a_2}\right)^3 P_2. \qquad \text{(Eqn 7.29)}$$
10. For planets in first-order mean motion resonant orbits with periods in the ratio $j + 1 : j$, the TTVs may be several minutes for an exterior perturbing planet of terrestrial mass. Thus terrestrial planets may be detected via the transit timing of Jupiter-mass exoplanets. The TTV is reasonably well fitted by the approximation
$$\delta t_2 \approx \frac{P_2}{4.5j} \frac{M_1}{M_1 + M_2}, \qquad \text{(Eqn 7.30)}$$
where M_1 is the mass of the perturbing planet and M_2 is the mass of the transiting planet. The libration period for the TTV is
$$P_{\text{lib}} \approx 0.5 j^{-4/3} \mu_2^{-2/3} P_2, \qquad \text{(Eqn 7.31)}$$
where μ_2 is the mass of the transiting planet divided by the total mass of the system.
11. Exomoons may exist and would cause TTVs and transit duration variations. The maximum amplitude of the TTV is
$$\delta t_M \approx \frac{a_M M_M}{2^{1/2} M_P} \frac{P}{2\pi a}, \qquad \text{(Eqn 7.32)}$$
and the maximum variation in the transit duration is
$$\delta T_M \approx \left(\frac{a}{a_M}\right)^{1/2} \frac{M_M}{[M_P(M_P + M_*)]^{1/2}} \frac{T_{\text{dur}}}{2^{1/2}}, \qquad \text{(Eqn 7.33)}$$
where T_{dur} is the unperturbed duration of the transit.

12. Exomoons can exist only in stable orbits that are inside an exoplanet's Hill sphere, of radius

$$R_{\rm H} = a \left(\frac{M_{\rm P}}{3M_*} \right)^{1/3}, \qquad \text{(Eqn 7.24)}$$

and outside its Roche limit. Giant planets in a star's habitable zone may have terrestrial-mass exomoons in stable orbits. Such exomoons would produce transit variation signatures of order 1 minute.

13. Exotrojans may exist at an exoplanet's L4 and L5 points. Terrestrial-mass exotrojans co-orbiting with Jupiter-mass exoplanets could also give rise to TTVs with amplitudes of order 1 minute. The TTV due to an exotrojan is

$$\delta t \approx \mu_{\rm T} P \frac{\Delta\psi(t_i)}{2\pi}, \qquad \text{(Eqn 7.34)}$$

where $\mu_{\rm T}$ is the mass of the Trojan divided by the total mass of the system, and $\Delta\psi(t_i)$ is the angular displacement (in radians) of the exotrojan from the L4 or L5 point at the time of the ith transit. The typical libration timescale of the exotrojans is estimated as

$$P_{\rm lib} \approx \left(\frac{4}{27\mu_{\rm P}} \right)^{1/2} P, \qquad \text{(Eqn 7.35)}$$

where $\mu_{\rm P}$ is the mass of the planet divided by the total mass of the system.

14. Gas giant planets form beyond the snowline. Hot Jupiters have subsequently migrated into the inner parts of the system. It is possible that terrestrial exoplanets could form exterior to giant exoplanets, after they have migrated inwards.

Chapter 8 Brave new worlds

Introduction

This chapter is dedicated to exploring the prospects for future discoveries in the area of exoplanets, focusing on the possibilities for finding habitable planets outside our own Solar System. It is intentionally short, not because there is a shortage of material to cover, but rather because there are so many possibilities. Research in this area is progressing at an astounding pace, thus predictions about the future are likely to rapidly become stale. We begin by discussing the Kepler mission, then look at two complementary strategies for identifying transiting terrestrial planets. Then we move on to discuss the characterization of transiting terrestrial planets, focusing on the anticipated James Webb Space Telescope's contribution. The second half of the chapter focuses on habitability and gives a very brief summary of the prospects for discovering the signatures of extraterrestrial life. We conclude with some remarks about the importance of the transiting exoplanets for planetary astrophysics, and finally place the subject in a much broader context.

8.1 Future searches for transiting planets

8.1.1 Kepler: the future begins now!

With apologies to Buckaroo Banzai and Yoyodyne.

The Kepler mission, mentioned in Chapter 6, was designed to detect transiting Earth-sized planets. An example of the superb signal-to-noise ratio data that it produces was shown in Figure 6.6. In addition to a photometric precision of better than 0.01%, the detection of transits by an Earth-like planet in the habitable zone of a star like the Sun requires a long time-base of observations. Kepler will stare at the same field, making measurements of the same 156 000 stars for over three and a half years. Of these stars, about 5000 are similar to the Sun and bright enough for Kepler to reach the precision required to detect a transiting Earth.

Exercise 8.1 If all 5000 suitable stars in the Kepler field are analogues of the Sun, and each one of them hosts a single planet with radius $R_\oplus$ orbiting at 1 AU from the host star, how many transiting exo-Earths will Kepler discover? Comment on how many exo-Earths you expect Kepler to find if you do not assume that each star hosts a single planet with radius $R_\oplus$ orbiting at 1 AU. ■

Kepler seems bound to make discoveries of Earth-sized exoplanets, and may find super-Earths with $M_P > 6\,M_\oplus$ within the habitable zones of their host stars, if they are common enough. By the time you read this, these discoveries may have been announced.

The drawback of Kepler's strategy is that the field of view is small, at least in comparison with the wide-field surveys discussed in Chapter 2. This means that only a small fraction of the brightest stars in the sky will be observed, so on average the stars discovered to host transiting planets will be fairly distant, and thus appear relatively dim. Kepler has a less than 2% chance of discovering a transiting planet within 20 parsecs. The apparent brightness of the host star (and the planet itself) crucially determines the quality of the data that it is possible to gather on the system; as we saw in Chapters 5 and 6, the brightest known exoplanet host stars dominate the follow-up studies of planet atmospheres. The majority of the transiting exoplanets found by the wide-field surveys such as WASP and HAT have host stars with $V \lesssim 12$, and only the very brightest of the Kepler host stars will match this. Studies of these relatively distant (and therefore faint) Kepler planets will obviously fail to reveal the level of detail possible for those with host stars with $V \sim 8$.

8.1.2 Wide-field surveys in space

To find the brightest transiting Earth analogues, ESA is considering a mission called PLATO: a space observatory that will use an array of small telescopes and CCD detectors to extend the wide-field survey concept into space. PLATO will attain the precision required to detect terrestrial planet transits around main sequence stars with masses exceeding the Sun's and will detect these transits around some of the brightest stars in the sky. If Earth-like planets within the habitable zones of their host stars are common, PLATO should identify some nearby, bright examples. If PLATO is selected by ESA, it will be launched no earlier than 2017.

As we learned in Chapter 3, once a high-quality transit light curve has been obtained, the dominant source of uncertainty in the determination of exoplanet parameters is the uncertainty in the stellar parameters. The PLATO mission aims to address this by improving the characterization of the host stars via **asteroseismology**. Stars, including the Sun, oscillate due to the presence of low-amplitude sound waves. The frequencies of the standing waves depend on the interior structure of the star and are revealed by subtle photometric variability, with amplitudes of a few parts per million. Asteroseismology is the process of observing this variability and using it to infer the stellar structure.

PLATO must conduct very long observations in a stable environment, with as few interruptions as possible. This means that low Earth orbits are unsuitable. PLATO, if launched, will be placed in a large Lissajous orbit around the L2 point, as shown in Figure 8.1. This orbit avoids the high radiation levels and high scattered light contamination in low Earth orbits, and produces good radio contact with ground stations, so PLATO can send data regularly and frequently.

A design concept similar to PLATO was considered by NASA in 2009. The Transiting Exoplanet Survey Satellite (TESS) failed, however, to be selected as a NASA small explorer mission. A third proposed mission, LEAVITT, would scan 70% of the sky every week, performing multicolour photometry that would allow astrophysical mimics to be identified. LEAVITT, if approved, would be a NASA medium-class explorer.

Irrespective of whether any of these particular missions fly, it is very likely that within the next decade or two, a mission of this type will be launched. On this timescale, therefore, we are likely to have identified examples of transiting exo-Earths orbiting within the habitable zones of relatively nearby bright stars. Figure 8.2 quantifies this likelihood. It shows that PLATO should find several dozen habitable transiting terrestrial planets with $M_{\rm P} < 3\,{\rm M}_\oplus$, and about 20 of these around stars brighter than $V = 11$. In contrast, Kepler is not expected to

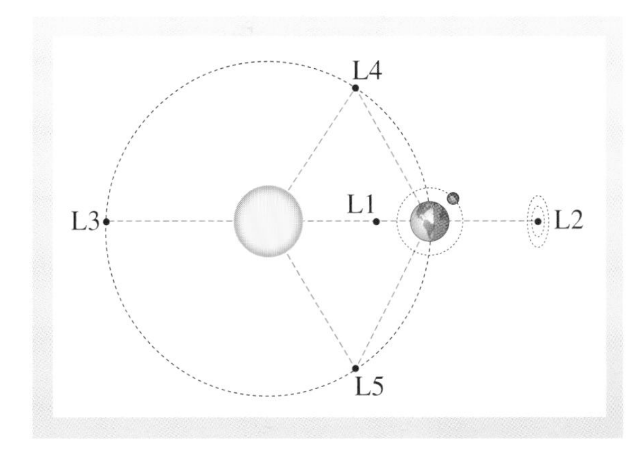

Figure 8.1 The planned orbits of PLATO and the James Webb Space Telescope (JWST) around Earth's L2 point. PLATO has a larger orbit around L2, which minimizes eclipses of Earth-bound ground stations by the Moon. JWST has a smaller orbit, which allows the light shield to shade the telescope from the Earth, Moon and Sun. The diagram is schematic: objects are not drawn to scale.

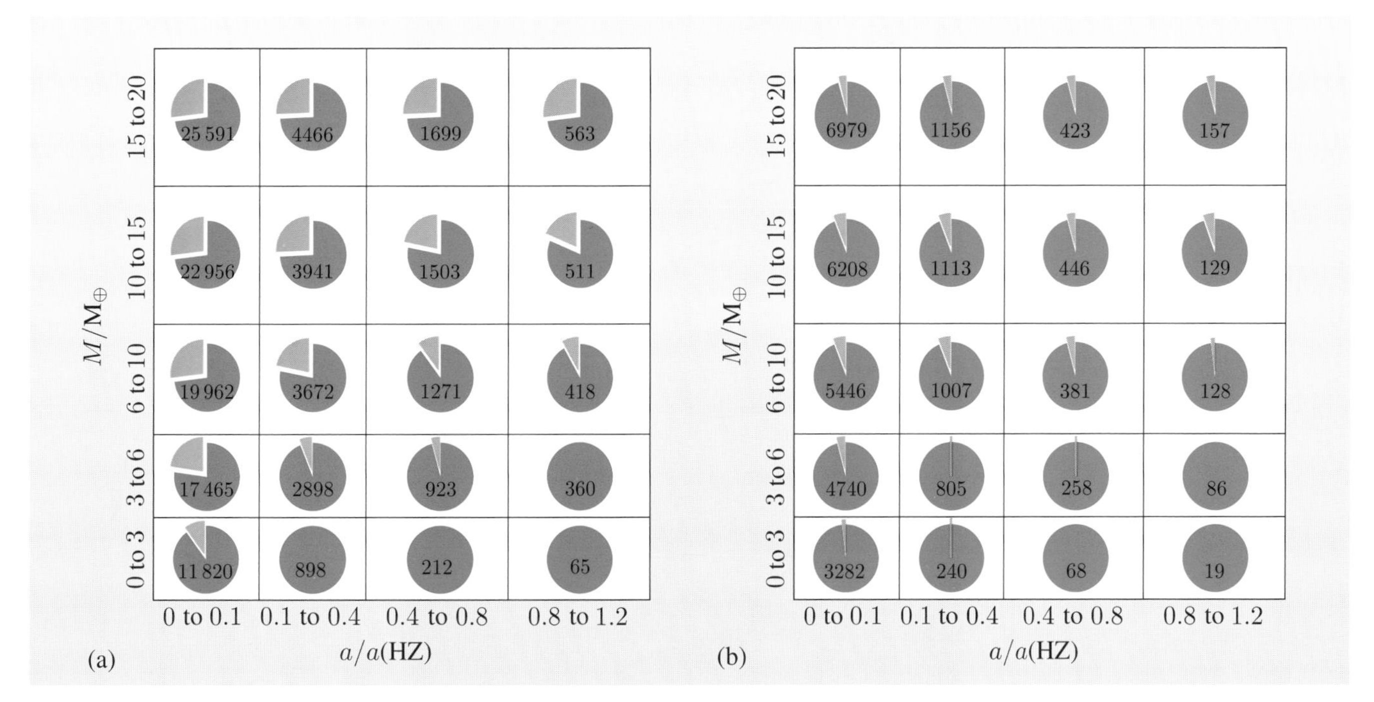

Figure 8.2 Probabilities of finding transiting terrestrial planets in the habitable zone for PLATO and Kepler. (a) Estimated numbers of detected transiting terrestrial planets as a function of mass and orbital separation normalized to the location of the habitable zone. The numbers within the blue discs give the estimations for PLATO, while the orange sectors indicate the relative numbers detected by Kepler. Kepler is not expected to find any planets less massive than $6\,M_\oplus$ within the habitable zone. (b) As (a) but for stars brighter than $V = 11$.

find any transiting planets less massive than $6\,M_\oplus$ within the habitable zone. The calculations behind Figure 8.2 assume that each star has only a single, randomly oriented planet within the mass-separation parameter space shown, and take account of all the necessary steps required to confirm the transit candidates. The discovery of a habitable exoplanet would arguably be one of the most exciting results in the history of science. George Ricker, leader of the TESS project, said:

> Decades, or even centuries after the TESS survey is completed, the new planetary systems it discovers will continue to be studied because they are both nearby and bright. In fact, when starships transporting colonists first depart the solar system, they may well be headed toward a TESS-discovered planet as their new home.

8.1.3 MEarth: transiting exoplanets around nearby M dwarf stars

MEarth is pronounced 'mirth'.

In contrast to the wide-field survey strategy, the MEarth survey targets individual nearby M dwarf stars using 0.4 m robotic telescopes. Being Earth-bound, these telescopes cannot match the photometric precision that is possible from space, but the small size of M dwarf stars means that a transiting planet of radius $2\,R_\oplus$ would produce a 0.5% deep transit. Figure 8.3 shows one of the MEarth telescopes.

Figure 8.3 One of the 0.4 m MEarth telescopes.

The first MEarth discovery was announced in the paper 'A super-Earth transiting a nearby low-mass star' by David Charbonneau et al. (2009, *Nature*, **462**, 891). This planet transits GJ 1214, a star of spectral class M4.5V, which has a distance of 13 parsecs and no previously discovered planets.

- What is the name of this first nearby transiting super-Earth?
- Since the star is GJ 1214 and there were no previously known planets in the system, this planet is named GJ 1214 b (see Chapter 1).

The mass of GJ 1214 b is $6.55\,M_\oplus$, and its radius is $2.68\,R_\oplus$. Its host star is quite faint in the V band, $V = 14.67$, despite its proximity, because low-mass M dwarf stars are intrinsically dim. The main sequence's steeply falling optical luminosity as stellar mass decreases is the reason why none of the host stars of transiting planets identified by wide-field surveys have spectral types as late as GJ 1214. GJ 1214 is by far the lowest-mass known transiting planet host star; the transit host with second lowest mass, GJ 436, is three times as massive (March 2010). As we saw in Chapter 1, most nearby stars are isolated M dwarfs. This is because the **initial mass function**, which describes how the number of stars formed in a volume of space varies with mass, is a steeply declining function of mass. As we discussed in Chapter 4, we do not yet know how the planeticity of stars varies as a function of stellar mass, nor do we know how the distribution of these planets in orbital period and planet mass varies with stellar mass. These are things that we should discover in the next decade (or two). The answers to these questions will determine whether George Ricker's first space colonists are most likely to embark on a journey to an MEarth-discovered planet or to a planet discovered by a wide-field survey in space.

Exercise 8.2 (a) Calculate the density of GJ 1214 b both in SI units and in terms of $\rho_\oplus$, and write a sentence comparing its density to that of the Earth.

(b) Calculate the acceleration due to gravity on the surface of GJ 1214 b, and compare it to that on the surface of the Earth.

(c) Comment on the suitability of GJ 1214 b for human occupation in the light of your answer in part (b). For example, would it appear to be possible to play golf on the surface of this planet? ■

During the Apollo 14 Moon landing, Alan Shepard fitted a smuggled golf club head to the handle of a lunar sample collection device and launched some golf balls. They are still there!

GJ 1214 b has a significantly lower density than Earth. In fact, it has a lower density than expected for a planet composed purely of water. Its discoverers suggest that it may be composed primarily of water enshrouded by an outer envelope of hydrogen and helium. Despite the rather encouraging value of surface gravity for GJ 1214 b, it doesn't look like a good home for an extraterrestrial golf course.

8.2 Characterization of terrestrial transiting exoplanets

One reason for wanting to identify the closest terrestrial transiting planets is because these will give us the best chance to empirically address the question: 'Is the Earth unique?' Don Pollacco, who leads the PLATO science consortium, said:

> PLATO surveys will be sensitive to Earth-sized planets in the *habitable zones* of late type stars ranging from F to M type. These will be prized targets for atmospheric analysis and will be the best targets for biomarker searches.

8.2.1 Spectroscopy with the James Webb Space Telescope

The facility that will lead in the characterization of these exo-Earths is the James Webb Space Telescope (JWST). As we saw in Chapters 5 and 6, the Hubble Space Telescope (HST) and the Spitzer Space Telescope (SST) have provided many of the most spectacular studies of the hot Jupiter exoplanets. These two observatories are nearing the ends of their missions, and NASA will follow them with the ambitious JWST, a space telescope with a $25\,m^2$ collecting area, which is 50 times the collecting area of SST. JWST is scheduled for launch in 2014, and will orbit around Earth's L2 point, as shown in Figure 8.1. JWST is optimized for the infrared, and will make observations in the wavelength interval 0.6–28 μm. As warm objects emit in the infrared, JWST must be kept cool, which means that it must be shielded from the radiation from the Sun, Earth and Moon that would otherwise heat the telescope and instruments. JWST has a light shield to block this radiation, but it also, obviously, blocks access to the sky for the purposes of observing. To minimize the angle subtended by the light shield, the telescope needs the Sun, Earth and Moon to be all in the same direction. The best place to accomplish this is the Earth's L2 point.

As we have seen throughout this book, the characterization of hot Jupiter exoplanets generally demands the extraction of a small differential signal from observations of bright stars. The characterization of terrestrial exoplanets will be even more demanding. JWST's orbit around L2 should allow it to make these high dynamic range observations, free from the systematic effects caused, for example, by HST's regular transitions from sunlight to Earth-shadow. JWST will be able to perform high-precision transmission and emergent spectroscopy, and will produce high-precision light curves. This should allow the techniques discussed in Chapters 5 and 6 to be applied successfully to transiting terrestrial exoplanets.

While we obviously do not know what we will find when we observe planets that we have not yet discovered, we can make educated guesses. Figure 8.4 shows the Earth's transmission and reflection spectra as determined from analysis of Earthshine reflected from the Moon. These data indicate the spectra that we might expect from Earth-twin exoplanets. As we saw in Figure 1.8, the contrast ratio between the Earth and the Sun is more favourable at longer wavelengths than shown in Figure 8.4, and JWST can take advantage of this.

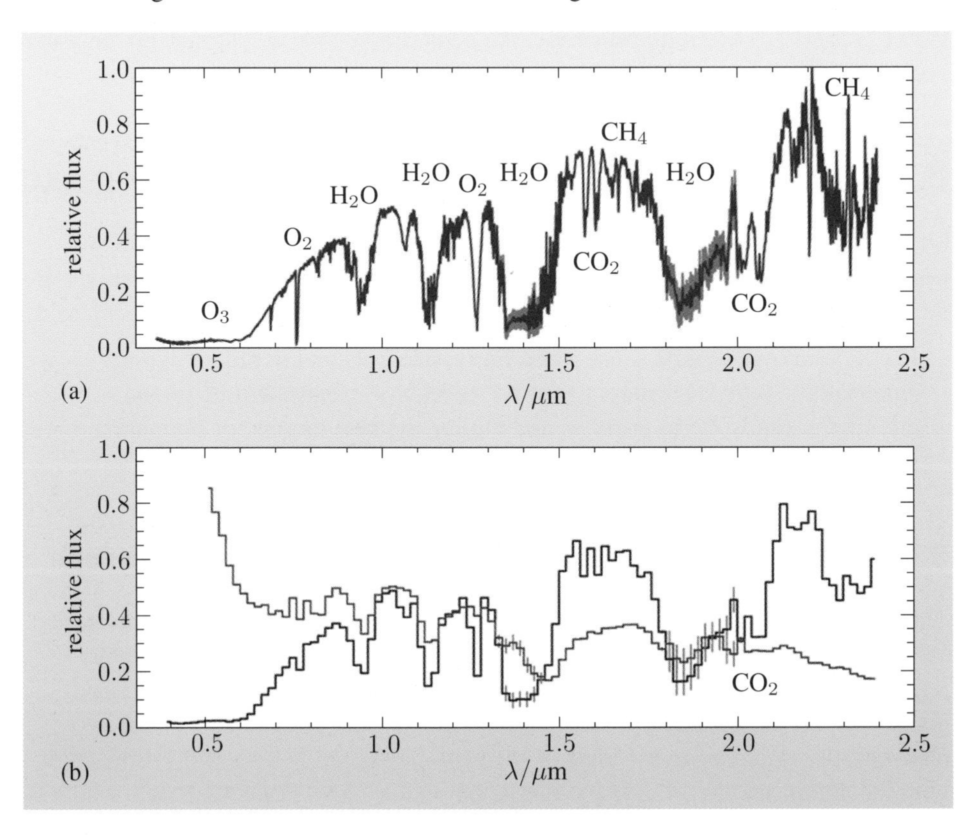

Figure 8.4 Spectra of the Earth's atmosphere derived from analysis of Earthshine reflected from the Moon. (a) The Earth's transmission spectrum is dominated by Rayleigh scattering, causing the strong rise in the continuum level towards longer wavelengths. There are strong absorption features caused by molecules in the Earth's atmosphere. (b) The black histogram shows the Earth's transmission spectrum degraded to a resolution of $0.02\,\mu$m. The blue line indicates the Earth's reflection spectrum. The steep increase in the latter towards the shortest wavelengths is due to Rayleigh reflectance.

Figures 8.5 and 8.6 respectively show model transmission spectra and emergent spectra for the newly-discovered transiting super-Earth GJ 1214 b (see Subsection 8.1.3).

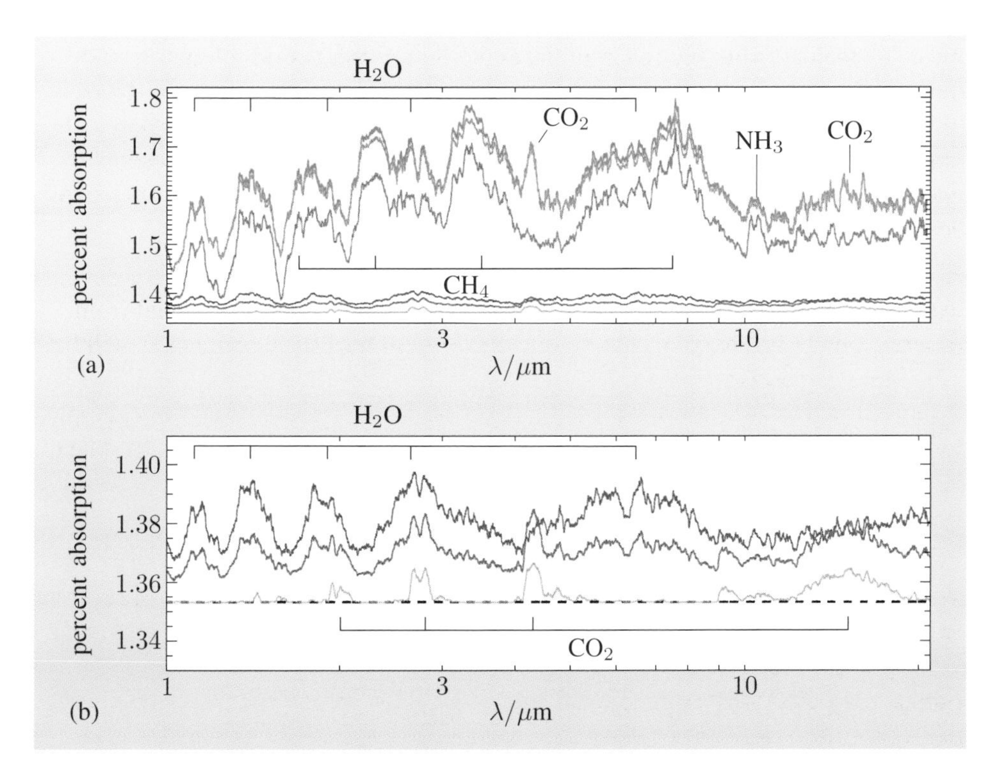

Figure 8.5 Model transmission spectra for the newly-discovered transiting super-Earth GJ 1214 b, assuming a variety of atmospheric compositions.

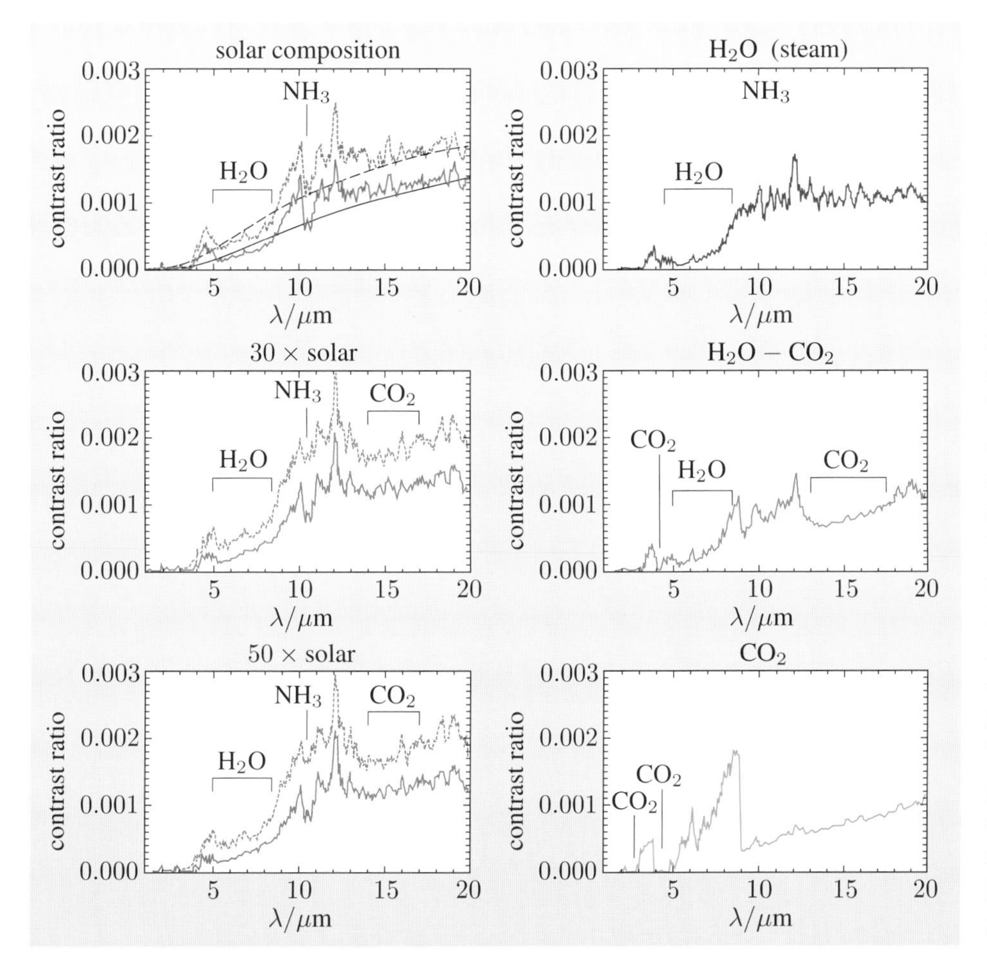

Figure 8.6 Model emergent spectra for the newly-discovered transiting super-Earth GJ 1214 b, assuming the same set of possible atmospheric compositions as in Figure 8.5. The vertical axes give the contrast ratio, i.e. planet emergent flux as a fraction of the stellar flux, while the horizontal axes give wavelength in microns. The solid coloured lines are models with efficient heat redistribution, while the dashed lines are models where heat redistribution is inefficient. The black lines in the first panel are the contrast ratios corresponding to both the star and the planet emitting as black bodies with the planet's T_{day} at 555 K and 660 K.

These models are consistent with the (limited) data available on this planet, and indicate the sort of features that JWST might detect. The next step in anticipating the results that JWST might produce is to pass the model spectrum through a simulator that takes into account the distance of the observing target, the overall quantum efficiency of the telescope and instrument, and any other sources of noise. This is, in essence, a sophisticated version of the signal-to-noise calculations that we discussed in Chapter 2, performed individually for each observed wavelength. Such simulations yield an expected observed spectrum, along with the anticipated error spectrum. An example of this is shown in Figure 8.7, where the blue line shows the model transmission spectrum, and the black histogram indicates the spectrum that might be expected from JWST after 28 hours of observing time.

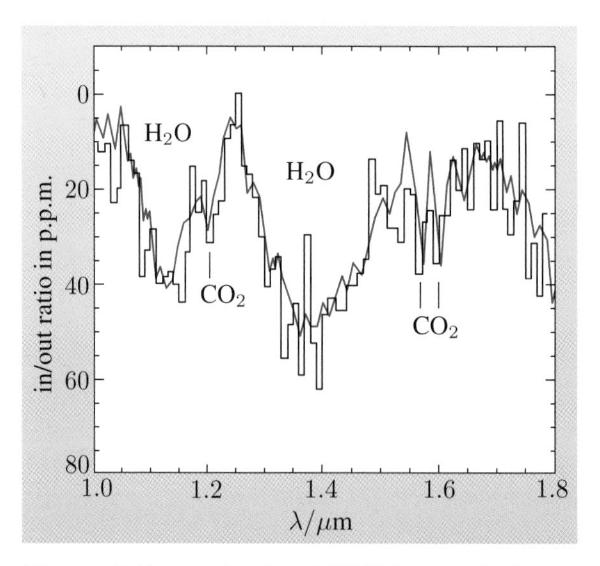

Figure 8.7 A simulated JWST transmission spectrum of an ocean planet orbiting around an M3V star.

It is worth pointing out that the example in Figure 8.7 is in many ways at the most favourable end of expectations: it assumes an ocean planet with a large atmospheric scale-height; the strength of transmission spectroscopy features is highly dependent on the atmospheric scale-height. It assumes an M3V star; for larger main sequence stars, the transit depths will be significantly smaller, rendering the transmission spectroscopy signals more difficult to detect. It assumes that the host star's infrared apparent brightness is $J = 8$, almost two magnitudes brighter than GJ 1214.

8.2.2 Confirmation of terrestrial transiting planet candidates

Before the extremely scarce resource of JWST observing time is invested in spectroscopy of an exoplanet, it must be confirmed as a bona fide planet. For terrestrial planets, detection of the stellar reflex radial velocity signal is a challenge, especially if they are in the habitable zone of a star like the Sun. As we saw in the solution to Exercise 5.5, the Rossiter–McLaughlin effect for a transiting Earth–Sun system is perhaps the strongest signal. It seems likely that the confirmation of any terrestrial transiting planets identified by wide-field surveys in space will rely on their Rossiter–McLaughlin signals.

Exercise 8.3 In this exercise we return to the viewpoint of the hypothetical extraterrestrial astronomers whom we considered in Chapter 1.

(a) Imagine the hypothetical case of the Sun having only a single planet, Earth. Calculate the ratio of the Sun's orbital reflex velocity, A_{RV}, to the amplitude of the Rossiter–McLaughlin effect, A_S. You may assume that the Sun's spin period is 25 days, that the inclination of the Sun's spin axis to the normal to the ecliptic is $7.15°$, and that the Earth is observed to transit with an impact parameter of 0.

(b) Now imagine a slightly different hypothetical case of the Sun having its entire complement of 8 planets, but with orbits oriented so that if viewed from a position in the ecliptic plane, only the Earth transits the Sun. Discuss the relative merits of the reflex radial velocity curve and the Rossiter–McLaughlin effect for confirming the presence of the Earth and measuring the Earth's characteristics. ■

One of the important steps in confirming the hot Jupiter transiting planets is to measure their mass. Without a measurement of the planet mass, it is impossible to know whether the object transiting the host star is a gas giant planet or giant planet-sized star: both brown dwarfs and white dwarfs can have radii similar to those of giant planets. Only the most massive white dwarfs have comparable size to terrestrial planets, so even though the reflex radial velocity signal of an exo-Earth may be difficult to detect, it should be straightforward to rule out the presence of a white dwarf.

- ● What is the approximate mass of the most massive white dwarf?
- ❍ The most massive white dwarfs have masses approaching the Chandrasekhar mass, $\sim 1.4\,M_{\odot}$.

The presence of a white dwarf with $R \sim R_{\oplus}$ could easily be ruled out with a couple of radial velocity measurements at precisions similar to those used to confirm hot Jupiter transiting planets.

8.3 Life in the Universe

The previous section established that it is very likely that the next decade or two will bring discoveries of transiting terrestrial planets in the habitable zones of their stars. This leads to the question of enormous significance: *Is there life elsewhere in the Universe?* Entire books have been and are being dedicated to the discussion of this topic, so we will limit the discussion here. The study of the Universe's potential for life is the relatively new science of **astrobiology**. One definition of it is as follows:

> Astrobiology seeks to understand the origin of the building blocks of life, how these biogenic compounds combine to create life, how life affects — and is affected by the environment from which it arose, and finally, whether and how life expands beyond its planet of origin.
>
> (From the website www.astrobiology.com, March 2010)

The discovery of the first exoplanets around main sequence stars made it abundantly clear that it is dangerously limiting to generalize from an example of one. Before the discovery of 51 Peg b, it was assumed that all planetary systems were arranged like our own Solar System. The existence of hot Jupiter exoplanets came as a complete surprise. In considering the question of life elsewhere in the Universe, it is probably impossible to fully free ourselves of the preconceptions that we carry as a result of our own perspective. Life on Earth is widespread, ubiquitous and of dazzling variety. Life has adapted to environments with almost the entire range of physical conditions present on the Earth. This does tend to suggest that, once established, life is reasonably robust and adaptable.

One of the immediate difficulties in addressing the questions of astrobiology is the definition of life itself. Any definition adopted must be general enough to avoid excluding examples of life that may be fundamentally different from Earthly life. We may assume, as we do in astronomy, that the laws of physics remain the same in different locations throughout the Universe. This implies that any alien life that we might encounter is likely to be composed of substances built from the elements of the periodic table and constrained by the chemistry of these elements and more complex entities containing them. Science deals with the specific, rather than with vague and ill-defined questions; one productive line of research in astrobiology is to examine Earthly life in the context of the physical environments elsewhere in the Galaxy. Earthly life is predicated on DNA and RNA, and it is useful to examine the effects of extreme conditions on these molecules: in these studies, DNA and RNA can be regarded as proxies for other complex molecules that might play important roles in alien life. Similarly, Earthly life uses photosynthesis to harness the energy of the Sun; while the biochemical pathways employed by alien life may differ, the energy needs of alien organisms are likely to be of a similar order of magnitude to those of Earthly life.

The probability of life existing elsewhere in the Universe is linked to the time that it took for life to arise on the Earth. If life arising is a probable event, then there may be many examples of alien life, and life would have arisen on Earth quickly. Conversely, if the origin of life required the coincidental occurrence of several unlikely events, then it may have occurred only here on Earth. In this case we would expect a long time to elapse before this unlikely happening. In fact, geological evidence suggests that life arose on Earth rather quickly (in geological terms) after the formation of the Earth's crust and very promptly after the **late heavy bombardment** phase in the Solar System's history, during which the Earth and other surviving bodies suffered frequent impacts from meteorites. There is evidence that photosynthesis was prevalent on Earth 3.5 Gyr ago, while the late heavy bombardment finished about 3.8 Gyr ago, i.e. only 300 Myr earlier. Thus life has existed on Earth for almost its entire history, as indicated schematically in Figure 8.8. This suggests that either the Earth is a very special place, or life may be commonplace.

The complex multicellular organisms that spring to mind when we visualize 'life' are all eukaryotes. As Figure 8.8 indicates, these did not arise until relatively late in the history of Earth. The first multicellular organisms do not appear in the fossil record until 580 Myr ago. This suggests that perhaps, while simple single-celled life arises readily, more 'advanced' life forms are less likely to arise. Figure 8.9 shows the tree of Earthly life, with the degree of genetic difference indicated by

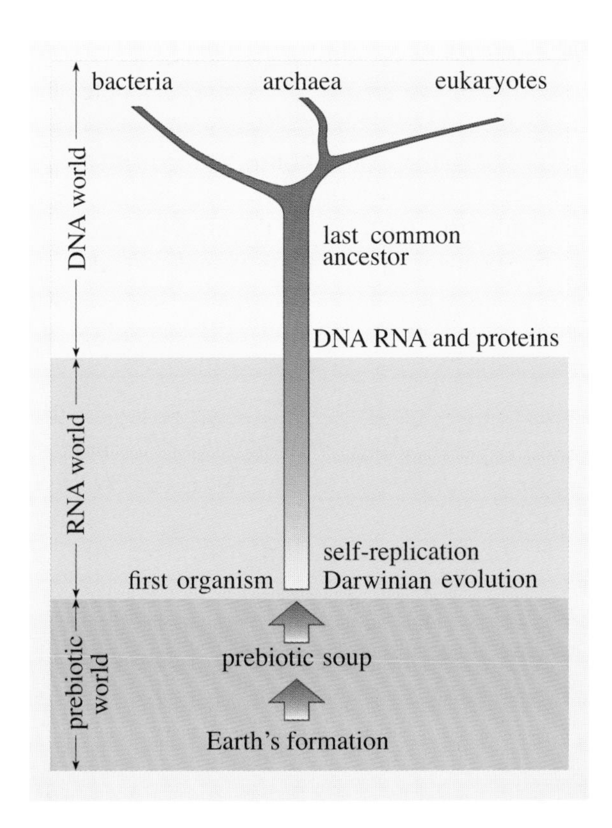

Figure 8.8 A schematic diagram indicating the history of life on Earth. The archaea are single-celled organisms whose structure resembles that of bacteria, but which are more closely genetically related to eukaryotes. Eukaryotes are organisms whose cells have more complex structures within the cell membrane; all animals, fungi and plants are eukaryotes.

the length of lines separating the different branches. The root of the tree is surrounded by heat-loving organisms, which suggests that the earliest Earthly forms of life were thermophilic. Examples of such **thermophiles** are still living on Earth today: bacteria are found, for example, living in deep-sea hydrothermal vents known as **black smokers** (Figure 8.10), and on the surface of burning coals near hot springs in Yellowstone Park.

Not only did multicellular organisms arrive late, they are arguably minor players in the present-day tableaux of life on Earth. Bacteria exist in every imaginable Earthly niche, and are numerically by far the most common life forms on Earth today. There are more bacteria in a single human's intestine than there are human beings on Earth! While it is difficult to be certain, it also seems likely that bacteria constitute a larger biomass on present-day Earth than archaea and eukaryotes combined. This seems intuitively unlikely when large trees in temperate regions are such a common feature of everyday life, but bacteria are common in soil and in the oceans, and can even live within rocks. We are unaware of bacteria simply because they are microscopic.

While recognizing the dangers of generalizing from our experiences on Earth, we can speculate based on the information above. It seems likely that alien life, if it exists, will be akin to bacteria. The complex multicellular organisms that ultimately evolved into human beings arrived late, and may have been the result of extremely unlikely circumstances. Our presence here constitutes the ultimate selection effect: Earth could be the only planet in the Galaxy, or indeed in a much larger region of space, where events conspired to give rise to multicellular organisms. Having said that, the developments in astronomy over the last 15 years have revealed that our Universe is more like that depicted in Star Trek than the author would previously have guessed.

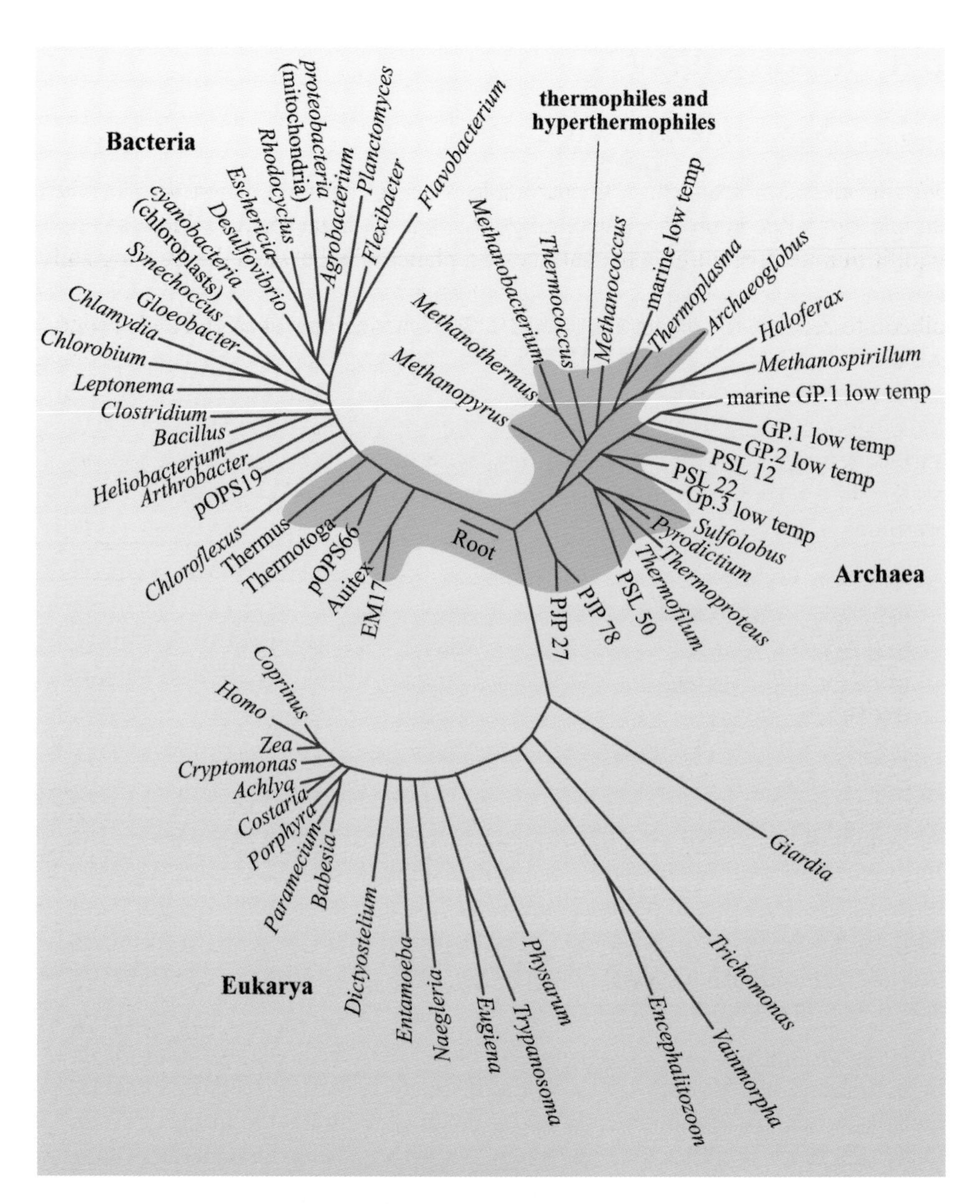

Figure 8.9 A schematic family tree for life on Earth.

Figure 8.10 A black smoker: a deep-sea hot-water vent.

8.4 Habitability

8.4.1 The habitable zone

We introduced the concept of the habitable zone in Chapter 7. There we used the simplest possible approach, calculating the range of distances for which the equilibrium temperature on the surface of a planet falls between the freezing and boiling points for water. This was clearly rather too simple: first, we set the albedo to zero, so the planet absorbed 100% of the incident starlight, and second, we did not consider the role of the planet's atmosphere in providing a **greenhouse effect**. For the Solar System's terrestrial planets, even when the correct value for the albedo is used, the equilibrium temperature is not the actual value of the surface temperature. For Mars the discrepancy is small, for the Earth it is over 30 K, and for Venus the discrepancy is over 500 K. The reason for this is the warming provided by the greenhouse effect, which is illustrated in Figure 8.11.

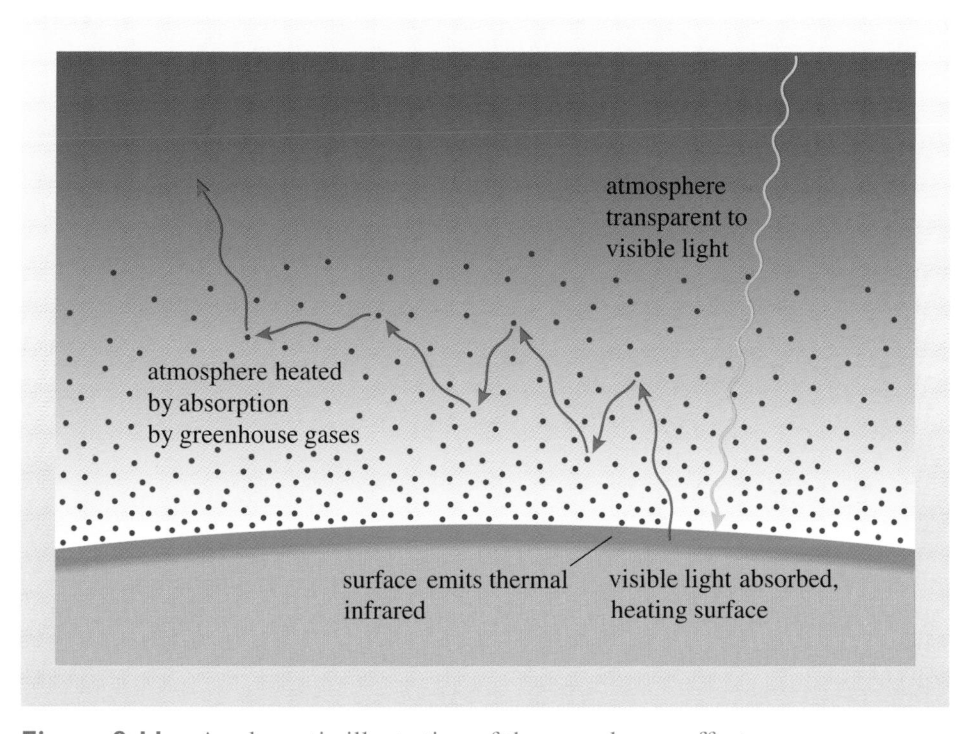

Figure 8.11 A schematic illustration of the greenhouse effect.

Sunlight is absorbed by the planet's surface, heating it so that it emits in the infrared. We assumed in our calculation of the equilibrium temperature (see the derivation of Equation 5.5) that this infrared radiation will escape efficiently. This is true only if it is not absorbed by the planet's atmosphere. Venus's atmosphere is optically thick, and the atmosphere absorbs much of the energy emitted by the planet's surface. The Earth's atmosphere also absorbs, as we see in Figure 8.4a, and this causes Earth to be warmer than its equilibrium temperature.

Much of the absorption in Earth's atmosphere is a result of CO_2 and CH_4, hence these are known as 'greenhouse gases'.

A seminal paper by Kasting, Whitmire and Reynolds, 'Habitable zones around main sequence stars' (1993, *Icarus*, **101**, 108), examined the habitable zone for Earth-like planets with atmospheres composed of carbon dioxide, water and nitrogen. Their basic premise is that liquid water is required for life. Water appears to be the best solvent for life to emerge in: it has a large dipole moment, which means that it readily forms hydrogen bonds, it can stabilize macromolecules, and it can play a role in orienting molecules.

Kasting, Whitmire and Reynolds found that the outer edge of the habitable zone is determined by the formation of CO_2 clouds, which cool the surface by increasing the albedo and changing the temperature gradient. The inner edge of the habitable zone is governed by water being broken down by **photolysis**, i.e. decomposition caused by stellar photons, and the resulting hydrogen escaping into interplanetary space. They found a habitable zone around the present Sun ranging from 0.95 AU to 1.37 AU. While on the main sequence, the Sun will brighten as the core contracts as a result of hydrogen being fused to helium. Consequently, the habitable zone moves outwards with time. Over the lifetime of the Sun to date, the **continuously habitable zone** is between 0.95 AU and 1.15 AU. Earth, of course, orbits within this region.

As we saw in Chapter 7, the habitable zone is wider for higher-mass main sequence stars. However, Kasting, Whitmire and Reynolds point out that the widths are approximately the same if distance is expressed on a logarithmic scale. A logarithmic distance scale is probably the most appropriate scale to use here since the planets in our own Solar System are spaced approximately logarithmically and because the distance at which planets form around a star is probably related to the star's mass. The width of the continuously habitable zone depends on the timespan used. Since lower-mass main sequence stars evolve more slowly, for any particular timespan they have a wider continuously habitable zone (in logarithmic space) than higher-mass main sequence stars.

8.4.2 What makes a habitable planet?

This question is comprehensively addressed in the paper of this title by Lammer et al. (2009, *Annual Review of Astronomy and Astrophysics*, **17**, 181). Here we give only a quick summary of some of the points that they discuss. The habitable zone provides only a partial description of the prerequisites for Earthly life. A planet bathed in gamma rays, for example, would not support human life. To be habitable, a planet needs to offer environments that are below some threshold flux of damaging electromagnetic radiation, and that are not bathed in energetic cosmic ray particles. On Earth the atmosphere provides protection from damaging electromagnetic radiation. The atmosphere itself is protected from erosion by the solar wind by the Earth's magnetic field. Thus an intrinsic magnetic field appears to be an important factor in making a planet habitable. This intrinsic magnetic field is generated by a **magnetic dynamo**, i.e. the motion of conductive fluids within the planet. In the case of the Earth it is the liquid iron outer core that generates the magnetic field. In turn, this magnetic dynamo may rely on the presence of large amounts of water, solidification of the inner core, and plate tectonics.

Plate tectonics also appears to be necessary in other ways for Earthly life. Plate tectonics is a complicated process that continuously recycles the Earth's crust and produces a continuous input of CO_2 into the atmosphere. Plate tectonic activity is driven by the transfer of internal heat, so can be maintained only while the planet retains sufficient heat. In their early histories, Mars, the Moon and Mercury probably had plate tectonic activity, but being smaller than Earth, they cooled more rapidly and plate tectonics ceased. Plate tectonics combined with liquid water probably provided the environment for the first Earthly life: somewhere similar to the black smoker shown in Figure 8.10. It seems rather odd that such an other-worldly place could be considered our ultimate ancestral home.

A minimum planetary size seems necessary for habitability. Low-mass planets like Mars are prone to losing their protective atmosphere, especially when they suffer impacts. They have a lower heat content and thus become too cool to maintain plate tectonic activity. In contrast, super-Earths maintain plate tectonic activity more readily than Earth-mass planets, for example having a less stringent requirement for the presence of liquid water.

The habitable zone described in Subsection 8.4.1 considers only the surface conditions. Some life forms, for example **chemolithotrophic bacteria** that obtain their energy from the oxidation of inorganic (non-carbon) compounds, thrive below the surface so long as liquid water is available.

8.4.3 Class I, II, III and IV habitable planets

Lammer et al. propose a classification of habitable planets into four habitat types. **Class I habitats** are bodies on which the stellar and geophysical conditions lead to Earth-analogue planets. In these habitats complex multicellular life may arise, though as we noted in Section 8.3, this may not happen frequently. **Class II habitats** are bodies where life may evolve, but with stellar and/or geophysical conditions that differ from those of Class I habitats. Mars and Venus are members of this class: the long-term conditions are unsuitable for the development of 'advanced' life forms. **Class III habitats** are bodies that harbour sub-surface water oceans in direct contact with a silicate-rich core. The Jovian moon Europa is probably the only Class III habitat in the Solar System. Conditions on Europa may be similar to those in which it is suggested that life arose in Earth's black smokers. Finally, **Class IV habitats** have sub-surface water that does not interact with the silicate-rich core. Saturn's moons Titan and Enceladus are possible examples of this class. It is possible that life may have emerged on or in Titan; the current conditions may be compatible with some examples of Earthly life.

8.5 Habitable worlds and biomarkers

It seems likely that habitable planets might be found soon. This is not at all the same thing as saying that we are likely to find planets inhabited by technologically advanced civilizations. Since what we observe on the Earth is subject to the ultimate selection effect, namely that we are here to observe and think about it, it is impossible to generalize from this lone example. The early origin of bacterial life on the Earth makes it relatively likely that this may have originated elsewhere. If life exists elsewhere, it is possible, though by no means guaranteed, that we may be able to detect evidence of its presence. We discuss these **biomarkers** next. A review of this topic is given by des Marais et al. in their paper 'Remote sensing of planetary properties and biosignatures on extrasolar terrestrial planets' (2002, *Astrobiology*, **2**, 153).

We know that Earth is teeming with life, while we have detected no signs of life on Mars and Venus. Figure 8.12 shows the infrared spectra of these three planets; the prominent ozone (O_3) feature between 9 μm and 10 μm distinguishes Earth's spectrum from those of Venus and Mars. This feature is an example of a biomarker: the bulk of the ozone in Earth's atmosphere is produced as a result of oxygen liberated by photosynthesis. O_2 and O_3 are probably the most reliable biosignature gases: for a Class I habitat, free O_2 or O_3 is probably a reliable indicator of life. O_3 is easier to detect at low oxygen abundance, especially in the ultraviolet, where it is a very strong absorber.

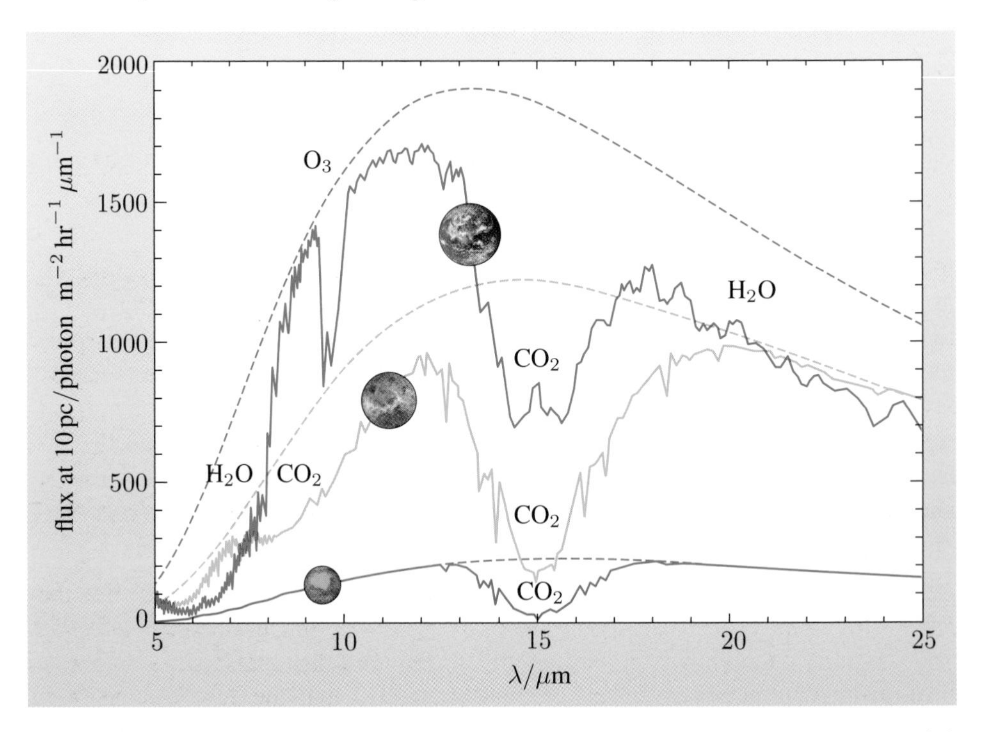

Figure 8.12 The emergent spectra of the Solar System's three largest terrestrial planets. The prominent ozone feature between 9 μm and 10 μm distinguishes Earth's spectrum from those of Venus and Mars.

Water, H_2O, is not a biomarker, but strong water features may indicate that the planet is subject to a 'moist' greenhouse effect. There is only a very limited range of conditions where this can occur and the planet remains habitable. Carbon dioxide, CO_2, is required for Earthly photosynthesis and other important biochemical processes. Its presence in the atmosphere is not, however, a clear biomarker, as Figure 8.12 makes abundantly clear. Methane, CH_4, is produced in abundance by cows, and early in the Earth's history was produced primarily by methanogenic bacteria. However, methane is also produced non-biologically by outgassing associated with volcanism.

The detection of a planet's biosphere from its spectrum is by no means straightforward. The Earth's spectrum changes significantly as the cloud cover varies, and the emergent spectrum is determined by the surface as well as the atmosphere. This leads to the idea of detecting the spectral signature of surface vegetation on an exoplanet. In particular, a **red edge** in the near infrared arises because the reflectance of chlorophyll changes rapidly with wavelength. Unfortunately, the chlorophyll red edge would be very difficult to detect in the presence of clouds: unless the surface properties are known (for an exoplanet they probably would not be), it is not possible to disentangle the signatures of clouds and the red edge. Furthermore, the red edge is probably not a universal feature.

Nonetheless, we are at the beginning of what promises to be a very interesting time. The first spectral observations of terrestrial exoplanets seem likely to reveal a diverse array of possible planet characteristics. This will allow us to begin to develop general scenarios for planet evolution, and to set our own planet in context.

8.6 Conclusions

Exoplanet astrophysics is an extremely rich, exciting, intellectually and philosophically stimulating field. In this book we focused on transiting planets, because knowing the radius of an exoplanet allows much more definite interpretation of its emission. Transiting planets also offer opportunities to apply the techniques discussed in Chapters 5 and 6, thereby directly detecting atomic, molecular and ionic species present in their atmospheres, learning about the temperature profiles in their atmospheres, and probing the alignment between the orbital and stellar angular momenta. This is a young field, and many developments are yet to come: the next big thing looks likely to be direct imaging of exoplanets. It is worth bearing in mind, however, that the whole of stellar astrophysics is underpinned by the interpretation of data on eclipsing binary stars. The transiting exoplanets are likely to continue to play the analogous role in planetary astrophysics.

We cannot hope to match the prose at the conclusion of the novel that inspired the title of this chapter, but we can provide a much more optimistic sentiment. This book has aimed to give you an introduction to the remarkable discoveries that we have made about planetary systems around stars other than the Sun. It is hugely inspiring that human ingenuity has made these discoveries possible a mere two centuries after the industrial revolution. The power of rational thought to predict, analyze and understand our Universe is prodigious. The exploration of the new worlds that we have found promises to be a huge adventure, which perhaps may ultimately include exploration in the old-fashioned sense as well as the scientific sense. If any of this is to happen, however, we need to use our collective ingenuity to understand and repair the effects that our industrial activity and our burgeoning population are having on our own planet. Pure science activities like the study of exoplanets may appear to be a distraction from this imperative goal, but perhaps some of the surprising new knowledge generated by exoplanetary science will provide insight into the climate evolution of the Earth. Ultimately this may be the only thing that matters: the author's favourite planet is definitely the Earth.

Summary of Chapter 8

1. Kepler will find transiting terrestrial planets.
2. The MEarth survey has identified a transiting super-Earth, GJ 1214 b, around a nearby low-mass M dwarf star. This planet may be composed primarily of water.

3. Wide-field surveys in space, such as ESA's PLATO mission, are likely to find dozens of transiting terrestrial planets within the habitable zones of nearby bright stars.
4. The Rossiter–McLaughlin effect will provide confirmation of transiting terrestrial planets, and the James Webb Space Telescope will play a key role in characterizing these new worlds.
5. Alien life, if it exists, seems most likely to resemble Earthly bacteria.
6. Because stars brighten while on the main sequence, the continuously habitable zone is smaller than the instantaneous habitable zone.
7. A magnetic field, a protective atmosphere, plate tectonics and a minimum size may all be prerequisites for habitability.
8. Habitable bodies can be divided into four classes. Earth is the prototype of Class I. Mars and Venus exemplify Class II: planets with long-term conditions unsuitable to the development of advanced life. Europa and Titan, bodies with sub-surface water (respectively) in contact or not in contact with a silicate-rich core, are examples of Class III and IV habitats.
9. Spectroscopic detection of life via biomarkers, most importantly O_2 and O_3, may be possible.

References and further reading

We intentionally avoided cluttering the text with numerous references to the primary literature. In this appendix we give references to selected sources and some relevant further reading. This is not intended to be an exhaustive list. Where a reference is relevant to more than one chapter, it is listed here in the under the first relevant chapter.

Chapter 1

1. Hatzes, A. P. et al. (2000) 'Evidence for a Long-Period Planet Orbiting ϵ Eridani', *ApJLett*, **544**, L145. Discusses evidence for the nearest known exoplanet.
2. Chauvin, G., Lagrange, A-M., Dumas, C., Zuckerman, B., Mouillet, D., Song, I., Beuzit, J-L. and Lowrance, P. (2004) 'A giant planet candidate near a young brown dwarf. Direct VLT/NACO observations using IR wavefront sensing', *A&A*, **425**, L29. The source of Figure 1.7, an infrared image of an exoplanet around a brown dwarf.
3. Kalas, P., Graham, J. R., Chiang, E., Fitzgerald, M. P., Clampin, M., Kite, E. S., Stapelfeldt, K., Marois, C. and Krist, J. (2008) 'Optical Images of an Exosolar Planet 25 Light-Years from Earth', *Science*, **322**, 1345. The discovery of Fomalhaut b.
4. Marois, C., Macintosh, B., Barman, T., Zuckerman, B., Song, I., Patience, J., Lafreniere, D. and Doyon, R. (2008) 'Direct Imaging of Multiple Planets Orbiting the Star HR 8799', *Science* **322**, 1348. Discovery of three planets around HR 8799.
5. Beaugé, C., Ferraz-Mello, S. and Michtchenko, T. A. (2008) 'Planetary Masses and Orbital Parameters from Radial Velocity Measurements', chapter in *Extrasolar Planets: Formation, Detection and Dynamics* edited by Rudolf Dvorak, ISBN: 978-3-527-40671-5. This chapter gives some of the derivation of Equation 1.6, and we follow their treatment in parts of Section 1.3.
6. Brouwer, D. and Clemence G. M, (1961) *Methods of Celestial Mechanics*, Academic Press, NY.
7. Murray, C. D. and Dermott, S. F., (1999) *Solar System Dynamics*, CUP. This reference and the previous one provide the first-principles derivation of the underpinnings of Equation 1.6. Also relevant to the material in Chapter 7.
8. Bailey, J., Butler, R. P., Tinney, C. G., Jones, H. R. A., O'Toole, S., Carter, B. D. and Marcy, G. W. (2009) 'A Jupiter-like planet orbiting the nearby M dwarf GJ 832', *ApJ*, **690**, 743. A planet similar to Jupiter discovered by the radial velocity technique.
9. Mayor, M., Udry, S., Naef, D., Pepe, F., Queloz, D., Santos, N. C. and Burnet, M. (2004) 'The CORALIE survey for southern extra-solar planets XII. Orbital solutions for 16 extra-solar planets discovered with CORALIE', *A&A*, **415**, 391. Examples of fits to reflex radial velocity data. The source of Figures 1.16 and 1.17.
10. Sackett, P. D., 'Searching for Unseen Planets via Occultation and Microlensing', chapter in *Planets Outside the Solar System: Theory and*

Observations (*NATO Science Series C*), Springer ISBN-13: 978-0792357087. Also available as a preprint, arXiv:astro-ph/9811269v1. The underpinnings of Section 1.4, an introduction to the detection of planets by microlensing, and equations used in Chapters 2 and 3.

Chapter 2

1. Pont, F., Sasselov, D. and Holman, M. (eds) (2009) 'Transiting Planets. Proceedings of the IAU Symposium 253', CUP. A comprehensive collection of research contributions on the subject of this book.
2. Mazeh, T. et al. (2000) 'The Spectroscopic Orbit of the Planetary Companion Transiting HD 209458', *ApJLett*, **532**, L55. The HD 209458 b discovery paper, and the source of Figure 2.3.
3. Charbonneau, D., Brown, T. M., Latham, D. W. and Mayor, M. (2000) 'Detection of Planetary Transits Across a Sun-like Star', *ApJLett*, **529**, L45,
4. Henry, G. W., Marcy, G. W., Butler, R. P. and Vogt, S. S. (2000) 'A Transiting "51 Peg-like" Planet', *ApJLett*, **529**, L41.
   This reference and the previous one, published on consecutive pages of Astrophysical Journal Letters, independently announce the discovery of transits in HD 209458 b. The former paper presents the first complete transits knowingly observed; Hipparcos had previously observed transits of HD 209458 b but these were not noticed until later. The former paper is the source of Figure 2.4 and of the data in Figures 3.16 and 3.18.
5. Horne, K. (2003) 'Status and Prospects of Planetary Transit Searches: Hot Jupiters Galore', *ASP Conf. Ser.*, **294**, 361. The starting point for the material in Section 2.2.
6. Pollacco, D. et al. (2006) 'The WASP Project and the SuperWASP Cameras', *PASP*, **118**, 1407. Paper describing the issues involved in wide-field survey photometry and the initial solutions adopted by the WASP consortium.
7. Collier-Cameron, A. et al. (2007) 'Efficient identification of exoplanetary transit candidates from SuperWASP light curves', *MNRAS*, **380**, 1230. Paper describing the transit-search strategy adopted by the WASP consortium.

Chapter 3

1. Brown, T. M., Charbonneau, D., Gilliland, R. L., Noyes, R. W. and Burrows, A. (2000) 'Hubble Space Telescope Time-Series Photometry of the Transiting Planet of HD 209458', *ApJ*, **552**, 699. The first ultra-high precision transit light curve.
2. Morrill, J. S. and Korendyke, C. M. (2008) 'High-Resolution Center-to-Limb Variation of the Quiet Solar Spectrum near Mg II', *ApJ*, **687**, 646. A recent study of limb darkening in the Sun.
3. Van Hamme, W. (1993) 'New limb-darkening coefficients for modeling binary star light curves', *Astronomical Journal*, **106**, 2096. Calculations of limb-darkening using model stellar atmospheres.
4. Knutson, H. A., Charbonneau, D., Noyes, R. W., Brown, T. M. and Gilliland, R. L. (2007) 'Using Stellar Limb-Darkening to Refine the

Properties of HD 209458 b', *ApJ*, **655**, 564. An analysis of multicolour transit light curves with consideration of the stellar limb-darkening.

5. Southworth, J. (2008) 'Homogeneous studies of transiting extrasolar planets - I. Light-curve analyses', *MNRAS*, **386**, 1644
Southworth, J. (2009) 'Homogeneous studies of transiting extrasolar planets - II. Physical properties', *MNRAS*, **394**, 272.
Papers analysing a number of different transiting exoplanets using a uniform method. These papers were used in Exercise 3.4 and Figure 4.21.
6. Mandel, K. and Agol, E. (2002) 'Analytic Light Curves for Planetary Transit Searches', *ApJLett*, **580**, L171. Seminal paper giving the equations we derive in Sections 3.2.2 and 3.2.3.
7. Sing, D. K. (2010) 'Stellar limb-darkening coefficients for CoRot and Kepler', *A&A*, **510**, 21. An atlas of limb-darkening coefficients generated for use in analysis of transit light curves.

See also the list of references given with Table 3.1.

Chapter 4

1. Johnson, J. A., (2009) 'International Year of Astronomy Invited Review on Exoplanets', *PASP*, **121**, 309. Succinct recent review of the highlights of exoplanet research up to 2008. Source of Figure 4.8.
2. Ho, S. and Turner, E. L., (2010) 'The Posterior Distribution of sin(i) Values For Exoplanets With $M_T \sin(i)$ Determined From Radial Velocity Data', arXiv1003.4738
Paper considering the assertion that the relative bin populations in the mass distribution of radial velocity planets is unaffected by the unknown $\sin i$.
3. Udry S., (2008) 'Statistical Properties of Exoplanets', chapter in *Extrasolar Planets* (eds) Deeg, Antonio Belmonte and Aparicio, CUP, ISBN-10: 0521868084, ISBN-13: 978-0521868082. Comprehensive chapter which we drew on extensively in Chapter 4. The book is at a postgraduate level.
4. Leger et al. (2009) 'Transiting exoplanets from the CoRoT space mission VIII. CoRoT-7b: the first super-Earth with measured radius', *A&A*, **506**, 287. The paper announcing the discovery of the first known terrestrial exoplanet.
5. Barnes, R., Raymond, S. N., Greenberg, R., Jackson, B. and Kaib, N. A. (2010) 'CoRoT-7b: Super-Earth Or Super-Io', *ApJLett*, **709**, L95. Interesting paper discussing tidal heating of the first known rocky exoplanet.
6. Guillot, T. (2005) 'The Interiors of Giant Planets: Models and Outstanding Questions', *Ann. Rev. Earth and Plan. Sciences*, **33**, 493. Comprehensive review which formed the basis of Section 4.3.
7. Liu, X., Burrows, A. and Ibgui L. (2008) 'Theoretical Radii of Extrasolar Giant Planets: The Cases of TrES-4, XO-3b, and HAT-P-1b', *ApJ*, **687**, 1191,
and
Arras, P. and Bildsten, L. (2008) 'Thermal Structure and Radius Evolution of Irradiated Gas Giant Planets', *ApJ*, **650**, 394.
Calculations of the contraction of giant planets. The former considers variations in atmospheric opacity, the presence of an inner heavy-element

core, and possible heating due to orbital tidal dissipation. The latter focuses on strong irradiation and thermal structure. The sources of Figures 4.14, 4.16 and 4.17.

8. Sasselov, D. (2008) 'Astronomy: Extrasolar planets', *Nature*, **451**, 29. A short article outlining some of the most interesting issues in the exoplanet field. Figure 4.19 was adapted from here.
9. Charbonneau, D., Brown, T. M., Burrows, A. and Laughlin, G. (2007) 'When Extrasolar Planets Transit Their Parent Stars', chapter in *Protostars and Planets V* (eds) Reipurth, B., Jewitt, D. and Keil K. (University of Arizona Space Science Series) ISBN-10: 0816526540 ISBN-13: 978-0816526543 A concise research-level summary of much of the material in this book. The source of Figure 4.20.
10. Pont, F., (2009) 'Empirical evidence for tidal evolution in transiting planetary systems', *MNRAS*, **396**, 1789. An examination of the demographics of transiting planets for evidence of tidal effects. Also relevant to Chapter 7.
11. Mordasini, C., Alibert, Y., Benz, W. and Naef, D. (2009) 'Extrasolar planet population synthesis II. Statistical comparison with observations', *A&A*, **501**, 1161. One in a series of papers reporting exoplanet population synthesis using a core accretion planet formation model. This paper focuses on the comparison of the model's outputs with the observations.
12. Baraffe, I., Chabrier, G. and Barman, T. (2010) 'The physical properties of extra-solar planets', *Rep.Prog.Phys.* **73**, 1. An up-to-date review of much of the material in Chapter 4 at the research level. Also contains a discussion of atmospheres and biosignatures relevant to Chapters 5, 6 and 8, and a discussion of orbital and tidal issues relevant to Chapter 7.

Chapter 5

1. Smith, G. R. and Hunten, D. M. (1990) 'Study of planetary atmospheres by absorptive occultations', *Reviews of Geophysics*, **28**, 117. A review of the application of transmission spectroscopy to Solar System objects. Page limits precluded including some of these results as a counterpoint to the exoplanet transmission spectroscopy.
2. Charbonneau, D., Brown, T., Noyes, R. and Gilliland R. (2001) 'Detection of an Extrasolar Planet Atmosphere', *ApJ*, **568**, 377. Seminal paper applying transmission spectroscopy to HD 209458 b. The first half of Chapter 5 draws upon this paper extensively.
3. Lodders, K. (2004) 'Brown Dwarfs – Faint at Heart, Rich in Chemistry', *Science*, **303**, 323. An examination of the effect of temperature on cloud formation. This work was adapted for Fig 5.4.
4. Marley, M. S. (2008) 'The Atmospheres of Extrasolar Planets', eprint arXiv:0809.4664. To appear in the proceedings of the Les Houches Winter School "Physics and Astrophysics of Planetary Systems", (EDP Sciences: EAS Publications Series). A discussion of the concepts and processes required for understanding of exoplanet atmospheres.

5. Vidal-Madjar, A. et al. (2003) 'An Extended Upper Atmosphere around the Extrasolar Planet HD 209458 b', *Nature*, **422**, 143. The astonishing discovery of a very deep transit in Lyman α.
6. Pont, F., Knutson, H., Gilliland, R. L., Moutou, C. and Charbonneau, D. (2008) 'Detection of atmospheric haze on an extrasolar planet: the 0.55-1.05 μm transmission spectrum of HD 189733 b with the Hubble Space Telescope', *MNRAS*, **385**, 109,
   and
   Sing, D. K., Desert, J-M., Lecavelier Des Etangs, A., Ballester, G. E., Vidal-Madjar, A., Parmentier, V., Hebrard, G. and Henry, G. W. (2009) 'Transit spectrophotometry of the exoplanet HD 189733b. I. Searching for water but finding haze with HST NICMOS', *A&A*, **505**, 891. Optical and Near-infrared measurements suggesting that Rayleigh scattering from upper-atmospheric haze dominates the transmission spectrum. Sources for data shown in Figure 5.11.
7. Pont, F., Gilliland, R. L., Knutson, H., Holman, M. and Charbonneau, D. (2009) 'Transit infrared spectroscopy of the hot Neptune around GJ 436 with the Hubble Space Telescope', *MNRAS*, **393**, 6. Transmission spectroscopy of a hot Neptune.
8. Desert, J-., Lecavelier des Etangs, A., Hebrard, G., Sing, D. K., Ehrenreich, D., Ferlet, R. and Vidal-Madjar, A. (2009) 'Search for Carbon Monoxide in the Atmosphere of the Transiting Exoplanet HD 189733b', *ApJ*, **699**, 478. A study of the Spitzer transmission spectroscopy of HD 189733 b with the suggestion of the detection of CO in the atmosphere. Figure 5.12 was adapted from this paper.
9. Redfield, S., Endl, M., Cochran, W. D. and Koesterke, L. (2008) 'Sodium Absorption from the Exoplanetary Atmosphere of HD 189733 b Detected in the Optical Transmission Spectrum', *ApJ*, **673**, L87. The first ground-based transmission spectroscopy and the source of Figure 5.13.
10. Gaudi, B. S. and Winn, J. N. (2007) 'Prospects for the Characterization and Confirmation of Transiting Exoplanets via the Rossiter–McLaughlin Effect', *ApJ*, **655**, 550. A comprehensive discussion of the application of the Rossiter–McLaughlin Effect to exoplanet transits. The source for much of the second half of Chapter 5.
11. Fabrycky, D. C. and Winn, J. N. (2009) 'Exoplanetary Spin-Orbit Alignment: Results from the Ensemble of Rossiter–McLaughlin Observations', *ApJ*, **696**, 1230. A statistical interpretation of the exoplanet Rossiter–McLaughlin observations. Concludes that there may be two distinct modes of planet migration, leading to a well-aligned population and a distinct spin-orbit misaligned population.

Chapter 6

1. Cowan, N. B., Agol, E. and Charbonneau, D. (2007) 'Hot nights on extrasolar planets: mid-infrared phase variations of hot Jupiters', *MNRAS*, **379**, 641. Section 6.1 was based upon the 'toy model' introduced by this paper. The paper interprets Spitzer observations of three exoplanets within the context of this model.

2. Seager, S., Deming, D. and Valenti J. A. (2008) 'Transiting Exoplanets with JWST', in *Astrophysics in the Next Decade: The James Webb Space Telescope and Concurrent Facilities*, Thronson, H. A., Stiavelli, M., Tielens, A. (eds.) 2009, XII, 520 p., Hardcover ISBN: 978-1-4020-9456-9. Also available as preprint arXiv:0808.1913v1
   Anticipates the contributions of the James Webb Space Telescope to the field. Figure 6.1 was adapted from here. Highly relevant to Chapter 8.
3. Charbonneau, D., Allen, L. E., Megeath, S. T., Torres, G., Alonso, R., Brown, T. M., Gilliland, R. L., Latham, D. W., Mandushev, G., O'Donovan, F. T. and Sozzetti, A. (2005) 'Detection of Thermal Emission from an Extrasolar Planet', *ApJ*, **626**, 52,
   and
   Deming, D., Seager, S., Richardson, L. J. and Harrington, J. (2005) 'Infrared radiation from an extrasolar planet', *Nature*, **434**, 740.
   The first detections of thermal emission from exoplanets. Figure 6.4 derived from the joint press release associated with these papers.
4. Deming, D., Harrington, J., Seager, S. and Richardson, L. J. (2006) 'Strong Infrared Emission from the Extrasolar Planet HD 189733 b', *ApJ*, **644**, 560.
   The first absolutely unequivocal detection of thermal emission from an exoplanet. The source of Figure 6.5.
5. Deming, D., Harrington, J., Laughlin, G., Seager, S., Navarro, S. B., Bowman, W. C. and Horning, K. (2007) 'Spitzer Transit and Secondary Eclipse Photometry of GJ 436 b', *ApJLett*, **667**, 199. A study of the Neptune-sized planet GJ 436 b, deriving its radius from the transit and its brightness temperature from secondary eclipse. The secondary eclipse is not at orbital phase 0.5, indicating an eccentric orbit. The brightness temperature is high suggesting tidal heating. Also relevant to Chapter 7.
6. Borucki, W. J. et al. (2010) 'Kepler Planet-Detection Mission: Introduction and First Results', *Science*, **327**, 977. The first five exoplanets to be discovered by the Kepler mission.
7. Welsh, W. F., Orosz, J. A., Seager, S., Fortney, J. J., Jenkins, J., Rowe, J. F., Koch, D., Borucki, William J. (2010) 'The Discovery of Ellipsoidal Variations in the Kepler Light Curve of HAT-P-7', *ApJLett*, **713**, 145. The orbital light curve of a transiting exoplanet system showing ellipsoidal variations from the host star as well as the reflected light phase curve and secondary eclipse. The analysis of the data shown in Figure 6.6, plus a further 23 days of data.
8. Charbonneau, D., 1,2, Knutson, H. A., 1, Barman, T., 3, Allen, L. E., 1, Mayor, M., 4, Megeath, S. T., 5, Queloz, D., 4, and Udry, S. (2008) 'The Broadband Infrared Emission Spectrum of the Exoplanet HD 189733b', *ApJ*, **686**, 1341. Secondary eclipse measurements in five band passes, from 3.6 μm to 24 μm. The source of Figure 6.7.
9. Fortney, J. J., Lodders, K., Marley, M. S. and Freedman, R. S. (2008) 'A Unified Theory for the Atmospheres of the Hot and Very Hot Jupiters: Two Classes of Irradiated Atmospheres', *ApJ*, **678**, 1419. Paper suggesting the existence of pM and pL class planets.

10. Harrington, J., Hansen, B. M., Luszcz, S. H., Seager, S., Deming, D., Menou, K., Cho, J. Y-K., Richardson, L. J. (2006) 'The Phase-Dependent Infrared Brightness of the Extrasolar Planet υ *And b*', **314**, 623. The first measurement of the orbital phase dependence of the detected thermal emission from an exoplanet.
11. Knutson, H. A., Charbonneau, D., Allen, L. E., Fortney, J. J., Agol, E. Cowan, N. B., Showman, A. P., Cooper, C. S., Megeath, S. T., (2007) 'A map of the daynight contrast of the extrasolar planet HD 189733 b', *Nature*, **447**, 183. The first map of the emission from an exoplanet. The source of Figures 6.13 and 6.14.
12. Agol E., Cowan, N. B., Bushong, J., Knutson, H., Charbonneau, D., Deming D., Steffen, J. H. (2008) preprint 'Transits and secondary eclipses of HD 189733 with Spitzer', arXiv:0807.2434v1. Also in Proceedings of the IAU Symposium 253, book listed under Chapter 2. An analysis of Spitzer data on HD 189733 b looking for variations. Also relevant to Chapter 7.

Chapter 7

1. Rodriguez, A. and Ferraz-Mello, S. (2010) 'Tidal decay and circularization of the orbits of short-period planets', EAS Publications Series, **42**, 411. Derives the equations for the circularization of orbits in Section 7.2.1.
2. Barnes, J. W. (2007) 'Effects of orbital eccentricity on extrasolar planet transit detectability and light curves', *PASP*, **119**, 986,
and
Burke, C. J. (2008) 'Impact of orbital eccentricity on the detection of transiting extrasolar planets', *ApJ*, **679**, 1566.
Results from the latter paper are discussed in Section 7.2.3. The former paper is a useful precursor to the Burke paper.
3. Jones, B. W., Sleep, N., Underwood, D. (2006) 'Which exoplanetary systems could harbour habitable planets?', *Int. Jnl. Astrobiology*, **5(3)**, 251. Section 7.3.1 drew from this paper.
4. Batygin, K., Bodenheimer, P. and Laughlin, G. (2009) 'Determination of the Interior Structure of Transiting Planets in Multiple-Planet Systems', *ApJ*, **704**, L49. Interesting paper showing how the orbital dynamics in multiple planet systems can be used to infer the interior structure of the planets.
5. Raghavan, D. et al. (2006) 'Two Suns in The Sky: Stellar Multiplicity in Exoplanet Systems', *ApJ*, **646**, 523. Provides a list of planets in binary and triple star systems.
6. Agol, E. et al., (2005) 'On detecting terrestrial planets with timing of giant planet transits', *MNRAS*, **359**, 567. The source of much of Sections 7.4 and 7.5.
7. Kipping, D. (2009) 'Transit timing effects due to an exomoon', *MNRAS*, **392**, 181. The source for much of Section 7.6.
8. Ford, E. and Holman, M. (2007) 'Using transit timing observations to search for Trojans of transiting extrasolar planets', *ApJ*, **664**, L51. The source for much of Section 7.7.

9. Fogg, M. J. and Nelson, R. P. (2005) 'Oligarchic and giant impact growth of terrestrial planets in the presence of gas giant planet migration', *A&A*, **441**, 791. An example of the type of work referred to in Section 7.8.

Chapter 8

1. Ford, E. B. et al. (2009) 'From Discovery to Understanding: Principles for Maximizing Scientific Return on Exoplanet Research', Astro2010: The Astronomy and Astrophysics Decadal Survey, Science White Papers, no. 80. One of many papers submitted to the United States review of priorities. This paper outlines many of the most interesting anticipated results and makes a strong case for the close collaboration of observers and theorists.
2. Montgomery, R. and Laughlin, G. (2009) 'Formation and detection of Earth mass planets around low mass stars', *Icarus*, **202**,1. A paper which investigates the formation of terrestrial planets, and concludes a space-borne wide-field survey is the most likely way to discover them.
3. Charbonneau D. et al., (2009) 'A super-Earth transiting a nearby low-mass star', *Nature*, **462**, 891. A transiting terrestrial planet around the nearby M dwarf star GJ 1214.
4. Gilmour, I. and Sephton, M. (2003) *An Introduction to Astrobiology*, OU / CUP ISBN 0 521 83736 7. A much more comprehensive treatment of the material in Section 8.3, written at a less mathematically-demanding level than the present book. The source of Figures 8.8 and 8.9.
5. Kasting, Whitmire and Reynolds (1993) 'Habitable Zones around Main Sequence Stars', *Icarus*, **101**, 108. A classic paper.
6. Lammer et al. (2009) 'What makes a habitable planet?', *Ann. Rev. A&A*, **17**, 181. A comprehensive review paper which was the source of Sections 8.4.2 and 8.4.3. Not an easy read as it includes much material drawn from a variety of disciplines.
7. des Marais et al. (2002) 'Remote Sensing of Planetary Properties and Biosignatures on Extrasolar Terrestrial Planets', *Astrobiology*, **2**, 153. A review outlining how signatures of life on other planets may be detected by astronomical observations.
8. Rowe, J. et al. (2010) 'Kepler Observations of Transiting Hot Compact Objects', *ApJLett*, **713**, L150. A fascinating puzzle! What are these objects? Why are they so hot?

Appendix

Table A.1 Common SI unit conversions and derived units.

Quantity	Unit	Conversion
speed	m s^{-1}	
acceleration	m s^{-2}	
angular speed	rad s^{-1}	
angular acceleration	rad s^{-2}	
linear momentum	kg m s^{-1}	
angular momentum	$\text{kg m}^2\,\text{s}^{-1}$	
force	newton (N)	$1\,\text{N} = 1\,\text{kg m s}^{-2}$
energy	joule (J)	$1\,\text{J} = 1\,\text{N m} = 1\,\text{kg m}^2\,\text{s}^{-2}$
power	watt (W)	$1\,\text{W} = 1\,\text{J s}^{-1} = 1\,\text{kg m}^2\,\text{s}^{-3}$
pressure	pascal (Pa)	$1\,\text{Pa} = 1\,\text{N m}^{-2} = 1\,\text{kg m}^{-1}\,\text{s}^{-2}$
frequency	hertz (Hz)	$1\,\text{Hz} = 1\,\text{s}^{-1}$
charge	coulomb (C)	$1\,\text{C} = 1\,\text{A s}$
potential difference	volt (V)	$1\,\text{V} = 1\,\text{J C}^{-1} = 1\,\text{kg m}^2\,\text{s}^{-3}\,\text{A}^{-1}$
electric field	N C^{-1}	$1\,\text{N C}^{-1} = 1\,\text{V m}^{-1} = 1\,\text{kg m s}^{-3}\,\text{A}^{-1}$
magnetic field	tesla (T)	$1\,\text{T} = 1\,\text{N s m}^{-1}\,\text{C}^{-1} = 1\,\text{kg s}^{-2}\,\text{A}^{-1}$

Table A.2 Other unit conversions.

wavelength
1 nanometre (nm) $= 10\,\text{Å} = 10^{-9}\,\text{m}$
1 ångstrom $= 0.1\,\text{nm} = 10^{-10}\,\text{m}$

angular measure
$1° = 60\,\text{arcmin} = 3600\,\text{arcsec}$
$1° = 0.017\,45$ radian
1 radian $= 57.30°$

temperature
absolute zero: $0\,\text{K} = -273.15\,°\text{C}$
$0\,°\text{C} = 273.15\,\text{K}$

spectral flux density
1 jansky (Jy) $= 10^{-26}\,\text{W m}^{-2}\,\text{Hz}^{-1}$
$1\,\text{W m}^{-2}\,\text{Hz}^{-1} = 10^{26}\,\text{Jy}$

cgs units
$1\,\text{erg} = 10^{-7}\,\text{J}$
$1\,\text{dyne} = 10^{-5}\,\text{N}$
$1\,\text{gauss} = 10^{-4}\,\text{T}$
$1\,\text{emu} = 10\,\text{C}$

mass–energy equivalence
$1\,\text{kg} = 8.99 \times 10^{16}\,\text{J}/c^2$ (c in m s^{-1})
$1\,\text{kg} = 5.61 \times 10^{35}\,\text{eV}/c^2$ (c in m s^{-1})

distance
1 astronomical unit (AU) $= 1.496 \times 10^{11}\,\text{m}$
1 light-year (ly) $= 9.461 \times 10^{15}\,\text{m} = 0.307\,\text{pc}$
1 parsec (pc) $= 3.086 \times 10^{16}\,\text{m} = 3.26\,\text{ly}$

energy
$1\,\text{eV} = 1.602 \times 10^{-19}\,\text{J}$
$1\,\text{J} = 6.242 \times 10^{18}\,\text{eV}$

cross-sectional area
$1\,\text{barn} = 10^{-28}\,\text{m}^2$
$1\,\text{m}^2 = 10^{28}\,\text{barn}$

pressure
$1\,\text{bar} = 10^5\,\text{Pa}$
$1\,\text{Pa} = 10^{-5}\,\text{bar}$
1 atmosphere $= 1.013\,25\,\text{bar}$
1 atmosphere $= 1.013\,25 \times 10^5\,\text{Pa}$

Table A.3 Constants.

Name of constant	Symbol	SI value
Fundamental constants		
gravitational constant	G	$6.673 \times 10^{-11}\,\mathrm{N\,m^2\,kg^{-2}}$
Boltzmann's constant	k	$1.381 \times 10^{-23}\,\mathrm{J\,K^{-1}}$
speed of light in vacuum	c	$2.998 \times 10^{8}\,\mathrm{m\,s^{-1}}$
Planck's constant	h	$6.626 \times 10^{-34}\,\mathrm{J\,s}$
	$\hbar = h/2\pi$	$1.055 \times 10^{-34}\,\mathrm{J\,s}$
fine structure constant	$\alpha = e^2/4\pi\varepsilon_0\hbar c$	$1/137.0$
Stefan–Boltzmann constant	σ	$5.671 \times 10^{-8}\,\mathrm{J\,m^{-2}\,K^{-4}\,s^{-1}}$
Thomson cross-section	σ_{T}	$6.652 \times 10^{-29}\,\mathrm{m^2}$
permittivity of free space	ε_0	$8.854 \times 10^{-12}\,\mathrm{C^2\,N^{-1}\,m^{-2}}$
permeability of free space	μ_0	$4\pi \times 10^{-7}\,\mathrm{T\,m\,A^{-1}}$
Particle constants		
charge of proton	e	$1.602 \times 10^{-19}\,\mathrm{C}$
charge of electron	$-e$	$-1.602 \times 10^{-19}\,\mathrm{C}$
electron rest mass	m_{e}	$9.109 \times 10^{-31}\,\mathrm{kg}$ $= 0.511\,\mathrm{MeV}/c^2$
proton rest mass	m_{p}	$1.673 \times 10^{-27}\,\mathrm{kg}$ $= 938.3\,\mathrm{MeV}/c^2$
neutron rest mass	m_{n}	$1.675 \times 10^{-27}\,\mathrm{kg}$ $= 939.6\,\mathrm{MeV}/c^2$
atomic mass unit	u	$1.661 \times 10^{-27}\,\mathrm{kg}$
Astronomical constants		
mass of the Sun	$\mathrm{M}_{\odot}$	$1.99 \times 10^{30}\,\mathrm{kg}$
radius of the Sun	$\mathrm{R}_{\odot}$	$6.96 \times 10^{8}\,\mathrm{m}$
luminosity of the sun	$\mathrm{L}_{\odot}$	$3.83 \times 10^{26}\,\mathrm{W}$
mass of the Earth	$\mathrm{M}_{\oplus}$	$5.97 \times 10^{24}\,\mathrm{kg}$
radius of the Earth	$\mathrm{R}_{\oplus}$	$6.37 \times 10^{6}\,\mathrm{m}$
mass of Jupiter	$\mathrm{M_J}$	$1.90 \times 10^{27}\,\mathrm{kg}$
radius of Jupiter	$\mathrm{R_J}$	$7.15 \times 10^{7}\,\mathrm{m}$
astronomical unit	AU	$1.496 \times 10^{11}\,\mathrm{m}$
light-year	ly	$9.461 \times 10^{15}\,\mathrm{m}$
parsec	pc	$3.086 \times 10^{16}\,\mathrm{m}$
Hubble parameter	H_0	$(70.4 \pm 1.5)\,\mathrm{km\,s^{-1}\,Mpc^{-1}}$ $(2.28 \pm 0.05) \times 10^{-18}\,\mathrm{s^{-1}}$
age of Universe	t_0	$(13.73 \pm 0.15) \times 10^{9}$ years
current critical density	$\rho_{\mathrm{c},0}$	$(9.30 \pm 0.40) \times 10^{-27}\,\mathrm{kg\,m^{-}}$
current dark energy density	$\Omega_{\Lambda,0}$	$(73.2 \pm 1.8)\%$
current matter density	$\Omega_{\mathrm{m},0}$	$(26.8 \pm 1.8)\%$
current baryonic matter density	$\Omega_{\mathrm{b},0}$	$(4.4 \pm 0.2)\%$
current non-baryonic matter density	$\Omega_{\mathrm{c},0}$	$(22.3 \pm 0.9)\%$
current curvature density	$\Omega_{\mathrm{k},0}$	$(-1.4 \pm 1.7)\%$
current deceleration	q_0	-0.595 ± 0.025

Table A.4 Zero-points for magnitude to flux conversions.

Band	Central Wavelength /nm	Spectral flux density for magnitude 0.0 /W m^{-2} nm^{-1}
U	365	4.27×10^{-11}
B	440	6.61×10^{-11}
V	550	3.64×10^{-11}
R	700	1.74×10^{-11}
I	900	8.32×10^{-12}
J	1250	3.18×10^{-12}
H	1650	1.18×10^{-12}
K	2200	4.17×10^{-13}
L	3600	6.23×10^{-14}
M	4800	2.07×10^{-14}
N	10 200	1.23×10^{-15}

Solutions to exercises

Exercise 1.1 First we need to convert the distance to SI units:

$$\begin{aligned}68 \text{ light-years} &= 68 \times 9.461 \times 10^{15}\,\text{m}\\ &= 6.5 \times 10^{17}\,\text{m}.\end{aligned}$$

Then we can obtain the time taken by dividing the distance by the speed of travel:

$$\begin{aligned}\text{time} &= \frac{\text{distance}}{\text{speed}}\\ &= \frac{6.5 \times 10^{17}\,\text{m}}{12 \times 10^{3}\,\text{m}\,\text{s}^{-1}}\\ &= 5.4 \times 10^{13}\,\text{s}\\ &= \frac{5.4 \times 10^{13}\,\text{s}}{365.25\,\text{day}\,\text{yr}^{-1} \times (24 \times 3600)\,\text{s}\,\text{day}^{-1}}\\ &= 1.7 \times 10^{6}\,\text{yr} \approx 2\,\text{Myr}.\end{aligned}$$

(Myr indicates 10^6 yr, or a megayear.)

Thus the time taken for Pioneer 10 to reach Aldebaran is almost 2 million years.

Exercise 1.2 In interpreting Figure 1.8, we note that both axes are logarithmic.

(a) The most favourable contrast ratio occurs when the vertical distance between Jupiter's curve and the Sun's is minimized. The two curves are converging as they disappear off the right-hand side of the figure, so the most favourable contrast ratio occurs for wavelengths around (or greater than) $100\,\mu\text{m}$. The value of the contrast ratio at $100\,\mu\text{m}$ is approximately 10 000.

(b) The spectral energy distribution of Jupiter peaks at around $0.5\,\mu\text{m}$. This is very different (by a factor of around 200) to the wavelength of the most favourable contrast ratio. The reason for the discrepancy is that the Sun's spectral energy distribution also peaks at around $0.5\,\mu\text{m}$, and the second, thermal emission component of Jupiter's spectral energy distribution gives a more favourable contrast ratio.

(c) Wavelengths around $20\,\mu\text{m}$ are at the peak of Jupiter's thermal emission. Though the contrast ratio (almost 100 000) is less favourable than at longer wavelengths, the flux from Jupiter is over 20 times higher at $20\,\mu\text{m}$ than it is at $100\,\mu\text{m}$. There is limited value in having a favourable contrast ratio if the flux from both objects is immeasurably small!

Exercise 1.3 Kepler's third law in the form used for planetary orbits is

$$\frac{a^3}{P_{\text{orb}}^2} = \frac{G(M_* + M_\text{P})}{4\pi^2}. \qquad \text{(Eqn 1.1)}$$

To make the estimate, we will consider the star as a small mass in orbit around a much larger mass positioned at the centre of the Galaxy. We can therefore replace $M_* + M_\text{P}$ with M_{total}, where this is the mass of the Galaxy. We will use the Galactocentric distance as the value for a, and this will allow us to make an estimate of P_{orb}, the time taken for a complete orbit around the Galaxy. Thus we have

$$P_{\text{orb}} = \left(\frac{4\pi^2 a^3}{GM_{\text{total}}}\right)^{1/2}. \qquad \text{(S1.1)}$$

To use this we need to convert all quantities into SI units (which we can accomplish using the information in the Appendix):

$$M_{\text{total}} = 10^{12}\,\text{M}_\odot = 1.99 \times 10^{30} \times 10^{12}\,\text{kg} \approx 2 \times 10^{42}\,\text{kg},$$
$$a = 8\,\text{kpc} = 8 \times 10^3 \times 3.086 \times 10^{16}\,\text{m} \approx 2 \times 10^{20}\,\text{m}.$$

In each case we have retained only one significant figure as we are making a rough estimate. Thus we have

$$\begin{aligned}
P_{\text{orb}} &\approx \left(\frac{4\pi^2(2 \times 10^{20}\,\text{m})^3}{7 \times 10^{-11}\,\text{N}\,\text{m}^2\,\text{kg}^{-2} \times 2 \times 10^{42}\,\text{kg}}\right)^{1/2} \\
&\approx \left(\frac{4 \times 10 \times 8 \times 10^{60}\,\text{m}^3}{7 \times 2 \times 10^{31}\,\text{kg}\,\text{m}\,\text{s}^{-2}\,\text{m}^2\,\text{kg}^{-2}\,\text{kg}}\right)^{1/2} \\
&\approx \left(\frac{3 \times 10^{62}\,\text{m}^3}{1 \times 10^{32}\,\text{m}^3\,\text{s}^{-2}}\right)^{1/2} \\
&\approx \left(3 \times 10^{30}\,\text{s}^2\right)^{1/2} \\
&\approx 2 \times 10^{15}\,\text{s}.
\end{aligned}$$

It's difficult to grasp how long 10^{15} s is, so we will convert the answer to years:

$$\begin{aligned}
P_{\text{orb}} &\approx \frac{2 \times 10^{15}\,\text{s}}{3600\,\text{s}\,\text{h}^{-1} \times 24\,\text{h}\,\text{day}^{-1} \times 365.25\,\text{day}\,\text{yr}^{-1}} \\
&\approx 6 \times 10^7\,\text{yr}.
\end{aligned}$$

Thus we have deduced that it takes about 60 Myr for the Sun to complete its orbit around the Galaxy. (Note: this is a very rough answer due to the initial approximations and the accumulated rounding errors.)

Exercise 1.4 (a) Equation 1.12 has a simple dependence on i:

$$V \propto \sin i.$$

The function $\sin i$ has a maximum value of 1, when $i = 90°$; this corresponds to the line of sight to the system being exactly in the plane of the orbit, as shown in Figure S1.1a. The minimum value of the radial velocity corresponds to $\sin i = 0$, which occurs when $i = 0°$; this corresponds to the plane of the orbit coinciding with the plane of the sky as viewed by the observer, as shown in Figure S1.1b. The orbital velocities of the star and its planet are always orthogonal to the line of sight, and zero radial velocity variation is observed. At intermediate orientations, $0° < i < 90°$, a finite radial velocity variation with amplitude less than that of the true orbital velocity, $\boldsymbol{v}_*$, of the star is observed, as shown in Figure S1.1c: $V = |\boldsymbol{v}_*| \sin i$.

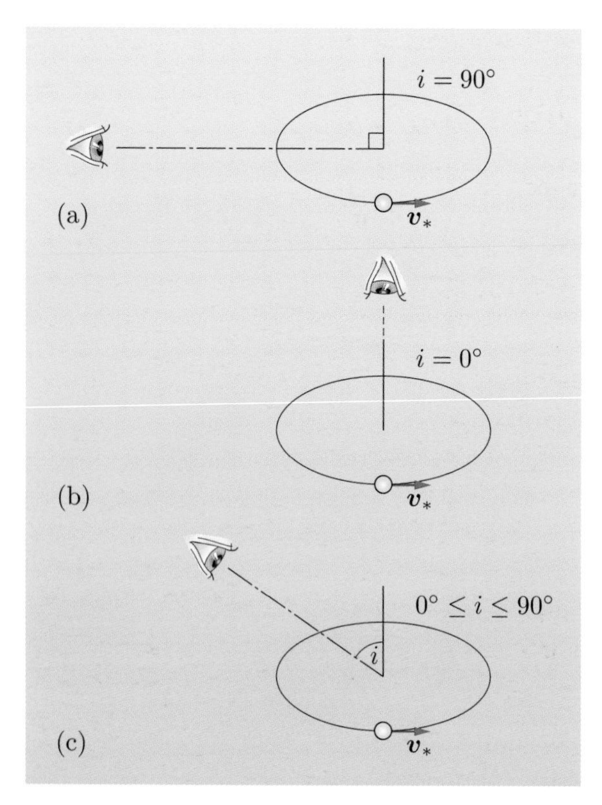

Figure S1.1 (a) An elliptical orbit viewed from an orbital inclination of $i = 90°$; the z-axis lies in the plane of the orbit. (b) The same orbit viewed from $i = 0°$; the plane of the orbit coincides with the plane of the sky, and there is no component of the orbital velocity in the direction towards or away from the observer. (c) At intermediate orientations, $0° < i < 90°$, only the component $v_* \sin i$ of the orbital motion is in the direction towards or away from the observer.

(b) The eccentricity appears twice in Equation 1.12. Inside the brackets in the numerator, it multiplies a constant ($\cos \omega_{\text{OP}}$, where ω_{OP} is a constant parameter of the orbit). The radial velocity variability is exclusively in the $\cos(\theta(t) + \omega_{\text{OP}})$ term, so this first occurrence of the eccentricity does not affect the radial velocity variations. The second appearance of the eccentricity is in the denominator of the terms that determine the amplitude multiplying the time-variable cosine term. Here it contributes to the $\sqrt{1 - e^2}$ term, which has the value 0 for $e = 1$ and the value 1 for $e = 0$. As the eccentricity approaches 1, this term in the denominator approaches 0, and therefore the amplitude of the radial velocity variation approaches infinity. As the eccentricity increases, the amplitude of the radial velocity variations increases.

(c) The observed radial velocity is given by Equation 1.12:

$$V(t) = V_{0,z} + \frac{2\pi a M_{\text{P}} \sin i}{(M_{\text{P}} + M_*) P \sqrt{1 - e^2}} \left(\cos(\theta(t) + \omega_{\text{OP}}) + e \cos \omega_{\text{OP}}\right).$$

The variable part of this is

$$\frac{2\pi a M_{\text{P}} \sin i}{(M_{\text{P}} + M_*) P \sqrt{1 - e^2}} \cos(\theta(t) + \omega_{\text{OP}}),$$

and $\cos(\theta(t) + \omega_{\text{OP}})$ varies cyclicly between -1 and $+1$. The radial velocity amplitude is therefore

$$A_{RV} = \frac{2\pi a M_P \sin i}{(M_P + M_*)P\sqrt{1-e^2}}. \qquad \text{(Eqn 1.13)}$$

For the specific case of Jupiter orbiting around the Sun, we substitute the appropriate subscripts:

$$A_{RV} = \frac{2\pi a \mathrm{M_J} \sin i}{(\mathrm{M_J} + \mathrm{M_\odot})\mathrm{P_J}\sqrt{1-\mathrm{e_J^2}}}. \qquad \text{(S1.2)}$$

The data were conveniently all given in SI units except for $\mathrm{P_J}$, which we must convert from years to seconds:

$$\begin{aligned}\mathrm{P_J} &= 12\,\mathrm{yr}\\ &= 12\,\mathrm{yr} \times 365.25\,\mathrm{days\,yr^{-1}} \times 24\,\mathrm{h\,day^{-1}} \times 60 \times 60\,\mathrm{s\,h^{-1}}\\ &= 3.8 \times 10^8\,\mathrm{s},\end{aligned}$$

where we have given our converted value to two significant figures, while performing the intermediate steps in the conversion to a higher precision.

Substituting values into Equation S1.2, we obtain

$$\begin{aligned}A_{RV} &= \frac{2\pi \times 8 \times 10^{11}\,\mathrm{m} \times 2 \times 10^{27}\,\mathrm{kg} \times \sin i}{2 \times 10^{30}\,\mathrm{kg} \times 3.8 \times 10^8\,\mathrm{s} \times \sqrt{1-0.0025}}\\ &= 13 \sin i\,\mathrm{m\,s^{-1}}.\end{aligned}$$

Exercise 1.5 (a) No. Equation 1.13 also contains a factor P in the denominator. Planetary orbits obey Kepler's third law, so $P^2 \propto a^3$. This means that $P \propto a^{3/2}$, so $A_{RV} \propto a^{1-3/2} = a^{-1/2}$. The radial velocity amplitude decreases as the planet's orbital semi-major axis increases.

(b) Planet mass appears in both the denominator and the numerator of Equation 1.13, giving a dependence

$$A_{RV} \propto \frac{M_P}{M_P + M_*}.$$

The quantity on the right-hand side will increase with M_P if M_* is held fixed. This proportionality also includes the only appearance of the stellar mass, M_*, in Equation 1.13. If M_P is held fixed, the right-hand side will decrease as M_* increases.

The eccentricity contributes to Equation 1.13 solely through the factor $\sqrt{1-e^2}$ in the denominator, so

$$A_{RV} \propto \frac{1}{\sqrt{1-e^2}}.$$

As e increases, $(1-e^2)$ decreases, so the right-hand side of the proportionality increases with increasing e.

We have already shown that the radial velocity amplitude decreases as the planet's orbital semi-major axis increases, so summarizing our analysis of Equation 1.13, we see that the radial velocity amplitude is highest for massive planets in close-in eccentric orbits around low-mass host stars.

Exercise 2.1 (a) Equation 1.21 is

$$\text{geometric transit probability} \approx \frac{R_*}{a}. \qquad \text{(Eqn 1.21)}$$

To use this we need a value for the semi-major axis, a. Kepler's third law (Equation 1.1) can be expressed as

$$a^3 = G(M_* + M_\mathrm{P})\frac{P^2}{4\pi^2}, \qquad \text{(S2.1)}$$

so the orbital semi-major axis is

$$a = (G(M_* + M_\mathrm{P}))^{1/3}\left(\frac{P}{2\pi}\right)^{2/3} \approx \left(GM_*\left(\frac{P}{2\pi}\right)^2\right)^{1/3}, \qquad \text{(S2.2)}$$

where the approximation lies in neglecting the planet's mass compared with that of the star. Since Jupiter's mass is $\sim 10^{-3}\,\mathrm{M}_\odot$ and other giant planets and stars have masses of the same order of magnitude, to the precision at which we are working this is a good approximation, so we will revert to using an equals sign in the subsequent working. In SI units, the quantities that we need to substitute into Equation S2.2 are

$$\begin{aligned}
G &= 6.67 \times 10^{-11}\,\mathrm{N\,m^2\,kg^{-2}},\\
M_* &= 1.12\,\mathrm{M}_\odot = 1.12 \times 1.99 \times 10^{30}\,\mathrm{kg} = 2.23 \times 10^{30}\,\mathrm{kg},\\
P &= 3.52\ \text{days} = 3.52 \times 24 \times 3600\,\mathrm{s} = 3.04 \times 10^5\,\mathrm{s}.
\end{aligned}$$

Consequently, we have

$$\begin{aligned}
a &= \left(6.67 \times 10^{-11} \times 2.23 \times 10^{30}\left(\frac{3.04 \times 10^5}{6.28}\right)^2 \mathrm{N\,m^2\,kg^{-1}\,s^2}\right)^{1/3}\\
&= \left(3.485 \times 10^{29}\,\mathrm{kg\,m\,s^{-2}\,m^2\,kg^{-1}\,s^2}\right)^{1/3}\\
&= \left(3.485 \times 10^{29}\,\mathrm{m^3}\right)^{1/3}\\
&= 7.04 \times 10^9\,\mathrm{m}. \qquad \text{(S2.3)}
\end{aligned}$$

We are given the radius, R_*, and simply need to convert this to metres:

$$R_* = 1.146\,\mathrm{R}_\odot = 1.146 \times 6.96 \times 10^8\,\mathrm{m} = 7.98 \times 10^8\,\mathrm{m}. \qquad \text{(S2.4)}$$

Consequently, substituting the values from Equations S2.3 and S2.4 into Equation 1.21, we have

$$\text{geometric transit probability} \approx \frac{R_*}{a} \approx \frac{7.98 \times 10^8\,\mathrm{m}}{7.04 \times 10^9\,\mathrm{m}} \approx 0.113.$$

The probability of a planet in a circular orbit like HD 209458 b's transiting from any random line of sight is approximately 11%, i.e. better than 1 in 10!

(b) The assumptions implicitly made by adopting Equation 1.21 are (i) that the orbit is randomly oriented, and (ii) that the orbit is circular. Since HD 209458 b was discovered by the radial velocity technique, whose sensitivity to a given planet decreases steadily as the orbital inclination decreases, the probability of HD 209458 b transiting was actually slightly higher than suggested by Equation 1.21. For a non-circular orbit, the planet spends time at a variety of distances, which will affect the probability of transiting. The probability depends on e and $\omega_{\rm OP}$ as well as a. A final subtlety for eccentric orbits is that the planet moves more quickly when it is closer to the star (as prescribed by Kepler's second law, or equivalently the conservation of angular momentum), so factoring in the finite observational coverage renders the transits slightly less likely to be caught. To quantitatively assess the relative importance of these three factors requires more information than is given in the question.

Exercise 2.2 (a) We have

$$A_{\rm RV} = \frac{2\pi a M_{\rm P} \sin i}{(M_{\rm P} + M_*)P\sqrt{1-e^2}} \qquad \text{(Eqn 1.13)}$$

and

$$a \approx \left(G(M_* + M_{\rm P}) \left(\frac{P}{2\pi}\right)^2 \right)^{1/3}. \qquad \text{(Eqn S2.1)}$$

Assuming a circular orbit, $e = 0$, and making the approximation $M_{\rm P} \ll M_*$, these become

$$A_{\rm RV} = \frac{2\pi a M_{\rm P} \sin i}{M_* P} \qquad \text{(S2.5)}$$

and

$$a = \left(GM_* \left(\frac{P}{2\pi}\right)^2 \right)^{1/3}. \qquad \text{(Eqn S2.2)}$$

Rearranging Equation S2.5, we have

$$M_{\rm P} \sin i = \frac{A_{\rm RV} M_* P}{2\pi a},$$

and substituting in for a, we obtain

$$\begin{aligned} M_{\rm P} \sin i &= \frac{A_{\rm RV} M_* P}{2\pi} \left(\frac{2\pi}{P}\right)^{2/3} \left(\frac{1}{GM_*}\right)^{1/3} \\ &= A_{\rm RV} \left(\frac{M_*^2 P}{2\pi G}\right)^{1/3}. \qquad \text{(S2.6)} \end{aligned}$$

$A_{\rm RV}$ and P are observables, and everything else on the right-hand side except for M_* is a constant, so this is the expression that we seek.

(b) The values for HD 209458 are

$$A_{\rm RV} = 84.67 \pm 0.70\,{\rm m\,s^{-1}},$$

$$M_* = 1.12\,{\rm M_\odot} = 1.12 \times 1.99 \times 10^{30}\,{\rm kg} = 2.23 \times 10^{30}\,{\rm kg},$$

$$P = 3.52\ {\rm days} = 3.52 \times 24 \times 3600\,{\rm s} = 3.04 \times 10^5\,{\rm s},$$

so we have

$$\begin{aligned}
M_{\rm P}\sin i &= A_{\rm RV}\left(\frac{M_*^2 P}{2\pi G}\right)^{1/3}\\
&= 84.67\,{\rm m\,s^{-1}}\left(\frac{(2.23\times10^{30})^2\,{\rm kg^2}\times3.04\times10^5\,{\rm s}}{6.28\times6.67\times10^{-11}\,{\rm N\,m^2\,kg^{-2}}}\right)^{1/3}\\
&= 84.67\times\left(3.609\times10^{75}\,\frac{{\rm kg^2\,s}}{{\rm kg\,m\,s^{-2}m^2\,kg^{-2}}}\right)^{1/3}\,{\rm m\,s^{-1}}\\
&= 84.67\times1.53\times10^{25}\left({\rm kg^3\,m^{-3}\,s^3}\right)^{1/3}\,{\rm m\,s^{-1}}\\
&= 1.30\times10^{27}\,{\rm kg}.
\end{aligned}$$

In the above, we explicitly kept the units of all the quantities when we substituted in, which allowed us to check that we (i) had got a dimensionally correct expression in part (a), and (ii) were using an appropriate choice of units for each quantity. If we had tried to use $M_* = 1.12\,{\rm M_\odot}$ in the expression, our final units would have been ${\rm M_\odot^{2/3}\,kg^{1/3}}$, alerting us that there was something amiss. Of course, assuming that the expression that we begin with is correct, if we use SI units throughout then we should always obtain an answer in SI units. Using this fact without working the units through wastes a valuable check of our working.

(c) We are given $i = 86.71^\circ \pm 0.05^\circ$, so $\sin i = 0.9984$. Consequently,

$$M_{\rm P} = \frac{1.30\times10^{27}\,{\rm kg}}{\sin i} = \frac{1.30\times10^{27}\,{\rm kg}}{0.9984} = 1.30\times10^{27}\,{\rm kg}.$$

Converting this to the other mass units requested:

$$M_{\rm P} = \frac{1.30\times10^{27}\,{\rm kg}}{1.90\times10^{27}\,{\rm kg\,M_J^{-1}}} = 0.684\,{\rm M_J}$$

and

$$M_{\rm P} = \frac{1.30\times10^{27}\,{\rm kg}}{1.99\times10^{30}\,{\rm kg\,M_\odot^{-1}}} = 6.53\times10^{-4}\,{\rm M_\odot}.$$

(d) Thus we can work out the ratio between $M_{\rm P}$ and M_* for the HD 209458 system:

$$\frac{M_{\rm P}}{M_*} = \frac{6.53\times10^{-4}\,{\rm M_\odot}}{1.12\,{\rm M_\odot}} = 5.83\times10^{-4}.$$

The planet's mass is a factor of 2 less than a thousandth that of the star. Since we have worked to a precision of only three significant figures, the approximation $M_{\rm P} \ll M_*$ was applicable.

(e) There are two basic approaches that could be used.

To tackle the problem *analytically*, the full versions of Kepler's third law (Equation 1.1) and Equation 1.13 would be used, and the algebra would need to carry through all the instances of $M_{\rm P}$. This would result in a much more complex expression, but it could still be solved.

To tackle the problem *iteratively*, we could replace the expression $(M_{\rm P} + M_*)$ with $M_{\rm total}$ in Kepler's third law (Equation 1.1) and Equation 1.13. Then the method in parts (a)–(d) could be used to evaluate M_P using the approximation $M_{\text{total}} \approx M_*$. This would be our first estimate for M_P, and we could then refine our estimate by using $M_{\text{total}} \approx M_* + M_P$ and calculating a new value for M_P. This technique should be repeated until, at the precision required, the new value of M_P does not differ from the value used to estimate it.

Exercise 2.3 (a) The bandpass is broad and centred on 700 nm, so we will use the fiducial flux density value for the R band, which is centred on 700 nm. Since the value of the sky brightness is already given in pixel units in Table 2.1, we can slightly simplify the procedure outlined in the text. The sky flux density per pixel is given by applying the standard conversion from magnitudes:

$$m_1 - m_2 = 2.5 \log_{10}(F_2/F_1), \tag{S2.7}$$

where $m_1 = 6.8$ (from Table 2.1), $m_2 = 0.0$, the fiducial magnitude, F_1 is the flux density corresponding to m_1, and $F_2 = 1.74 \times 10^{-11}\,\text{W m}^{-2}\,\text{nm}^{-1}$ (the R band value from Table A.4 in the Appendix). Making F_1 the subject of Equation S2.7, we have

$$\begin{aligned} F_1 &= F_2 \times 10^{-(m_1-m_2)/2.5} \\ &= 1.74 \times 10^{-11}\,\text{W m}^{-2}\,\text{nm}^{-1} \times 10^{-6.8/2.5} \\ &= 1.74 \times 10^{-11} \times 1.91 \times 10^{-3}\,\text{W m}^{-2}\,\text{nm}^{-1} \\ &= 3.32 \times 10^{-14}\,\text{W m}^{-2}\,\text{nm}^{-1}. \end{aligned}$$

We will take 700 nm to be the typical wavelength of the radiation. Thus the typical photon energy is

$$\begin{aligned} E_{\text{ph}} = h\nu &= \frac{hc}{\lambda} \\ &= \frac{6.63 \times 10^{-34}\,\text{J s} \times 3.00 \times 10^{8}\,\text{m s}^{-1}}{700 \times 10^{-9}\,\text{m}} = 2.84 \times 10^{-19}\,\text{J}. \end{aligned}$$

This allows us to convert from flux to photon rate, l_{sky}:

$$\begin{aligned} l_{\text{sky}} = \frac{F_1}{E_{\text{ph}}} &= \frac{3.32 \times 10^{-14}\,\text{W m}^{-2}\,\text{nm}^{-1}}{2.84 \times 10^{-19}\,\text{J}} \\ &= 1.169 \times 10^{5}\,\text{s}^{-1}\,\text{m}^{-2}\,\text{nm}^{-1}. \end{aligned}$$

We expect, therefore, $l_{\text{sky}} = 1.17 \times 10^5$ sky photons per second per square metre of telescope aperture per nanometre included in the bandpass to fall on each pixel of the PASS survey CCD.

Using the values that we are given: the telescope aperture is of diameter 2.5 cm, so the collecting area is $A = \frac{\pi}{4} \times 0.025^2\,\text{m}^2 = 4.91 \times 10^{-4}\,\text{m}^2$. The bandpass is wide ($\Delta\lambda = 300\,\text{nm}$), so the photon rate per pixel is

$$\begin{aligned} \frac{dn_{\text{sky}}}{dt} &= 1.169 \times 10^{5}\,\text{s}^{-1}\,\text{m}^{-2}\,\text{nm}^{-1} \times A \times \Delta\lambda \\ &= 1.169 \times 10^{5}\,\text{s}^{-1}\,\text{m}^{-2}\,\text{nm}^{-1} \times 4.91 \times 10^{-4}\,\text{m}^2 \times 300\,\text{nm} \\ &= 1.72 \times 10^{4}\,\text{s}^{-1}. \end{aligned}$$

We expect almost 20 000 sky photons per pixel per second for the PASS survey.

(b) The expected number of sky photons, n_{sky}, per pixel in a 10 s exposure is

$$n_{\text{sky}} = 1.72 \times 10^{4}\,\text{s}^{-1} \times 10\,\text{s} = 1.72 \times 10^{5},$$

or almost 200 000 sky photons per pixel per exposure.

Exercise 2.4 We have

$$F = \frac{L}{4\pi r^2}, \qquad \text{(Eqn 2.2)}$$

and we know that sources can be detected if $F \geq S$. This means that a source will be detected if

$$S \leq \frac{L}{4\pi r^2},$$

where we have substituted $S \leq F$ in Equation 2.2. The limiting distance will be when the flux exactly equals S, so that the limiting distance, d_{max}, corresponds to

$$S = \frac{L}{4\pi d_{\text{max}}^2}.$$

Making the distance the subject of the equation, this becomes

$$d_{\text{max}} = \left(\frac{L}{4\pi S}\right)^{1/2}, \qquad \text{(Eqn 2.3)}$$

as required

Exercise 2.5

(a) The arc length is $2a\theta$, where θ is the angle indicated in Figure S2.1. From this figure, we see that

$$\sin\theta = \frac{R_*}{a},$$

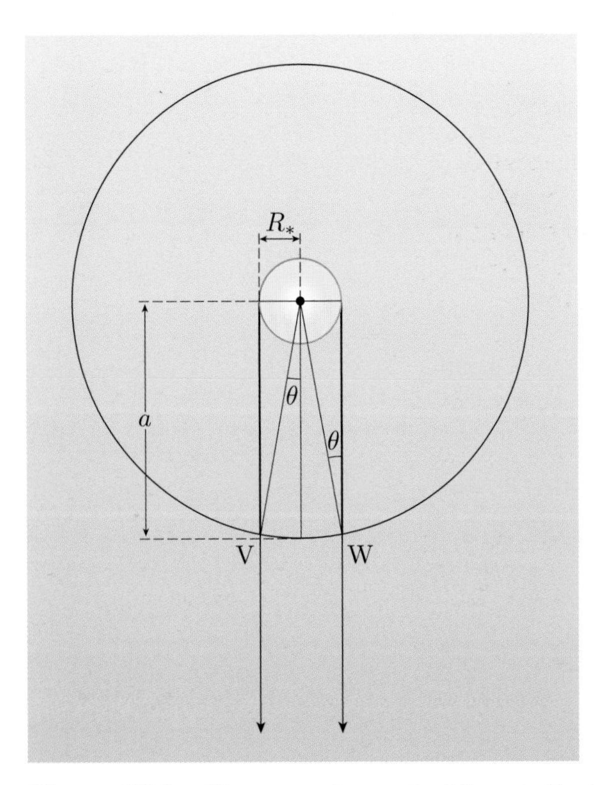

Figure S2.1 The two outer vertical lines indicate the path taken by light rays on each side of the star to a distant observer. The rays are parallel. Points V and W are the intersections of these lines with the planet's orbit.

so

$$\theta = \sin^{-1}\left(\frac{R_*}{a}\right).$$

Hence we have

$$\begin{aligned} T_{\text{dur}} &= P \times \frac{\text{length of arc from V to W}}{2\pi a} \\ &= P \times \frac{2a\sin^{-1}(R_*/a)}{2\pi a} \\ &= \frac{P}{\pi}\sin^{-1}\left(\frac{R_*}{a}\right). \end{aligned} \qquad \text{(S2.8)}$$

(b) The deviation between the exact and approximated values for T_{dur} is simply the difference between the right-hand sides of Equations 2.6 and S2.8:

$$\Delta T_{\text{dur}} = \frac{P}{\pi}\left(\sin^{-1}\left(\frac{R_*}{a}\right) - \frac{R_*}{a}\right). \qquad \text{(S2.9)}$$

The fractional deviation is simply ΔT_{dur} divided by T_{dur}, i.e.

$$\frac{\Delta T_{\text{dur}}}{T_{\text{dur}}} = \frac{\dfrac{P}{\pi}\left(\sin^{-1}\left(\dfrac{R_*}{a}\right) - \dfrac{R_*}{a}\right)}{\dfrac{P}{\pi}\sin^{-1}\left(\dfrac{R_*}{a}\right)} = 1 - \frac{\dfrac{R_*}{a}}{\sin^{-1}\left(\dfrac{R_*}{a}\right)}.$$

(c) As we saw in the solutions to Exercises 2.1 and 2.2, Kepler's third law tells us that

$$a = \left(\frac{P}{2\pi}\right)^{2/3}(GM_*)^{1/3}.$$

(d) To evaluate $\Delta T_{\text{dur}}/T_{\text{dur}}$, we need to use our expression for a for the specific values of $M_* = 1\,\text{M}_\odot$ and orbital period $P = 1$ day. Generally, it is advisable to substitute in expressions to obtain the answer algebraically, before substituting in numerical values. In this case, however, the quantity that we seek depends only on the ratio R_*/a, and the algebraic expression for a is significantly more complicated than this.

So, evaluating a with

$$\begin{aligned} P &= 1\text{ day} = 86\,400\,\text{s}, \\ M_* &= 1.99 \times 10^{30}\,\text{kg}, \\ G &= 6.67 \times 10^{-11}\,\text{N}\,\text{m}^2\,\text{kg}^{-2}, \end{aligned}$$

we have

$$\begin{aligned} a &= \left(\frac{P}{2\pi}\right)^{2/3}(GM_*)^{1/3} \\ &= \left(1.375 \times 10^4\,\text{s}\right)^{2/3}\left(1.327 \times 10^{20}\,\text{N}\,\text{m}^2\,\text{kg}^{-1}\right)^{1/3} \\ &= 5.740 \times 10^2 \times 5.101 \times 10^6\,\text{s}^{2/3}\,\text{N}^{1/3}\,\text{m}^{2/3}\,\text{kg}^{-1/3} \\ &= 2.928 \times 10^9\,\text{s}^{2/3}\,(\text{kg}\,\text{m}\,\text{s}^{-2})^{1/3}\,\text{m}^{2/3}\,\text{kg}^{-1/3} \\ &= 2.93 \times 10^9\,\text{m}. \end{aligned}$$

In SI units, $1\,R_\odot = 6.96 \times 10^8$ m, so

$$\frac{R_*}{a} = \frac{6.96 \times 10^8}{2.93 \times 10^9} = 0.238$$

and

$$\sin^{-1}\left(\frac{R_*}{a}\right) = \sin^{-1}(0.238) = 0.240,$$

so

$$\frac{\Delta T_{\rm dur}}{T_{\rm dur}} = 1 - \frac{0.238}{0.240} = 0.0083 \approx 1\%.$$

(e) Assuming that the star's properties remain constant, the semi-major axis will increase as $P^{2/3}$ with the increasing orbital period of the planet. The ratio R_*/a will therefore decrease as $P^{-2/3}$. The small angle approximation $\sin\theta \approx \theta$ becomes increasingly accurate as the angle, θ, decreases (where $\theta = R_*/a$ in this case). So for longer orbital periods, the approximation will be better, and the fractional deviation will be less than 1%. Demonstrating this numerically, for $P = 5.2$ days,

$$a(5.2\ \text{days}) = a(1\ \text{day}) \times \left(\frac{5.2}{1}\right)^{2/3} = 2.93 \times 10^9\ \text{m} \times 3.00 = 8.79 \times 10^9\ \text{m},$$

so

$$\frac{R_*}{a} = \frac{6.96 \times 10^8}{8.79 \times 10^9} = 0.0792$$

and

$$\sin^{-1}\left(\frac{R_*}{a}\right) = \sin^{-1}(0.0792) = 0.0793.$$

Consequently,

$$\frac{\Delta T_{\rm dur}}{T_{\rm dur}} = 1 - \frac{0.0792}{0.0793} \approx 0.1\%.$$

The approximation becomes ever more exact as the planet's orbital period is increased.

(f) The approximation used in Equation 2.6 is good to a precision of about 1% or better for the calculation of the durations of transits of planets in orbits $P > 1$ day around main sequence host stars. This precision should be sufficient for an estimate.

Exercise 2.6 We have

$$T_{\rm dur} \approx \frac{PR_*}{\pi a}. \qquad \text{(Eqn 2.6)}$$

Converting the values that we are given to metres and hours, we have

$$P = 1.3382 \times 24\,\text{h} = 32.12\,\text{h},$$

$$R_* = 1.15 \times 6.96 \times 10^8\ \text{m} = 8.00 \times 10^8\ \text{m},$$

$$a = 0.023 \times 1.5 \times 10^{11}\ \text{m} = 3.45 \times 10^9\ \text{m},$$

and substituting in these values gives

$$T_{\text{dur}} \approx \frac{32.12 \times 8.00 \times 10^8\,\text{h m}}{\pi \times 3.45 \times 10^9\,\text{m}} \approx 2.37\,\text{h}.$$

So the transit duration is 2.4 hours (to 2 s.f.).

Exercise 2.7 (a) Equation 2.34 tells us that the number of transiting planet discoveries predicted scales as $L_*^{3/2}$. If stars were more luminous, the signal received would be stronger, and the signal-to-noise ratio at any given distance would be better. This $N \propto L^{3/2}$ dependency is exactly the same as derived in Equation 2.5, and arises from the dependency of the survey volume on the luminosity of the sources (assuming a fixed flux threshold for detection).

(b) L_{SN} is the limiting signal-to-noise ratio and is proportional to the flux from the planet host star at the faint limit. This limiting host star brightness is an example of the limiting flux, S, in Equation 2.5. The number of planets is proportional to $L_{\text{SN}}^{-3/2}$ according to Equation 2.34, and consequently this equation expresses the same $N \propto S^{-3/2}$ relationship as Equation 2.5.

Exercise 2.8 Assuming, for simplicity, that the 16 cameras have been constantly operational since 2004, we can estimate the number of months of operation to date. At the time of writing (March 2009), this is about 5 years, or 60 months. Thus the number of data points is of the order of

$$\begin{aligned}\text{number of points} &= N_{\text{months}} \times n_{\text{cameras}} \times 5 \times 10^8 \\ &= 60 \times 16 \times 5 \times 10^8 \\ &= 4.8 \times 10^{11} \\ &= 5 \times 10^{11}\ (\text{to 1 s.f.}).\end{aligned}$$

Thus the archive needs to organize and store on the order of 10^{12} unique photometric data points.

Exercise 2.9 The minimum value of the ratio is almost zero: the value that occurs for a barely detectable grazing transit.

The maximum value will correspond to a central transit, i.e. impact parameter of $b = 0.0$. From the results of Exercise 2.5 we know that we can adopt the approximate expression for the transit duration

$$T_{\text{dur}}(b = 0.0) \approx \frac{PR_*}{\pi a}, \qquad \text{(Eqn 2.6)}$$

so the ratio that we must calculate is

$$\frac{T_{\text{dur}}(b = 0.0)}{P} \approx \frac{R_*}{\pi a}. \qquad \text{(S2.10)}$$

Substituting for a from Kepler's third law (Equation 1.1), i.e.

$$a = (G(M_* + M_{\text{P}}))^{1/3} \left(\frac{P}{2\pi}\right)^{2/3}$$

into Equation S2.10, we obtain

$$\begin{aligned}\frac{T_{\text{dur}}(b = 0.0)}{P} &\approx \frac{R_*}{\pi} \frac{2^{2/3}\pi^{2/3}}{P^{2/3}G^{1/3}M_*^{1/3}} \\ &\approx R_* \left(\frac{4}{\pi P^2 G M_*}\right)^{1/3}.\end{aligned}$$

From this we can see that for a given star, the ratio will be a maximum for the minimum value of the orbital period as

$$\frac{T_{\rm dur}(b=0.0)}{P} \propto P^{-2/3}.$$

At the time of writing (October 2009) the shortest-period known planet is CoRoT-7 b with $P = 0.85$ days. Adopting this period and the mass and radius of the Sun, we can obtain an estimate for the maximum value for the ratio. The input values are

$$P = 0.85\ \text{days} = 0.85 \times 24 \times 3600\,\text{s} = 7.3 \times 10^4\ \text{s},$$
$$R_* = 6.96 \times 10^8\ \text{m},$$
$$M_* = 1.99 \times 10^{30}\ \text{kg},$$

so we have

$$\begin{aligned}\frac{T_{\rm dur}(b=0.0)}{P} &\approx R_* \left(\frac{4}{\pi P^2 G M_*}\right)^{1/3} \\ &\approx 6.96 \times 10^8\ \text{m} \times \left(\frac{4}{\pi \times (7.3\times 10^4)^2 \times 6.67 \times 10^{-11} \times 1.99 \times 10^{30}\,\text{s}^2\,\text{N}\,\text{m}^2\,\text{kg}^{-2}\,\text{kg}}\right)^{1/3} \\ &\approx 0.085.\end{aligned}$$

The transit duration varies between 0% and 8.5% of a planet's orbital period.

Exercise 2.10 The radial velocity semi-amplitude is given by Equation 1.13 as

$$A_{\rm RV} = \frac{2\pi a M_{\rm P} \sin i}{(M_{\rm P} + M_*)P(1-e^2)^{1/2}},$$

where the semi-major axis is given by Kepler's third law (Equation 1.1) as

$$a = \left[\frac{GP^2(M_* + M_{\rm P})}{4\pi^2}\right]^{1/3}.$$

Substituting the second equation into the first, we have

$$A_{\rm RV} = \left(\frac{2\pi G}{P}\right)^{1/3} \frac{M_{\rm P}\sin i}{(M_* + M_{\rm P})^{2/3}(1-e^2)^{1/2}},$$

and rearranging to make $M_{\rm P}$ the subject of the equation, we get

$$M_{\rm P} = \frac{A_{\rm RV}(M_* + M_{\rm P})^{2/3}(1-e^2)^{1/2}}{\sin i}\left(\frac{P}{2\pi G}\right)^{1/3}.$$

Now, if we assume that the radial velocities measured in the two spectra represent the extremes of the reflex orbital motion of the star, the minimum orbital period of the putative planet must be 12 hours, and the minimum value of the radial velocity amplitude is given by the difference between the two radial velocity measurements. Since the star has the same spectral type as the Sun, clearly $M_* = \mathrm{M}_\odot$, and the mass term on the right-hand side has a minimum value of $M_*^{2/3}$. Also, we note that the maximum value of $\sin i$ is 1, so setting this at its limit will also give a minimum value for $M_{\rm P}$. For such a short orbital period, the orbit would circularize, so we may assume $e \sim 0$, and the equation therefore reduces to

$$M_{\rm P} \geq A_{\rm RV} M_*^{2/3} \left(\frac{P}{2\pi G}\right)^{1/3}.$$

So, if the observed radial velocity changes are due to the reflex motion caused by an unseen orbiting body, the minimum mass of this object is

$$\begin{aligned}
M_{\mathrm{P}} &\geq 10^4\,\mathrm{m\,s^{-1}} \times (1.99 \times 10^{30}\,\mathrm{kg})^{2/3} \times \left(\frac{12 \times 3600\,\mathrm{s}}{2\pi \times 6.673 \times 10^{-11}\,\mathrm{N\,m^2\,kg^{-2}}}\right)^{1/3} \\
&\geq 7.4 \times 10^{28}\,\mathrm{m\,s^{-1}\,kg^{2/3}\,s^{1/3}\,N^{-1/3}\,m^{-2/3}\,kg^{2/3}} \\
&\geq 7.4 \times 10^{28}\,\mathrm{m\,s^{-1}\,kg^{2/3}\,s^{1/3}\,(kg\,m\,s^{-2})^{-1/3}\,m^{-2/3}\,kg^{2/3}} \\
&\geq 7.4 \times 10^{28}\,\mathrm{kg} \\
&\geq 39\,\mathrm{M_J}.
\end{aligned}$$

Since the putative planet has a mass in excess of almost 40 times the mass of Jupiter (about $0.037\,\mathrm{M_\odot}$), this object may be a brown dwarf, but cannot be a planet, and so can be eliminated from further follow-up.

Exercise 3.1 (a) The semi-major axis of the Earth's orbit is $a = 1$ AU and the period of the Earth's orbit is $P = 1$ year. So, from Equation 3.1, the Earth's orbital speed is

$$\begin{aligned}
v &= \frac{2\pi a}{P} \\
&= \frac{2\pi \times 1\,\mathrm{AU}}{1\,\mathrm{year}} \\
&= \frac{2\pi \times 1.496 \times 10^{11}\,\mathrm{m}}{365.25 \times 24 \times 3600\,\mathrm{s}} \\
&= 2.98 \times 10^4\,\mathrm{m\,s^{-1}}.
\end{aligned}$$

So the orbital speed of the Earth is about $30\,\mathrm{km\,s^{-1}}$.

(b) Although this planet is at the same distance from its star as the Earth is from the Sun, the orbital periods will differ because the star has a mass different to that of the Sun. We can calculate the orbital period from Kepler's third law (Equation 1.1) as

$$\begin{aligned}
P &= \left(\frac{4\pi^2 a^3}{G(M_* + M_{\mathrm{P}})}\right)^{1/2} \\
&\approx \left(\frac{4\pi^2 a^3}{GM_*}\right)^{1/2} \\
&\approx \left(\frac{4\pi^2 \times (1.496 \times 10^{11}\,\mathrm{m})^3}{6.673 \times 10^{-11}\,\mathrm{N\,m^2\,kg^{-2}} \times 0.5 \times 1.99 \times 10^{30}\,\mathrm{kg}}\right)^{1/2} \\
&\approx 4.46 \times 10^7\,\mathrm{s}.
\end{aligned}$$

This orbital period is about 1.41 years. So, using Equation 3.1 once again, the orbital speed of the planet is

$$\begin{aligned}
v &= \frac{2\pi a}{P} = \frac{2\pi \times 1\,\mathrm{AU}}{4.46 \times 10^7\,\mathrm{s}} \\
&= \frac{2\pi \times 1.496 \times 10^{11}\,\mathrm{m}}{4.46 \times 10^7\,\mathrm{s}} = 2.11 \times 10^4\,\mathrm{m\,s^{-1}}.
\end{aligned}$$

So the orbital speed of the planet is about $21\,\mathrm{km\,s^{-1}}$.

Exercise 3.2 (a) We start from Equation 3.4:

$$T_{\text{dur}} = \frac{P}{\pi} \sin^{-1}\left(\frac{\sqrt{(R_* + R_{\text{P}})^2 - a^2 \cos^2 i}}{a}\right).$$

This can be rewritten as

$$T_{\text{dur}} = \frac{P}{\pi} \sin^{-1}\left[\left(\frac{R_*^2}{a^2} + \frac{2R_* R_{\text{P}}}{a^2} + \frac{R_{\text{P}}^2}{a^2} - \cos^2 i\right)^{1/2}\right].$$

Since $a \gg R_* \gg R_{\text{P}}$, the second and third terms inside the brackets are much smaller than the first, so

$$T_{\text{dur}} \approx \frac{P}{\pi} \sin^{-1}\left(\frac{R_*^2}{a^2} - \cos^2 i\right)^{1/2}.$$

Since the term in the argument of the inverse sine function will be small, and $\sin x \approx x$ for small values of x in radians, we have

$$T_{\text{dur}} \approx \frac{P}{\pi}\left(\frac{R_*^2}{a^2} - \cos^2 i\right)^{1/2},$$

as required.

(b) To use this equation to determine i, we need to first obtain a value for a. Kepler's third law tells us that

$$\frac{a^3}{P^2} \approx \frac{GM_*}{4\pi^2}.$$

Since we have a Sun-like star, the right-hand side is the same for the planet in the question and the Earth, so

$$\frac{a^3}{P^2} = \frac{\text{a}_\oplus^3}{\text{P}_\oplus^2},$$

where $\text{a}_\oplus$ and $\text{P}_\oplus$ are the semi-major axis and the orbital period of the Earth, respectively. Hence

$$a^3 = \text{a}_\oplus^3 \left(\frac{P}{\text{P}_\oplus}\right)^2,$$

thus

$$\begin{aligned} a &= \text{a}_\oplus \left(\frac{P}{\text{P}_\oplus}\right)^{2/3} \\ &= 1\,\text{AU}\left(\frac{6\text{ days}}{365.25\text{ days}}\right)^{2/3} \\ &= 1.496 \times 10^{11}\,\text{m} \times (0.0164)^{2/3} \\ &= 9.67 \times 10^{9}\,\text{m}. \end{aligned}$$

Now we must recast the equation to solve for i:

$$T_{\text{dur}} \approx \frac{P}{\pi}\left(\frac{R_*^2}{a^2} - \cos^2 i\right)^{1/2},$$

so

$$\frac{T_{\text{dur}}^2 \pi^2}{P^2} \approx \left(\frac{R_*}{a}\right)^2 - \cos^2 i,$$

thus

$$\cos^2 i \approx \left(\frac{R_*}{a}\right)^2 - \frac{T_{\text{dur}}^2 \pi^2}{P^2},$$

giving

$$\cos i \approx \left[\left(\frac{R_*}{a}\right)^2 - \left(\frac{T_{\text{dur}}\pi}{P}\right)^2\right]^{1/2}$$

and finally

$$i \approx \cos^{-1}\left[\left(\frac{R_*}{a}\right)^2 - \left(\frac{T_{\text{dur}}\pi}{P}\right)^2\right]^{1/2}.$$

Now, substituting in $R_* = 1\,\mathrm{R}_\odot = 6.96 \times 10^8\,\mathrm{m}$, $a = 9.67 \times 10^9\,\mathrm{m}$, $T_{\text{dur}} = 2$ hours and $P = 6$ days $= 144$ hours, we have

$$\begin{aligned} i &\approx \cos^{-1}\left[\left(\frac{6.96 \times 10^8\,\mathrm{m}}{9.67 \times 10^9\,\mathrm{m}}\right)^2 - \left(\frac{2\pi\,\mathrm{h}}{144\,\mathrm{h}}\right)^2\right]^{1/2} \\ &\approx \cos^{-1}\left[\left(\frac{6.96}{96.7}\right)^2 - \left(\frac{\pi}{72}\right)^2\right]^{1/2} \\ &\approx \cos^{-1}(0.0573) \\ &\approx 86.7^\circ. \end{aligned}$$

(c) If $T_{\text{dur}} = 4\,\mathrm{h}$ and all other parameters of the system are as above, then we have

$$\begin{aligned} i &\approx \cos^{-1}\left[\left(\frac{6.96 \times 10^8\,\mathrm{m}}{9.67 \times 10^9\,\mathrm{m}}\right)^2 - \left(\frac{4\pi\,\mathrm{h}}{144\,\mathrm{h}}\right)^2\right]^{1/2} \\ &\approx \cos^{-1}\left[\left(\frac{6.96}{96.7}\right)^2 - \left(\frac{\pi}{36}\right)^2\right]^{1/2} \\ &\approx \cos^{-1}(-0.002\,43)^{1/2}. \end{aligned}$$

Since the square root of a negative number is not a real number, this equation is invalid. That is, the transit duration is too long to be possible in such a system. We would be forced to conclude that the measurements are erroneous or that the candidate is an astrophysical mimic: probably it is either a blended eclipsing binary or a grazing eclipse binary.

Exercise 3.3 (a) The signal is $\Delta F = 0.0164F$, where we have used the depth as given in the caption of Figure 3.5. The noise is $1.1 \times 10^{-4}F$. So

$$\frac{\text{Signal}}{\text{Noise}} = \frac{0.0164F}{1.1 \times 10^{-4}F} = 149.$$

(b) The curvature in the transit floor causes a drop from approximately $0.9870F$ at second contact to approximately $0.9835F$ at mid-transit. The 'signal' level is therefore $(0.9870 - 0.9835)F = 0.0035F$.

If the scatter was more than about 0.3% (or $3 \times 10^{-3}F$), it would be difficult to detect the curvature.

Exercise 3.4 As the light emerging from close to the limb of the planet has $\gamma = 80°$, so $\mu = \cos\gamma = 0.174$.

The linear limb darkening law (Equation 3.9) gives

$$\frac{I(\mu)}{I(1)} = 1 - u(1 - \mu),$$

so

$$\begin{aligned}\frac{I(0.174)}{I(1)} &= 1 - 0.215 \times (1 - 0.174)\\ &= 0.82.\end{aligned}$$

The logarithmic limb darkening law (Equation 3.10) gives

$$\frac{I(\mu)}{I(1)} = 1 - u_l(1 - \mu) - \nu_l\mu\ln\mu,$$

so

$$\begin{aligned}\frac{I(0.174)}{I(1)} &= 1 - 0.14 \times (1 - 0.174) + 0.12 \times 0.174 \times \ln 0.174\\ &= 1 - 0.116 + 0.036\\ &= 0.92.\end{aligned}$$

The quadratic limb darkening law (Equation 3.11) gives

$$\frac{I(\mu)}{I(1)} = 1 - u_\mathrm{q}(1 - \mu) - \nu_\mathrm{q}(1 - \mu)^2,$$

so

$$\begin{aligned}\frac{I(0.174)}{I(1)} &= 1 - 0.29 \times (1 - 0.174) + 0.13 \times (1 - 0.174)^2\\ &= 1 - 0.240 + 0.089\\ &= 0.85.\end{aligned}$$

Clearly, these three limb darkening prescriptions give markedly different amounts of limb darkening. The form of limb darkening law adopted *does* make a difference.

Exercise 3.5 The impact parameter is $b = 3\,\mathrm{R}_\odot/4$, and the inclination angle is $i = 86.5°$, so the semi-major axis of the orbit is

$$a = b/\cos i = 3\,\mathrm{R}_\odot/4\cos 86.5° = 12.3\,\mathrm{R}_\odot$$

or

$$a = 12.3 \times 6.96 \times 10^8\,\mathrm{m} = 8.56 \times 10^9\,\mathrm{m}.$$

Using Kepler's third law (Equation 1.1), the period of the orbit is

$$P_{\text{orb}} = \left(\frac{4\pi^2 a^3}{GM_*}\right)^{1/2}$$

$$= \left(\frac{4\pi^2 \times (8.56 \times 10^9\,\text{m})^3}{6.673 \times 10^{-11}\,\text{N}\,\text{m}^2\,\text{kg}^{-2} \times 1.99 \times 10^{30}\,\text{kg}}\right)^{1/2}$$

$$= 4.32 \times 10^5\,\text{s}.$$

The orbital period is therefore 120 hours (or 5 days).

1 hour before or after mid-transit therefore corresponds to a phase angle of

$$\pm\omega t = \frac{\pm 2\pi t}{P_{\text{orb}}} = \frac{\pm(2\pi \times 1\,\text{h})}{120\,\text{h}} = \pm 0.052\ \text{radians}.$$

We therefore need to work out the position of the planet in three cases:
(i) $\omega t = 2\pi - 0.052 = 6.231$ radians, (ii) $\omega t = 0$ radians, (iii) $\omega t = 0.052$ radians.

(i) In this case, the horizontal component of the planet's position, with respect to the centre of the star's disc, is

$$a \sin \omega t = 12.3\,\text{R}_\odot \times \sin(6.231\ \text{radians}) = -0.64\,\text{R}_\odot.$$

The vertical component of the planet's position, with respect to the centre of the star's disc, is

$$a \cos i \cos \omega t = 12.3\,\text{R}_\odot \times \cos 86.5^\circ \times \cos(6.231\ \text{radians}) = 0.75\,\text{R}_\odot.$$

(ii) In this case, the horizontal component of the planet's position, with respect to the centre of the star's disc, is

$$a \sin \omega t = 12.3\,\text{R}_\odot \times \sin(0\ \text{radians}) = 0.$$

The vertical component of the planet's position, with respect to the centre of the star's disc, is

$$a \cos i \cos \omega t = 12.3\,\text{R}_\odot \times \cos 86.5^\circ \times \cos(0\ \text{radians}) = 0.75\,\text{R}_\odot.$$

(iii) In this case, the horizontal component of the planet's position, with respect to the centre of the star's disc, is

$$a \sin \omega t = 12.3\,\text{R}_\odot \times \sin(0.052\ \text{radians}) = 0.64\,\text{R}_\odot.$$

The vertical component of the planet's position, with respect to the centre of the star's disc, is

$$a \cos i \cos \omega t = 12.3\,\text{R}_\odot \times \cos 86.5^\circ \times \cos(0.052\ \text{radians}) = 0.75\,\text{R}_\odot.$$

The curvature of the locus of the planet is too small to notice when working at this level of precision, and the locus of the transit is as shown in Figure S3.1.

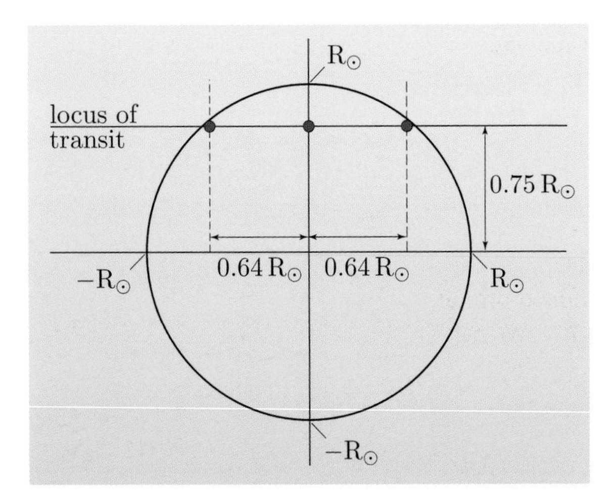

Figure S3.1 The locus of the planetary transit.

Exercise 3.6 (a) Since we are told that the star has negligible limb darkening, we may use the relationships appropriate to a stellar disc of uniform brightness. The observed flux is given by

$$\begin{aligned} F(t) &= F_{\text{unocculted}} - \Delta F \\ &= \pi I_0 R_*^2 - I_0\, A_{\text{e}}(t) \\ &= I_0\left(\pi R_*^2 - A_{\text{e}}(t)\right), \end{aligned} \tag{S3.1}$$

where

$$A_{\text{e}} = \begin{cases} 0 & \text{if } 1+p<\xi, \\ R_*^2\left(p^2\alpha_1 + \alpha_2 - \dfrac{\sqrt{4\xi^2-(1+\xi^2-p^2)^2}}{2}\right) & \text{if } 1-p<\xi\leq 1+p, \\ \pi p^2 R_*^2 & \text{if } 1-p\geq\xi, \end{cases} \tag{Eqn 3.27}$$

with

$$p = \frac{R_{\text{P}}}{R_*}, \quad \cos\alpha_1 = \frac{p^2+\xi^2-1}{2\xi p}, \quad \cos\alpha_2 = \frac{1+\xi^2-p^2}{2\xi},$$
$$\xi = \frac{a}{R_*}\left(\sin^2\omega t + \cos^2 i\cos^2\omega t\right)^{1/2} \quad \text{and} \quad \omega = \frac{2\pi}{P_{\text{orb}}}.$$

(b) For $p = 0.1$ and $\xi = 0.2$, clearly $1 - p \geq \xi$, so the third case of Equation 3.27 is appropriate, which corresponds to the planet falling entirely within the stellar disc. In this case, $A_{\text{e}} = \pi p^2 R_*^2 = \pi R_{\text{P}}^2$, so the change in flux is $\Delta F = I_0 A_{\text{e}} = I_0 \pi R_{\text{P}}^2$. The relative change in flux is therefore

$$\frac{\Delta F}{F} = \frac{I_0\pi R_{\text{P}}^2}{I_0\pi R_*^2} = \frac{R_{\text{P}}^2}{R_*^2},$$

as required.

(c) The orbital inclination, i, is known from the transit duration, T_{dur}. First contact, at time $t = t_1$, occurs when $\xi_1 = 1 + p$, and second contact, at time $t = t_2$, occurs when $\xi_2 = 1 - p$. Knowing these two times, and knowing the transit duration, $T_{\rm dur}$, and the orbital period, P, we can evaluate ξ for the two contact points:

$$\xi_1 = 1 + p = \frac{a}{R_*}\left(\sin^2 \omega t_1 + \cos^2 i \cos^2 \omega t_1\right)^{1/2} = \frac{a}{R_*}\Gamma_1,$$

$$\xi_2 = 1 - p = \frac{a}{R_*}\left(\sin^2 \omega t_2 + \cos^2 i \cos^2 \omega t_2\right)^{1/2} = \frac{a}{R_*}\Gamma_2,$$

where Γ_1 and Γ_2 have been introduced as shorthand for the known, but complicated, expressions involving the measured times and the determined orbital inclination. Consequently, we have two expressions relating p, a and R_*. We can therefore treat these as simultaneous equations, eliminate a, and make p the subject of the resulting equation:

$$p = \frac{\Gamma_1 - \Gamma_2}{\Gamma_1 + \Gamma_2}. \qquad \text{(S3.2)}$$

We recall that $p = R_{\rm P}/R_*$, so Equation S3.2 gives us the ratio of the star and planet radii in terms of the measured contact times and the inclination.

(d) The method illustrated in part (b) depends on measuring the depth of the transit, ΔF, and relating this to the fraction of the stellar flux occulted, which is assumed to be proportional to the area of the planet's disc. If the star is limb darkened, the definition of ΔF is complicated: if we use the depth at the centre of the transit, this will probably result in an erroneously large planet size. At mid-transit the occulted fraction of the total stellar flux is likely to be greater than $(R_{\rm P}/R_*)^2$ because the bright central regions of the stellar disc are being occulted. Limb darkening seriously affects the planet size deduced from this method.

On the other hand, the method illustrated in part (c) depends only on measuring the timing at which the contact points occur. So long as the second and third contacts can be clearly identified in the light curve, this method is unaffected by the limb darkening.

Exercise 3.7

(a) A large planet will give rise to a prolonged ingress and egress, and the transit will be relatively deep; a large impact factor means that the duration of the transit is relatively short; and very little limb darkening implies that the transit floor will be flat. A sketch of such a transit is shown in Figure S3.2a.

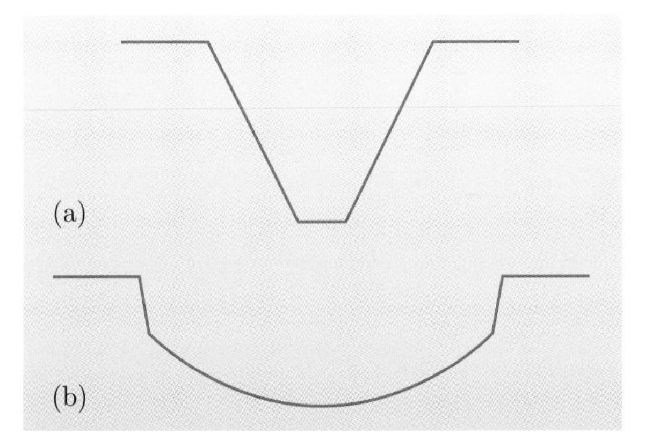

Figure S3.2 (a) A transit light curve for a large planet with a large impact factor and very little limb darkening. (b) A transit light curve for a small planet with a small impact factor and significant limb darkening.

(b) A small planet will give rise to a short ingress and egress, and the transit will be relatively shallow; a small impact factor means that the duration of the transit is relatively long; and significant limb darkening implies that the transit floor will be curved. A sketch of such a transit is shown in Figure S3.2b.

Exercise 4.1 We need to use Kepler's third law, but recognize that a planet in orbit around a solar-mass star with an orbital period of 1 year will have a semi-major axis of 1 AU. Hence we can write

$$\frac{a^3}{M_* P^2} = \frac{(1\,\mathrm{AU})^3}{1\,\mathrm{M}_\odot \times (1\,\mathrm{yr})^2}.$$

(a) (i) In this case, the orbital period is given by

$$\begin{aligned}
P &= \left(\frac{a^3}{M_*} \times \frac{1\,\mathrm{M}_\odot \times (1\,\mathrm{yr})^2}{(1\,\mathrm{AU})^3}\right)^{1/2} \\
&= \left(\frac{(1\,\mathrm{AU})^3}{0.7\,\mathrm{M}_\odot} \times \frac{1\,\mathrm{M}_\odot \times (1\,\mathrm{yr})^2}{(1\,\mathrm{AU})^3}\right)^{1/2} \\
&= \left(\frac{1}{0.7}\right)^{1/2} \mathrm{yr} = 1.195\,\mathrm{yr} = 437\ \mathrm{days}.
\end{aligned}$$

(ii) In this case, the orbital period is given by

$$\begin{aligned}
P &= \left(\frac{a^3}{M_*} \times \frac{1\,\mathrm{M}_\odot \times (1\,\mathrm{yr})^2}{(1\,\mathrm{AU})^3}\right)^{1/2} \\
&= \left(\frac{(1\,\mathrm{AU})^3}{1.5\,\mathrm{M}_\odot} \times \frac{1\,\mathrm{M}_\odot \times (1\,\mathrm{yr})^2}{(1\,\mathrm{AU})^3}\right)^{1/2} \\
&= \left(\frac{1}{1.5}\right)^{1/2} \mathrm{yr} = 0.816\,\mathrm{yr} = 298\ \mathrm{days}.
\end{aligned}$$

(b) (i) In this case, the semi-major axis of the orbit is given by

$$\begin{aligned}
a &= \left(M_* P^2 \times \frac{(1\,\mathrm{AU})^3}{1\,\mathrm{M}_\odot \times (1\,\mathrm{yr})^2}\right)^{1/3} \\
&= \left(0.7\,\mathrm{M}_\odot \times (500\ \mathrm{days})^2 \times \frac{(1\,\mathrm{AU})^3}{1\,\mathrm{M}_\odot \times (1\,\mathrm{yr})^2}\right)^{1/3} \\
&= (0.7)^{1/3} \times \left(\frac{500}{365.25}\right)^{2/3} \mathrm{AU} = 1.095\,\mathrm{AU}.
\end{aligned}$$

(ii) In this case, the semi-major axis of the orbit is given by

$$\begin{aligned}
a &= \left(M_* P^2 \times \frac{(1\,\mathrm{AU})^3}{1\,\mathrm{M}_\odot \times (1\,\mathrm{yr})^2}\right)^{1/3} \\
&= \left(1.5\,\mathrm{M}_\odot \times (500\ \mathrm{days})^2 \times \frac{(1\,\mathrm{AU})^3}{1\,\mathrm{M}_\odot \times (1\,\mathrm{yr})^2}\right)^{1/3} \\
&= (1.5)^{1/3} \times \left(\frac{500}{365.25}\right)^{2/3} \mathrm{AU} = 1.411\,\mathrm{AU}.
\end{aligned}$$

(c) (i) For a planet in a 1 AU orbit around a late M type star with $M_* = 0.2\,\mathrm{M}_\odot$, the orbital period would be

$$\begin{aligned}
P &= \left(\frac{a^3}{M_*} \times \frac{1\,\mathrm{M}_\odot \times (1\,\mathrm{yr})^2}{(1\,\mathrm{AU})^3}\right)^{1/2} \\
&= \left(\frac{(1\,\mathrm{AU})^3}{0.2\,\mathrm{M}_\odot} \times \frac{1\,\mathrm{M}_\odot \times (1\,\mathrm{yr})^2}{(1\,\mathrm{AU})^3}\right)^{1/2} \\
&= \left(\frac{1}{0.2}\right)^{1/2}\,\mathrm{yr} = 2.236\,\mathrm{yr} = 817\ \mathrm{days}.
\end{aligned}$$

(ii) For a planet in a 500-day orbit around a late M type star with $M_* = 0.2\,\mathrm{M}_\odot$, the semi-major axis would be

$$\begin{aligned}
a &= \left(M_* P^2 \times \frac{(1\,\mathrm{AU})^3}{1\,\mathrm{M}_\odot \times (1\,\mathrm{yr})^2}\right)^{1/3} \\
&= \left(0.2\,\mathrm{M}_\odot \times (500\ \mathrm{days})^2 \times \frac{(1\,\mathrm{AU})^3}{1\,\mathrm{M}_\odot \times (1\,\mathrm{yr})^2}\right)^{1/3} \\
&= (0.2)^{1/3} \times \left(\frac{500}{365.25}\right)^{2/3}\,\mathrm{AU} = 0.721\,\mathrm{AU}.
\end{aligned}$$

(d) Exoplanets have been discovered around stars with a range of masses, from late M type with $M_* \sim 0.2\,\mathrm{M}_\odot$ to stars of mass significantly greater than that of the Sun. A planetary orbit with a given semi-major axis can therefore correspond to a range of orbital periods, spanning a factor of almost 3 in the examples above. Similarly, a planetary orbit with a given period can correspond to a range of orbital semi-major axes, spanning a factor of around 2 in the examples above.

Exercise 4.2 (a) Figure 1.4 in the box on 'The nearest stars and planets' shows that of the roughly 330 known stars within 10 pc, almost 240 are M stars, i.e. about 72% of the stars (including the white dwarfs) are M stars. Of the A–M spectral type stars, about 80% are M stars. We are not told how many of these stars are main sequence stars, and we will assume that they all are; this is justified because stars spend most of their lifetimes on the main sequence. Reading from Figure 4.8c, the fraction of M stars hosting one or more planets is $\sim 2 \pm 1\%$. Thus we expect the number of M stars within 10 pc hosting RV planets to be

$$N_{\mathrm{P,M}} \approx 240 \times (0.02 \pm 0.01) \approx 4.8 \pm 2.4.$$

Similarly, there are 71 F, G, K stars within 10 pc, and reading off the graph, the fraction of these hosting planets is $\sim 4.1 \pm 0.6\%$. Thus we expect the number of F, G, K stars hosting RV planets to be

$$N_{\mathrm{P,FGK}} \approx 71 \times (0.041 \pm 0.006) \approx 2.9 \pm 0.4.$$

Finally, there are 4 A stars within 10 pc in the RECONS census, and for the highest-mass F stars and the A stars, Figure 4.8c suggests that $\sim 9 \pm 3\%$ host giant planets. Thus

$$N_{\mathrm{P,A}} \approx 4 \times (0.09 \pm 0.03) \approx 0.36 \pm 0.12 \approx 0.4 \pm 0.1.$$

Gathering this together, the total number of systems hosting RV-detectable giant planets within 10 pc is expected to be about

$$N_{\mathrm{P,RV}} \approx (4.8 \pm 2.4) + (2.9 \pm 0.4) + (0.4 \pm 0.1) \approx 8.1 \pm 2.9.$$

Thus the fraction of stars within 10 pc hosting RV-detectable planets is expected to be

$$\text{fraction hosting planets} = \frac{\text{number hosting planets}}{\text{total number}} \approx \frac{8.1 \pm 2.9}{330} \approx 0.02 \pm 0.009.$$

(b) Over the mass range shown in Figure 4.8c, i.e. $0.1–1.9\,M_\odot$, the main sequence luminosity increases from $10^{-3}\,L_\odot$ to $10\,L_\odot$, i.e. by a factor of 10^4. Equation 2.2 tells us that the flux from a source decreases as the inverse square of the distance, and consequently the distance at which a star of luminosity L is brighter than V $\sim$ 10 varies as

$$d_{\text{max}} = \left(\frac{L}{4\pi S_{10}}\right)^{1/2}, \qquad \text{(Eqn 2.3)}$$

where we have used the notation S_{10} for the flux corresponding to magnitude V $\sim$ 10. Thus the most massive A stars in the range are visible at distances about 100 times the limiting distance for the least massive M stars in the range. Since the volume of space included within a limiting distance d_{max} is given by

$$\text{volume} \propto d_{\text{max}}^3,$$

the magnitude limit includes A stars in a volume 10^6 times the volume for M stars. If we assume that our local volume is representative of the relative numbers of A stars and M stars, then Figure 1.4 suggests that the ratio of A stars to M stars will be roughly

$$\frac{N_{\text{A}}}{N_{\text{M}}} \approx \frac{4 \times 10^6}{239} \approx 10^4. \qquad \text{(S4.1)}$$

Thus the magnitude-limited sample is strongly biased in favour of the more luminous, more massive stars. The number that we obtained in Equation S4.1 is an overestimate of the ratio of the numbers in the lowest-mass bin and the highest-mass bin in Figure 4.8c since we used the extremes ($M_* = 0.1\,M_\odot$ and $M_* = 1.9\,M_\odot$) rather than the average masses within the bins for the estimate, but it illustrates the point.

Since the magnitude-limited sample is dominated by the highest-mass bin, we expect that the percentage of stars with RV-detectable planets will be more or less the value for that bin, i.e. we expect that the percentage of stars with planets will be $\sim 9 \pm 3\%$, where we have read the number from Figure 4.8c.

(c) The two estimates are different. The answer in part (a) is effectively the fraction for the lowest-mass bin in Figure 4.8c because the volume-limited sample is dominated by these lowest-mass stars. The answer in part (b) is effectively the fraction for the highest-mass bin in Figure 4.8c because the flux-limited sample is dominated by these luminous high-mass stars.

(d) The answer in part (a) suggests that we should expect roughly 8 RV-detectable planets within 10 pc. Figure 1.4 shows 18 planets, but this includes the 8 in the Solar System, so it shows 10 exoplanets. This is consistent with our estimate, which is not surprising as both Figure 1.4 and Figure 4.8c draw on the known census of exoplanets. We should expect more exoplanets to be discovered within 10 pc, however, as the RV precision of available instrumentation improves, and longer time bases are sampled. These factors will allow previously undiscovered exoplanets to be detected.

Exercise 4.3 The reflex RV amplitude is given by Equation 1.13 as

$$A_{\rm RV} = \frac{2\pi a M_{\rm P} \sin i}{(M_{\rm P} + M_*)P\sqrt{1-e^2}}.$$

In all the cases here, the planetary mass is negligible compared to that of the star, so $M_{\rm P} + M_* \approx M_*$.

Hence for GJ 581 b, the reflex RV amplitude of the star is

$$A_{\rm RV,b} = \frac{2\pi \times (0.041 \times 1.496 \times 10^{11}\,{\rm m}) \times (15.64 \times 5.97 \times 10^{24}\,{\rm kg})}{(0.31 \times 1.99 \times 10^{30}\,{\rm kg}) \times (5.369 \times 24 \times 3600\,{\rm s})}$$
$$= 12.6\,{\rm m\,s^{-1}}.$$

Similarly, for GJ 581 c

$$A_{\rm RV,c} = \frac{2\pi \times (0.07 \times 1.496 \times 10^{11}\,{\rm m}) \times (5.36 \times 5.97 \times 10^{24}\,{\rm kg})}{(0.31 \times 1.99 \times 10^{30}\,{\rm kg}) \times (12.929 \times 24 \times 3600\,{\rm s}) \times (1-0.17)^{1/2}}$$
$$= 3.4\,{\rm m\,s^{-1}},$$

for GJ 581 d

$$A_{\rm RV,d} = \frac{2\pi \times (0.22 \times 1.496 \times 10^{11}\,{\rm m}) \times (7.09 \times 5.97 \times 10^{24}\,{\rm kg})}{(0.31 \times 1.99 \times 10^{30}\,{\rm kg}) \times (66.8 \times 24 \times 3600\,{\rm s}) \times (1-0.38)^{1/2}}$$
$$= 3.1\,{\rm m\,s^{-1}},$$

and for GJ 581 e

$$A_{\rm RV,e} = \frac{2\pi \times (0.03 \times 1.496 \times 10^{11}\,{\rm m}) \times (1.94 \times 5.97 \times 10^{24}\,{\rm kg})}{(0.31 \times 1.99 \times 10^{30}\,{\rm kg}) \times (3.149 \times 24 \times 3600\,{\rm s})}$$
$$= 1.9\,{\rm m\,s^{-1}}.$$

If, at a given time, the contributions to the reflex RV amplitude from each planet are all in phase with each other, then the maximum total amplitude that may be observed is simply $(12.6 + 3.4 + 3.1 + 1.9)\,{\rm m\,s^{-1}} = 21\,{\rm m\,s^{-1}}$.

Exercise 4.4 In Worked Example 4.1 we calculated that the pressure in the core of Jupiter is approximately 10^{13} Pa, or 10^8 bar. So Figure 4.12 indicates that the core temperature of Jupiter is roughly $\log(T/{\rm K}) = 4.4$, or $T = 25\,000$ K. The degeneracy parameter is $\theta = 0.03$. We can therefore rearrange Equation 4.18 to give the number density of electrons as

$$n = \left(\frac{2\pi mkT_{\rm F}}{h^2}\right)^{3/2}$$
$$= \left(\frac{2\pi mkT}{\theta h^2}\right)^{3/2}$$
$$= \left(\frac{2\pi \times (9.109 \times 10^{-31}\,{\rm kg}) \times (1.381 \times 10^{-23}\,{\rm J\,K^{-1}}) \times (25\,000\,{\rm K})}{0.03 \times (6.626 \times 10^{-34}\,{\rm J\,s})^2}\right)^{3/2}$$
$$= 1.8 \times 10^{30}\,({\rm kg\,J^{-1}\,s^{-2}})^{3/2}$$
$$= 1.8 \times 10^{30}\,({\rm kg\,kg^{-1}\,m^{-2}\,s^{2}\,s^{-2}})^{3/2}$$
$$= 1.8 \times 10^{30}\,{\rm m^{-3}}.$$

Exercise 4.5 (a) The energy radiated away is $E_{\text{rad}} = L\,\Delta t$, and the change in gravitational energy is $\Delta E_{\text{GR}} = (\mathrm{d}E_{\text{GR}}/\mathrm{d}t)\,\Delta t$. We have

$$L = -\frac{\zeta - 1}{\zeta}\frac{\mathrm{d}E_{\text{GR}}}{\mathrm{d}t}, \qquad \text{(Eqn 4.25)}$$

so

$$E_{\text{rad}} = L\,\Delta t = -\frac{\zeta - 1}{\zeta}\frac{\mathrm{d}E_{\text{GR}}}{\mathrm{d}t}\,\Delta t = -\frac{\zeta - 1}{\zeta}\,\Delta E_{\text{GR}}, \qquad \text{(S4.2)}$$

which is the general expression that we require.

(b) In the case of an ideal diatomic gas, we know that $\zeta = 3.2$, so using Equation S4.2 we have

$$E_{\text{rad}} = -\frac{\zeta - 1}{\zeta}\,\Delta E_{\text{GR}} = -\frac{2.2}{3.2}\,\Delta E_{\text{GR}} = -0.69\,\Delta E_{\text{GR}}.$$

(c) As the planet contracts, its gravitational energy decreases, and as calculated above, around 70% of this energy can be radiated away. The remainder of the 'lost' gravitational energy goes to increasing the internal energy of the material of which the planet is composed. Hence conservation of energy is maintained.

(d) The temperature, pressure and density of the H_2 gas during the collapse must all increase. As noted above, the internal energy of the gas will increase, and this will raise the temperature. Since the same amount of gas is contained within a smaller volume, an increase in temperature implies an increase in pressure, via the ideal gas law $PV = nkT$. Similarly, since the mass is constant but the volume is decreasing, the density of the gas must also increase.

Exercise 5.1 (a) Using Equation 5.5 we have

$$T_{\text{eq},\oplus} = \frac{1}{2}\left[\frac{(1 - \mathrm{A}_\oplus)\,\mathrm{L}_\odot}{\sigma\pi\mathrm{a}_\oplus^2}\right]^{1/4},$$

and we are told that $\mathrm{A}_\oplus = 0.30$. The other constants that we require are

$$\mathrm{L}_\odot = 3.83 \times 10^{26}\,\mathrm{J\,s^{-1}},$$

$$\mathrm{a}_\oplus = 1\,\mathrm{AU} = 1.496 \times 10^{11}\,\mathrm{m},$$

$$\sigma = 5.671 \times 10^{-8}\,\mathrm{J\,m^{-2}\,K^{-4}\,s^{-1}}.$$

Substituting in, we have

$$\begin{aligned} T_{\text{eq},\oplus} &= \frac{1}{2}\left[\frac{(1 - 0.3)3.83 \times 10^{26}\,\mathrm{J\,s^{-1}}}{5.67 \times 10^{-8}\,\mathrm{J\,m^{-2}\,K^{-4}\,s^{-1}} \times 3.14 \times (1.50 \times 10^{11}\,\mathrm{m})^2}\right]^{1/4} \\ &= \frac{1}{2}\left[\frac{0.7 \times 3.83 \times 10^{26}\,\mathrm{K}^4}{5.67 \times 10^{-8} \times 3.14 \times (1.50 \times 10^{11})^2}\right]^{1/4} \\ &= 255\,\mathrm{K}, \end{aligned}$$

where we have restricted ourselves to three significant figures in the calculations as we were given $\mathrm{A}_\oplus$ to only two significant figures.

(b) The temperature that we have calculated for the Earth is about $-19°\mathrm{C}$, which is in reasonable agreement with the actual temperature that we experience here on Earth. It is, of course, slightly cooler than the actual temperature, but there are a number of assumptions that are not completely valid. For example, the Earth does not have a uniform temperature and it does not radiate as a perfect black body.

Exercise 5.2 We have

$$T_{\text{eq}} = \frac{1}{2}\left[\frac{(1-A)L_*}{\sigma\pi a^2}\right]^{1/4}. \qquad \text{(Eqn 5.5)}$$

Since we are just asked for an approximate numerical value, we will not consider the error range that we have been given for L_*, and will simply adopt $L_* \approx 1.61\,\text{L}_\odot$. Substituting in the values that we are given, therefore, we have

$$\begin{aligned}
T_{\text{eq}} &\approx \frac{1}{2}\left[\frac{1.61 \times 3.83 \times 10^{26}\,\text{J}\,\text{s}^{-1}}{5.67 \times 10^{-8}\,\text{J}\,\text{m}^{-2}\,\text{K}^{-4}\,\text{s}^{-1} \times 3.14(0.0471)^2(1.50 \times 10^{11}\,\text{m})^2}\right]^{1/4}(1-A)^{1/4} \\
&\approx \tfrac{1}{2}\left[6.98 \times 10^{13}\,\text{K}^4\right]^{1/4}(1-A)^{1/4} \\
&\approx \tfrac{1}{2}\,[2890]\,(1-A)^{1/4}\,\text{K} \\
&\approx 1400(1-A)^{1/4}\,\text{K},
\end{aligned}$$

as required.

Exercise 5.3 (a) Referring to the figure, the transit depth is around 15%, i.e. $\Delta F/F = 0.15$.

(b) We know that the transit depth is related to the ratio of the stellar and planetary radii by

$$\frac{\Delta F}{F} = \frac{R_{\text{P}}^2}{R_*^2}. \qquad \text{(Eqn 1.18)}$$

Consequently, the transit depth of HD 209458 b at the wavelength of Lyman α implies that

$$\frac{R_{\text{P}}^2}{R_*^2} = 0.15$$

and so

$$\begin{aligned}
R_{\text{P}} &= \sqrt{0.15}R_* \\
&= 0.387 \times 1.15 \times 6.96 \times 10^8\,\text{m} \\
&= 3.1 \times 10^8\,\text{m} \\
&\approx \tfrac{1}{2}\,\text{R}_\odot \\
&\approx 4\,\text{R}_\text{J}.
\end{aligned}$$

This inferred value for R_{P} is approximately three times bigger than the currently accepted value for the radius of the planet, as it must be since the transit depth in Lyman α is approximately ten times deeper than the depth in the optical continuum. As we saw in Section 4.5, no known giant planets have radii this big, nor do models predict planetary radii of this size (cf. Section 4.4).

(c) The occulted area of the star in Lyman α is far larger than the area of the planet's disc as inferred from the optical continuum. This suggests that the planet has a large cloud of hydrogen surrounding it, so the area that absorbs in the spectral lines of hydrogen is far larger than the area of the planet's disc. The cloud of hydrogen will be transparent except at the wavelengths absorbed by hydrogen atoms. The most obvious explanation for the existence of this cloud surrounding the planet is that it is being evaporated off the surface of the planet by the intense irradiation of the nearby host star.

Exercise 5.4 (a) Equation 1.13 is

$$A_{\mathrm{RV}} = \frac{2\pi a M_{\mathrm{P}} \sin i}{(M_{\mathrm{P}} + M_*)P\sqrt{1-e^2}}.$$

Comparing this with the expression that we require, it is clear that we need to eliminate a from the equation. We can do this by using Kepler's third law

$$\frac{a^3}{P^2} = \frac{G(M_* + M_{\mathrm{P}})}{4\pi^2}, \qquad \text{(Eqn 1.1)}$$

which yields

$$a = \left(\frac{P^2 G(M_* + M_{\mathrm{P}})}{4\pi^2}\right)^{1/3},$$

and substituting this into Equation 1.13, we obtain

$$A_{\mathrm{RV}} = \frac{2\pi}{P}\,\frac{P^{2/3}G^{1/3}(M_* + M_{\mathrm{P}})^{1/3}}{2^{2/3}\pi^{2/3}}\,\frac{M_{\mathrm{P}}\sin i(1-e^2)^{-1/2}}{M_{\mathrm{P}} + M_*}.$$

Collecting terms,

$$A_{\mathrm{RV}} = \left(\frac{2\pi G}{P}\right)^{1/3} \frac{M_{\mathrm{P}} \sin i}{(M_{\mathrm{P}} + M_*)^{2/3}} \left(1-e^2\right)^{-1/2},$$

as required.

(b) If $M_{\mathrm{P}} \ll M_*$ and $e \approx 0.0$, then the expression for A_{RV} simplifies to

$$A_{\mathrm{RV}} \approx \left(\frac{2\pi G}{P}\right)^{1/3} \frac{M_{\mathrm{P}} \sin i}{M_*^{2/3}}.$$

We need to express each of the variables P, M_{P} and M_* as a normalized quantity. To do this we need to multiply and divide through by the appropriate quantities:

$$\begin{aligned}
A_{\mathrm{RV}} &\approx (2\pi G)^{1/3} \left(\frac{P}{\mathrm{yr}}\right)^{-1/3} \left(\frac{M_{\mathrm{P}}\sin i}{\mathrm{M}_\oplus}\right) \left(\frac{M_*}{\mathrm{M}_\odot}\right)^{-2/3} \frac{\mathrm{M}_\oplus}{\mathrm{yr}^{1/3}\,\mathrm{M}_\odot^{2/3}} \\
&\approx \left(\frac{2\pi G}{\mathrm{yr}\,\mathrm{M}_\odot^2}\right)^{1/3} \mathrm{M}_\oplus \left(\frac{P}{\mathrm{yr}}\right)^{-1/3} \left(\frac{M_{\mathrm{P}}\sin i}{\mathrm{M}_\oplus}\right) \left(\frac{M_*}{\mathrm{M}_\odot}\right)^{-2/3} \\
&\approx \left(\frac{2\pi \times 6.67 \times 10^{-11}\,\mathrm{N\,m^2\,kg^{-2}}}{365.25 \times 24 \times 3600\,\mathrm{s} \times (1.99 \times 10^{30})^2\,\mathrm{kg}^2}\right)^{1/3} \\
&\qquad \times 5.97 \times 10^{24}\,\mathrm{kg} \left(\frac{P}{\mathrm{yr}}\right)^{-1/3} \left(\frac{M_{\mathrm{P}}\sin i}{\mathrm{M}_\oplus}\right) \left(\frac{M_*}{\mathrm{M}_\odot}\right)^{-2/3} \\
&\approx \left(\frac{3.353 \times 10^{-78}\,\mathrm{kg\,m\,s^{-2}\,m^2\,kg^{-2}}}{\mathrm{s\,kg^2}}\right)^{1/3} \\
&\qquad \times 5.97 \times 10^{24}\,\mathrm{kg} \left(\frac{P}{\mathrm{yr}}\right)^{-1/3} \left(\frac{M_{\mathrm{P}}\sin i}{\mathrm{M}_\oplus}\right) \left(\frac{M_*}{\mathrm{M}_\odot}\right)^{-2/3} \\
&\approx 0.0894\,\mathrm{m\,s^{-1}} \left(\frac{P}{\mathrm{yr}}\right)^{-1/3} \left(\frac{M_{\mathrm{P}}\sin i}{\mathrm{M}_\oplus}\right) \left(\frac{M_*}{\mathrm{M}_\odot}\right)^{-2/3}.
\end{aligned}$$

Exercise 5.5 (a) We have

$$A_S = V_S \sin i_S \left(\frac{R_P^2}{R_*^2 - R_P^2} \right), \qquad \text{(Eqn 5.27)}$$

and for an exo-Earth $R_P \ll R_*$, so

$$A_S \approx V_S \sin i_S \, \frac{R_P^2}{R_*^2}.$$

We are required to express this in terms of normalized variables, so again we multiply and divide through by the appropriate quantities:

$$\begin{aligned} A_S &\approx \left(\frac{\mathrm{R}_\oplus}{\mathrm{R}_\odot}\right)^2 \times 5\times 10^3\,\mathrm{m\,s^{-1}} \left(\frac{V_S \sin i_S}{5\,\mathrm{km\,s^{-1}}}\right) \left(\frac{R_P}{\mathrm{R}_\oplus}\right)^2 \left(\frac{R_*}{\mathrm{R}_\odot}\right)^{-2} \\ &\approx \left(\frac{6.37\times 10^6\,\mathrm{m}}{6.96\times 10^8\,\mathrm{m}}\right)^2 \times 5\times 10^3\,\mathrm{m\,s^{-1}} \times \left(\frac{V_S \sin i_S}{5\,\mathrm{km\,s^{-1}}}\right) \left(\frac{R_P}{\mathrm{R}_\oplus}\right)^2 \left(\frac{R_*}{\mathrm{R}_\odot}\right)^{-2} \\ &\approx 0.42\,\mathrm{m\,s^{-1}} \times \left(\frac{V_S \sin i_S}{5\,\mathrm{km\,s^{-1}}}\right) \left(\frac{R_P}{\mathrm{R}_\oplus}\right)^2 \left(\frac{R_*}{\mathrm{R}_\odot}\right)^{-2}, \end{aligned}$$

as required.

(b) We are asked to express the ratio A_S/A_{RV} for an exo-Earth in terms of normalized parameters. We already obtained a suitable expression for A_{RV} in our work for Exercise 5.4:

$$A_{RV} \approx 0.0894\,\mathrm{m\,s^{-1}} \left(\frac{P}{\mathrm{yr}}\right)^{-1/3} \left(\frac{M_P \sin i}{\mathrm{M}_\oplus}\right) \left(\frac{M_*}{\mathrm{M}_\odot}\right)^{-2/3}, \qquad \text{(Eqn 5.17)}$$

which we can divide into the expression that we just obtained for A_S:

$$\begin{aligned} \frac{A_S}{A_{RV}} &\approx \frac{0.42\,\mathrm{m\,s^{-1}}}{0.0894\,\mathrm{m\,s^{-1}}} \left(\frac{V_S \sin i_S}{5\,\mathrm{km\,s^{-1}}}\right) \left(\frac{R_P}{\mathrm{R}_\oplus}\right)^2 \left(\frac{R_*}{\mathrm{R}_\odot}\right)^{-2} \left(\frac{P}{\mathrm{yr}}\right)^{1/3} \left(\frac{M_P \sin i}{\mathrm{M}_\oplus}\right)^{-1} \left(\frac{M_*}{\mathrm{M}_\odot}\right)^{2/3} \\ &\approx 4.69 \left(\frac{V_S \sin i_S}{5\,\mathrm{km\,s^{-1}}}\right) \left(\frac{R_P}{\mathrm{R}_\oplus}\right)^2 \left(\frac{M_P \sin i}{\mathrm{M}_\oplus}\right)^{-1} \left(\frac{P}{\mathrm{yr}}\right)^{1/3} \left(\frac{\rho_*}{\rho_\odot}\right)^{2/3}, \end{aligned}$$

where we gathered the MR^{-3} terms for the star in question and the Sun to form the ratio of densities.

(c) Exo-Earths will exhibit transits of depth $\Delta F \sim 10^{-4}F$ once a year. Performing photometry of the required precision over several years will be challenging, so this is not a terribly promising prospect. The reflex radial velocity amplitude of exo-Earths is $\sim$10 cm s^{-1}, which is beyond the capabilities of the best currently available instruments (November 2009), and may be less than the typical intrinsic radial velocity variability of main sequence stars (cf. Figure 4.1). It is possible that exo-Earths may be detected by their radial velocity variations, as the intrinsic stellar variability can probably be characterized and largely removed, but this will be challenging. The amplitude of the Rossiter–McLaughlin effect is about 5 times greater than the reflex radial velocity amplitude, and at 0.42 m s^{-1} it is within the capabilities of the best current instrumentation, e.g. HARPS.

Exercise 6.1 Before we begin any calculations, we will express the sizes and distances that we are given in Table 6.1 in a uniform set of units. We note that the sizes and distances always appear in the relevant equations as ratios of two lengths: they must do so, as neither temperature nor secondary eclipse depth has dimensions of length. Consequently, it doesn't really matter which unit we choose to express all the lengths in, but the simplest thing to do is to convert all three lengths to SI units, as the conversion factors in each case are given in the Appendix. Applying the conversion factors, we obtain Table S6.1.

Table S6.1 Selected parameters of HD 189733 b and its host star for use in Exercises 6.1 and 6.3.

Quantity	Value	Units	Value	Units
a	0.030 99	AU	4.64×10^{9}	m
R_{P}	1.14	R_{J}	8.15×10^{7}	m
R_{*}	0.788	$\text{R}_{\odot}$	5.48×10^{8}	m
T_{eff}	5000	K		
$T_{\text{bright}}(16\,\mu\text{m})$	4315	K		

The first step in calculating the secondary eclipse depth, $\Delta F_{\text{SE}}/F$, is to calculate the day side temperature of the planet, T_{day}. The appropriate equation is

$$T_{\text{day}}^{4} = (1 - P)(1 - A)\,\frac{R_{*}^{2}}{2a^{2}}\,T_{\text{eff}}^{4}, \qquad \text{(Eqn 6.4)}$$

and we are told to assume that no heat is redistributed to the night side of the planet, so $P = 0$, and that $A = 0.05$. First rearranging Equation 6.4, then substituting these values and the appropriate values from Table S6.1, we obtain

$$\left(\frac{T_{\text{day}}}{T_{\text{eff}}}\right)^{4} = (1 - P)(1 - A)\,\frac{R_{*}^{2}}{2a^{2}}$$

so

$$\begin{aligned}\left(\frac{T_{\text{day}}}{5000\,\text{K}}\right)^{4} &= (1)(0.95)\,\frac{(5.48 \times 10^{8}\,\text{m})^{2}}{2 \times (4.64 \times 10^{9}\,\text{m})^{2}}\\ &= 0.95 \times 0.006\,97\\ &= 6.63 \times 10^{-3}\end{aligned}$$

thus

$$\begin{aligned}T_{\text{day}} &= 5000\,\text{K} \times (6.63 \times 10^{-3})^{1/4}\\ &= 5000\,\text{K} \times 0.285\\ &= 1427\,\text{K}.\end{aligned}$$

We are not given any information on the star HD 189733 beyond the radius, the effective temperature, and the brightness temperature at 16 μm. We are told to assume that the star emits as a black body (though we know that it doesn't because the brightness temperature at 16 μm differs from the effective temperature) to estimate the fractional depth of the secondary eclipse. Without further information, we cannot comment on how justified we are in making this approximation for the stellar flux at 24 μm.

The simplest form of the equations holds if we can use the Rayleigh–Jeans law rather than the unapproximated Planck function. To assess whether this is a good approximation, we need to evaluate $hc/(k\,T_{\rm day})$ and compare this to the wavelength in question, $\lambda_{\rm c} = 24\,\mu{\rm m}$. Using our value of $T_{\rm day}$, we have

$$\begin{aligned}\frac{hc}{k\,T_{\rm day}} &= \frac{6.626\times10^{-34}\,{\rm J\,s}\times 2.998\times10^{8}\,{\rm m\,s^{-1}}}{1.381\times10^{-23}\,{\rm J\,K^{-1}}\times 1427\,{\rm K}}\\ &= 1.01\times10^{-5}\,{\rm m}\\ &= 10.1\,\mu{\rm m}.\end{aligned}\qquad\text{(S6.1)}$$

The essence of the Rayleigh–Jeans law is the replacement of $\exp(hc/(\lambda k\,T_{\rm day})) - 1$ with $hc/(\lambda k\,T_{\rm day})$. With the value that we obtained in Equation S6.1, the Rayleigh–Jeans approximation uses

$$\frac{hc}{\lambda k\,T_{\rm day}} \approx \frac{10\,\mu{\rm m}}{24\,\mu{\rm m}} \approx 0.42,$$

while the full Planck function would use

$$\exp\left(\frac{hc}{\lambda k\,T_{\rm day}}\right) - 1 = 1.52 - 1 = 0.52.$$

Thus we can see that the approximation is good to within 20%, which is probably adequate given that we have already approximated both the planet and star as black body emitters. We will use the Rayleigh–Jeans law, but note that this is perhaps overestimating the planet flux by 20%.

Adopting Equation 6.15, which is appropriate to the Rayleigh–Jeans regime and the black body approximations discussed above, we have

$$\begin{aligned}\frac{\Delta F_{\rm SE}}{F} &\approx \left[\frac{(1-P)(1-A)}{2a^2}\right]^{1/4}\frac{R_{\rm P}^2}{R_*^{3/2}}\\ &\approx \left[\frac{(1)(0.95)}{2(4.64\times10^{9}\,{\rm m})^2}\right]^{1/4}\frac{(8.15\times10^{7}\,{\rm m})^2}{(5.48\times10^{8}\,{\rm m})^{3/2}}\\ &\approx 0.0063,\end{aligned}$$

i.e. the secondary eclipse depth is predicted to be 0.6%.

Exercise 6.2

(a) We have

$$\Delta f_{{\rm P},\lambda} = \left(\frac{R_{\rm P}}{d}\right)^2\left[B_\lambda(T_{\rm day}) - B_\lambda(T_{\rm night})\right],\qquad\text{(Eqn 6.12)}$$

which gives us the amplitude $\Delta f_{{\rm P},\lambda}$ in flux units, *assuming that the orbit is edge-on*. If the orbit is not edge-on, we will always see part of both the day side and the night side hemispheres. The amplitude given in Equation 6.12 is a maximum value, attained only for exactly edge-on orbits. Having noted this caveat, to obtain contrast units we simply divide through by the total flux of the system:

$$\text{peak to peak amplitude in contrast units} = \frac{\Delta f_{{\rm P},\lambda}}{f_{*,\lambda} + f_{{\rm P},\lambda}}.$$

Since $f_{*,\lambda} \gg f_{{\rm P},\lambda}$ we can make the approximation

$$\text{peak to peak amplitude in contrast units} \approx \frac{\Delta f_{{\rm P},\lambda}}{f_{*,\lambda}}.$$

To find the amplitude in a bandpass centred on wavelength λ_c, we need to integrate over the bandpass (just as we did in Worked Example 6.2):

$$\frac{\Delta F_P}{F} = \frac{\int_{\lambda_l}^{\lambda_u} \Delta f_{P,\lambda} \, Q_\lambda \, d\lambda}{\int_{\lambda_l}^{\lambda_u} f_{*,\lambda} \, Q_\lambda \, d\lambda},$$

where, as in Worked Example 6.2, Q_λ is the weighting as a function of wavelength. If we assume that we can take the fluxes at the central wavelengths as proportional to the integrated flux over the bandpass, then

$$\begin{aligned}\frac{\Delta F_P}{F} &= \frac{\Delta f_{P,\lambda_c}}{f_{*,\lambda_c}} \\ &= \left(\frac{R_P}{d}\right)^2 \left[B_{\lambda_c}(T_{\text{day}}) - B_{\lambda_c}(T_{\text{night}})\right] \times \left(\frac{d}{R_*}\right)^2 \frac{1}{B_{\lambda_c}(T_{\text{bright}})} \\ &= \frac{B_{\lambda_c}(T_{\text{day}}) - B_{\lambda_c}(T_{\text{night}})}{B_{\lambda_c}(T_{\text{bright}})} \left(\frac{R_P}{R_*}\right)^2,\end{aligned}$$

which is the expression that we were asked for.

(b) For a transiting hot Jupiter, the orbit is within a few degrees of edge-on, so Equation 6.18 gives the amplitude of the phase function more or less exactly. Since the system is transiting, the quantity R_P/R_* can be determined empirically from the transit depth. The flux from the star can be measured, and consequently the only unknowns in Equation 6.18 are the day side and night side fluxes. The amplitude of the phase function in flux units can also be measured, for use in Equation 6.12. Assuming that the distance to the star is known, e.g. from the star's spectral type and apparent brightness, we have two equations and everything but the day side and night side fluxes is empirically known. In the case of a transiting planet, therefore, it is possible to deduce the temperature difference between the day side and night side hemispheres, so long as we assume that both hemispheres emit as black bodies.

(c) In the case of a hot Jupiter planet that does not transit, the orbital orientation will be unknown, and the full amplitude as described in Equation 6.18 would not be seen. The ratio R_P/R_* would be unconstrained, so with these unknown factors it would not be possible to deduce the temperature difference between the two hemispheres.

Exercise 6.3 (a) In Figure 6.5 the base of the secondary eclipse is one quarter of the way up from 0.994 to 0.996 on the vertical axis. Therefore the value of the relative intensity during the secondary eclipse is 0.9945. The values shown are normalized so that the out-of-eclipse level is 1.0000. The deficit during the secondary eclipse is therefore

$$\Delta F_{\text{SE}} = (1.0000 - 0.9945)F = 0.0055F.$$

In contrast units, therefore,

$$\frac{\Delta F_{\text{SE}}}{F} = 0.0055.$$

The secondary eclipse depth is just over one half of one per cent.

(b) In terms of the full Planck function, the secondary eclipse depth is given by

$$\frac{\Delta F_{\text{SE}}}{F} \approx p_{\lambda_c} \left(\frac{R_P}{a}\right)^2 + \frac{B_{\lambda_c}(T_{\text{day}})}{B_{\lambda_c}(T_{\text{bright}})} \left(\frac{R_P}{R_*}\right)^2, \qquad \text{(Eqn 6.13)}$$

and we are told that we can ignore the reflection component, so we have

$$\frac{\Delta F_{\mathrm{SE}}}{F} = \frac{B_{\lambda_c}(T_{\mathrm{day}})}{B_{\lambda_c}(T_{\mathrm{bright}})}\left(\frac{R_{\mathrm{P}}}{R_*}\right)^2,$$

where we have reverted to an equality, but note that we have approximated the integral over the bandpass and neglected the reflection term. The full Planck function is rather unwieldy, but we can simplify it by noting that we can express the equation with a ratio of Planck functions as the subject:

$$\frac{B_\lambda(T_{\mathrm{day}})}{B_\lambda(T_{\mathrm{bright}})} = \frac{\Delta F_{\mathrm{SE}}}{F}\left(\frac{R_*}{R_{\mathrm{P}}}\right)^2.$$

Substituting in for the Planck function, this becomes

$$\frac{\exp(hc/(\lambda_c k\,T_{\mathrm{bright}})) - 1}{\exp(hc/(\lambda_c k\,T_{\mathrm{day}})) - 1} = \frac{\Delta F_{\mathrm{SE}}}{F}\left(\frac{R_*}{R_{\mathrm{P}}}\right)^2, \qquad \text{(Eqn 6.19)}$$

which is the first expression that we were asked to derive. Manipulating this to isolate T_{day}, we have

$$\left(\exp\left(\frac{hc}{\lambda_c k\,T_{\mathrm{bright}}}\right) - 1\right)\frac{F}{\Delta F_{\mathrm{SE}}}\left(\frac{R_{\mathrm{P}}}{R_*}\right)^2 = \exp\left(\frac{hc}{\lambda_c k\,T_{\mathrm{day}}}\right) - 1$$

so

$$\left(\exp\left(\frac{hc}{\lambda_c k\,T_{\mathrm{bright}}}\right) - 1\right)\frac{F}{\Delta F_{\mathrm{SE}}}\left(\frac{R_{\mathrm{P}}}{R_*}\right)^2 + 1 = \exp\left(\frac{hc}{\lambda_c kT_{\mathrm{day}}}\right)$$

thus

$$\log_e\left[\left(\exp\left(\frac{hc}{\lambda_c k\,T_{\mathrm{bright}}}\right) - 1\right)\frac{F}{\Delta F_{\mathrm{SE}}}\left(\frac{R_{\mathrm{P}}}{R_*}\right)^2 + 1\right] = \frac{hc}{\lambda_c k\,T_{\mathrm{day}}}.$$

Making T_{day} the subject of the equation, we have

$$T_{\mathrm{day}} = \frac{hc}{\lambda_c k}\left[\log_e\left[\left(\exp\left(\frac{hc}{\lambda_c k\,T_{\mathrm{bright}}}\right) - 1\right)\frac{F}{\Delta F_{\mathrm{SE}}}\left(\frac{R_{\mathrm{P}}}{R_*}\right)^2 + 1\right]\right]^{-1}. \qquad \text{(Eqn 6.20)}$$

This equation, the second that we were asked to derive, gives us a (complicated) expression for T_{day}.

(c) Either of the expressions in part (b) can be used. Since Equation 6.20 has T_{day} as its subject, we will use that one. The expression is complicated, so we will break down the evaluation using the specific values for the 16 μm secondary eclipse of HD 189733 b. First we note that the quantity $hc/\lambda_c k$ appears twice, so we will evaluate that. The wavelength is the central wavelength, λ_c, of the bandpass is 16 μm. Consequently,

$$\frac{hc}{\lambda_c k} = \frac{6.626\times10^{-34}\,\mathrm{J\,s}\times2.998\times10^{8}\,\mathrm{m\,s^{-1}}}{16\times10^{-6}\,\mathrm{m}\times1.381\times10^{-23}\,\mathrm{J\,K^{-1}}}$$
$$= 899.0\,\mathrm{K}.$$

Similarly, using the values in the question and Table S6.1, we can evaluate

$$\frac{F}{\Delta F_{\mathrm{SE}}}\left(\frac{R_{\mathrm{P}}}{R_*}\right)^2 = \frac{1}{0.0055}\left(\frac{8.15\times10^{7}\,\mathrm{m}}{5.48\times10^{8}\,\mathrm{m}}\right)^2$$
$$= 4.0 \quad \text{(to 2 s.f.)},$$

where we quote the result to only two significant figures because our estimate of $\Delta F_{\rm SE}$ from the eclipse depth is no better than this. The exponential function on the right-hand side is

$$\exp\left(\frac{hc}{\lambda_{\rm c} k\, T_{\rm bright}}\right) = \exp\left(\frac{899.0\,{\rm K}}{4315\,{\rm K}}\right) = \exp(0.20835) = 1.2316.$$

Substituting these values into Equation 6.20, we have

$$\begin{aligned} T_{\rm day} &= 899.0\,{\rm K}\left[\log_{\rm e}\left[(1.2316-1)\times 4.0+1\right]\right]^{-1} \\ &= 899.0\,{\rm K}\left[\log_{\rm e}\left[0.9265+1\right]\right]^{-1} \\ &= 1.525\times 899.0\,{\rm K} \\ &= 1371\,{\rm K} = 1400\,{\rm K} \quad \text{(to 2 s.f.)}, \end{aligned}$$

where in the first line we used the convention that multiplication takes precedence to avoid yet another set of brackets surrounding the terms '$(1.2316-1)\times 4.0$'.

(d) The essence of the Rayleigh–Jeans law is to replace $\exp(hc/\lambda_{\rm c}kT)-1$ with $hc/\lambda_{\rm c}kT$. For the star, we have $T_{\rm bright}(16\,\mu{\rm m}) = 4315$, which means that

$$\frac{hc}{\lambda_{\rm c}kT} = 0.2084,$$

while

$$\exp\left(\frac{hc}{\lambda_{\rm c}kT}\right) - 1 = 0.2316.$$

These are in the same ballpark, but differ by more than 10%. Since we can read the most uncertain of the input quantities, $\Delta F_{\rm SE}$, from the graph at a better precision than this, it would probably not be justified to use the Rayleigh–Jeans law instead of the full Planck function to evaluate the $16\,\mu$m flux from the star HD 189733 b.

Comment: In Exercise 6.1 we made a bigger approximation, but in that exercise, we were making an estimate; here we are deriving results from state-of-the-art observations.

The planet has a day side temperature of 1400 K, which gives

$$\frac{hc}{\lambda_{\rm c}kT} = 0.6422,$$

while

$$\exp\left(\frac{hc}{\lambda_{\rm c}kT}\right) - 1 = 0.9006.$$

In this case it is clear that the Rayleigh–Jeans law would introduce a discrepancy of about 30% in the flux from the planet. It would not be a justifiable approximation.

(e) If the wavelength of the observation were shorter, the quantity $hc/\lambda_{\rm c}kT$ would become bigger, and the approximation, which is valid for $hc/\lambda_{\rm c}kT \ll 1$, would be worse. The Rayleigh–Jeans law is most applicable at long wavelengths.

(f) If the brightness temperature of the star were higher, the quantity $hc/\lambda_{\rm c}kT$ would become smaller, and the approximation, which is valid for $hc/\lambda_{\rm c}kT \ll 1$, would be better. The Rayleigh–Jeans law is most applicable at high temperatures.

Exercise 7.1 The magnitude of the force of gravity due to the Sun acting on the Earth is $F_\odot = G M_\odot M_\oplus/(1^2\,\mathrm{AU}^2)$, while the magnitude of the force of gravity due to Jupiter acting on the Earth is $F_J = G M_J M_\oplus/((5-1)^2\,\mathrm{AU}^2)$ when they are at their closest separation. The ratio of these forces is

$$\frac{F_\odot}{F_J} = \frac{M_\odot \times 16}{10^{-3}\,M_\odot \times 1} = 1.6 \times 10^4.$$

So the gravitational force due to the Sun is about sixteen thousand times stronger than the gravitational force due to Jupiter, and this is of course independent of the mass of the Earth.

Exercise 7.2 Using Equation 7.22,

$$\begin{aligned}\tau_{\mathrm{circ}} &= \frac{2}{21}\frac{Q_P}{k_{\mathrm{dP}}}\left(\frac{a^3}{GM_*}\right)^{1/2}\frac{M_P}{M_*}\left(\frac{a}{R_P}\right)^5 \\ &= \frac{2}{21} \times 10^5 \times \left(\frac{(7.63 \times 10^9\,\mathrm{m})^3}{(6.673 \times 10^{-11}\,\mathrm{N\,m^2\,kg^{-2}}) \times (2.39 \times 10^{30}\,\mathrm{kg})}\right)^{1/2} \\ &\quad \times \frac{9.31 \times 10^{26}\,\mathrm{kg}}{2.39 \times 10^{30}\,\mathrm{kg}} \times \left(\frac{7.63 \times 10^9\,\mathrm{m}}{1.24 \times 10^8\,\mathrm{m}}\right)^5 \\ &= 1.73 \times 10^{14}\,\mathrm{s},\end{aligned}$$

which is around 5 million years.

Exercise 7.3 The equilibrium temperature of a planet is given by

$$T_{\mathrm{eq}} = \frac{1}{2}\left(\frac{(1-A)L_*}{\sigma\pi a^2}\right)^{1/4}. \qquad \text{(Eqn 5.5)}$$

Rearranging this to make a the subject,

$$a = \frac{1}{(2T_{\mathrm{eq}})^2}\left(\frac{(1-A)L_*}{\sigma\pi}\right)^{1/2}.$$

(a) So, for the A type star, the inner boundary of the habitable zone, where $T_{\mathrm{eq}} = 373\,\mathrm{K}$, is given by

$$\begin{aligned}a &= \frac{1}{(2 \times 373\,\mathrm{K})^2}\left(\frac{(1-0) \times 10 \times 3.83 \times 10^{26}\,\mathrm{J\,s^{-1}}}{5.671 \times 10^{-8}\,\mathrm{J\,m^{-2}\,K^{-4}\,s^{-1}} \times \pi}\right)^{1/2} \\ &= 2.635 \times 10^{11}\,\mathrm{m} \\ &= 1.76\,\mathrm{AU}.\end{aligned}$$

Similarly, the outer boundary of the habitable zone for the A type star, where $T_{\mathrm{eq}} = 273\,\mathrm{K}$, is given by

$$\begin{aligned}a &= \frac{1}{(2 \times 273\,\mathrm{K})^2}\left(\frac{(1-0) \times 10 \times 3.83 \times 10^{26}\,\mathrm{J\,s^{-1}}}{5.671 \times 10^{-8}\,\mathrm{J\,m^{-2}\,K^{-4}\,s^{-1}} \times \pi}\right)^{1/2} \\ &= 4.918 \times 10^{11}\,\mathrm{m} \\ &= 3.29\,\mathrm{AU}.\end{aligned}$$

So the width of the habitable zone of the A type star is estimated as about 1.5 AU.

(b) Now, since $a \propto L_*^{1/2}$, and the luminosity of the K type star is 100 times smaller than that of the A type star, the inner and outer limits of the habitable zone for the K type star will be 10 times smaller than those for the A type star (all else being equal). As a result, the inner edge is at about 0.18 AU and the outer edge is at about 0.33 AU. The width of the habitable zone of the K type star is therefore estimated as about 0.15 AU.

Exercise 7.4 An exoplanet might exist in a stable orbit around any one of the six component stars (e.g. Figure S7.1a or b). Alternatively, an exoplanet could have a stable orbit around any one of the three binary pairs that make up the system (e.g. Figure S7.1c or d), or around the quadruple system that constitutes the 'visual binary' (Figure S7.1e). Finally, one might even have an exoplanet in a stable orbit around the entire sextuple system (Figure S7.1f), although such a planet would likely be only loosely bound to the system as it would be necessarily very distant from the centre of mass.

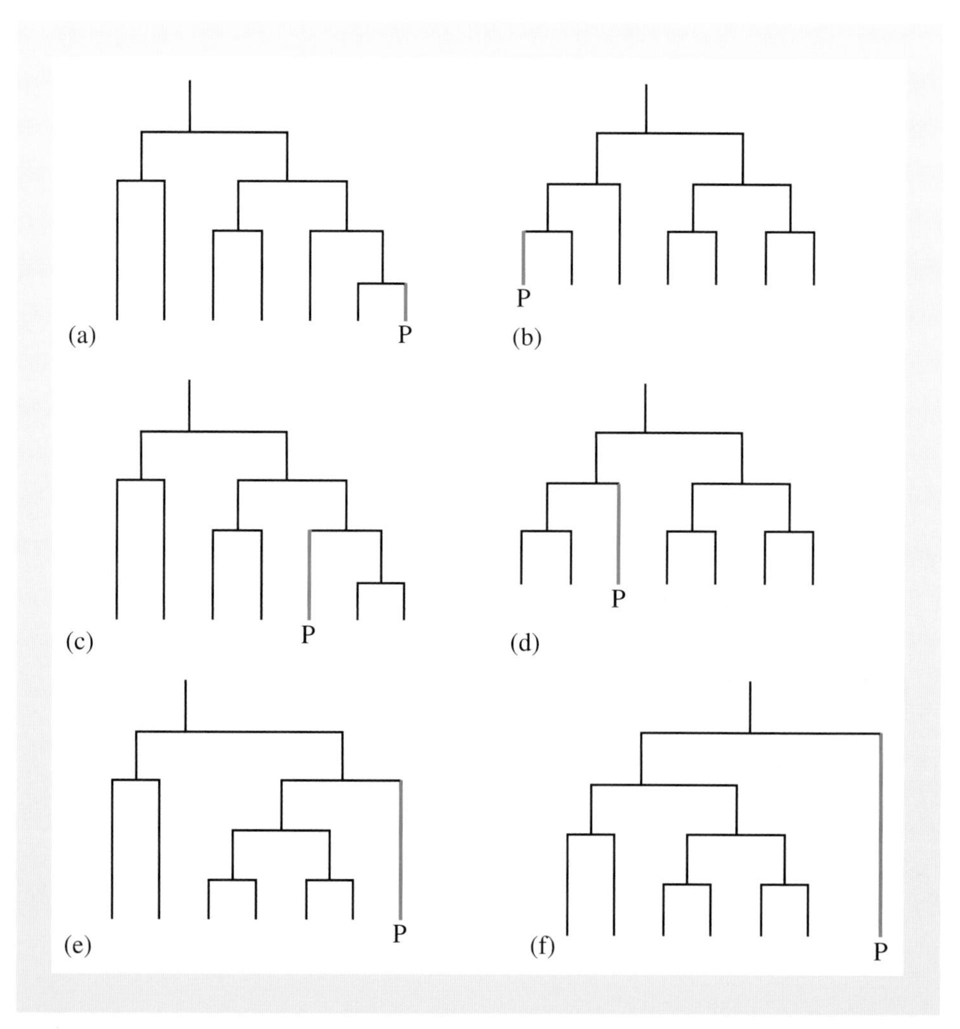

Figure S7.1 See solution to Exercise 7.4. 'P' indicates the possible location of a planet.

Exercise 7.5 (a) The maximum offset of the transit time occurs when the sine function has a value of ± 1. In this case, the offset is $|\delta t_2| \approx P_2 a_1 \mu_1 / 2\pi a_2$.

Now, the reduced mass is $\mu_1 \approx M_1/M_*$, which in this case is $\mu_1 \approx 3 \times 10^{-6}$. So

$$|\delta t_2| \approx \frac{30\text{ days} \times 0.072\text{ AU} \times 3 \times 10^{-6}}{2\pi \times 0.19\text{ AU}}.$$

The maximum offset is therefore 5.4×10^{-6} days or 0.47 seconds.

(b) If the mass of the inner planet were 10 times larger, the maximum offset in the transit time would also be 10 times larger, i.e. 4.7 seconds.

Exercise 7.6 The maximum transit timing offset will occur when the cosine term in the numerator of the right-hand side of Equation 7.28 is ± 1 and when the sine term in the denominator is 1. The timing offset is then

$$|\delta t_2| \approx \frac{P_2 \mu_1 a_1 \sqrt{1 - e_2^2}}{2\pi a_2 (1 - e_2)}.$$

(a) When $e_2 = 0.1$, this becomes

$$|\delta t_2| \approx \frac{30\text{ days} \times 3 \times 10^{-6} \times 0.072 \times \sqrt{0.99}}{2\pi \times 0.19 \times 0.9} \approx 6.0 \times 10^{-6}\text{ days} = 0.52\text{ s}.$$

(b) When $e_2 = 0.5$, this becomes

$$|\delta t_2| \approx \frac{30\text{ days} \times 3 \times 10^{-6} \times 0.072 \times \sqrt{0.75}}{2\pi \times 0.19 \times 0.5} \approx 9.4 \times 10^{-6}\text{ days} = 0.81\text{ s}.$$

Exercise 7.7 Using Equation 7.29, we first note that the reduced mass is $\mu_2 \approx M_2/M_* \approx 0.5 \times 10^{-3}$. The TTV is therefore

$$\begin{aligned}\delta t_1 &\approx (0.5 \times 10^{-3} \times 0.5 \times 100\text{ days}) \times \left(\frac{0.05}{0.42}\right)^3 \\ &\approx 4.2 \times 10^{-5}\text{ days} \\ &= 3.6\text{ seconds}.\end{aligned}$$

Exercise 7.8 (a) Using Equation 7.30, the maximum TTV is

$$\delta t_2 \approx \frac{12\text{ days}}{4.5 \times 3} \times \frac{3 \times 10^{-6}}{10^{-3}} \approx 2.7 \times 10^{-3}\text{ days} \approx 4\text{ minutes}.$$

(b) Using Equation 7.31, the libration period of this TTV is

$$P_{\text{lib}} \approx 0.5 \times 3^{-4/3} \times (10^{-3})^{-2/3} \times 12\text{ days} \approx 139\text{ days}.$$

So the maximum TTV would recur roughly every 11 or 12 transits.

Exercise 7.9 (a) Rearranging Equation 7.32, we have

$$\begin{aligned}a_{\text{M}} M_{\text{M}} &\approx \frac{2\pi a \times 2^{1/2} \times M_{\text{P}}\, \delta t_{\text{M}}}{P} \\ &\approx \frac{2 \times \pi \times 0.05\text{ AU} \times 2^{1/2} \times 10^{-3}\text{ M}_\odot \times 30\text{ s}}{4 \times 24 \times 3600\text{ s}} \\ &\approx 3.86 \times 10^{-8}\text{ AU M}_\odot.\end{aligned}$$

Rearranging Equation 7.33, we have

$$\begin{aligned} a_{\rm M}^{-1/2} M_{\rm M} &\approx \frac{2^{1/2}[M_{\rm P}(M_{\rm P}+M_*)]^{1/2}\,\delta T_{\rm M}}{T_{\rm dur}\,a^{1/2}} \\ &\approx \frac{2^{1/2}\times[10^{-3}\times(10^{-3}+1)]^{1/2}\,{\rm M}_\odot\times 15\,{\rm s}}{3\times 3600\,{\rm s}\times 0.05^{1/2}\,{\rm AU}^{1/2}} \\ &\approx 2.78\times 10^{-4}\,{\rm AU}^{-1/2}\,{\rm M}_\odot. \end{aligned}$$

Now, dividing the first of these by the second, we get

$$a_{\rm M}^{3/2} \approx 1.39\times 10^{-4}\,{\rm AU}^{3/2},$$

so

$$a_{\rm M} \approx 2.68\times 10^{-3}\,{\rm AU} = 4.0\times 10^{8}\,{\rm m}$$

(which is around four hundred thousand kilometres).

So, if the TTV and the transit duration variation are due to the presence of an exomoon, they imply that this moon must be orbiting fairly close to the planet. (The radius of Jupiter is about 7×10^7 m, and the radius of a hot Jupiter exoplanet may well be somewhat larger.)

Substituting this orbital radius back into the first equation,

$$M_{\rm M} \approx \frac{3.86\times 10^{-8}}{2.68\times 10^{-3}}\,{\rm M}_\odot \approx 1.44\times 10^{-5}\,{\rm M}_\odot,$$

which is about five times the mass of the Earth.

(b) Kepler's third law applied to the orbit of the putative exomoon around the exoplanet may be written as

$$\begin{aligned} P_{\rm M} &= \left(\frac{4\pi^2 a_{\rm M}^3}{G(M_{\rm P}+M_{\rm M})}\right)^{1/2} \\ &\approx \left(\frac{4\pi^2\times(4.0\times 10^8\,{\rm m})^3}{6.67\times 10^{-11}\,{\rm N\,m^2\,kg^{-2}}\times(10^{-3}+1.4\times 10^{-5})\times 1.99\times 10^{30}\,{\rm kg}}\right)^{1/2} \\ &\approx 1.37\times 10^{5}\,{\rm s}. \end{aligned}$$

The orbital period of the exomoon would therefore be about 38 hours.

Exercise 7.10 Using Equation 7.32,

$$\begin{aligned} \delta t_{\rm M} &\approx \frac{a_{\rm M} M_{\rm M}}{2^{1/2} M_{\rm P}}\,\frac{P}{2\pi a} \\ &\approx \frac{10^9\,{\rm m}\times 3\times 10^{-6}}{2^{1/2}\times 10^{-3}}\times\frac{46\times 24\times 3600\,{\rm s}}{2\pi\times 0.2\times 1.5\times 10^{11}\,{\rm m}} \\ &\approx 45\,{\rm s}. \end{aligned}$$

The maximum TTV is therefore less than 1 minute.

Using Equation 7.33,

$$\begin{aligned} \delta T_{\rm M} &\approx \left(\frac{a}{a_{\rm M}}\right)^{1/2}\frac{M_{\rm M}}{[M_{\rm P}(M_{\rm P}+M_*)]^{1/2}}\,\frac{T_{\rm dur}}{2^{1/2}} \\ &\approx \left(\frac{0.2\times 1.5\times 10^{11}\,{\rm m}}{10^9\,{\rm m}}\right)^{1/2}\times\frac{3\times 10^{-6}}{[10^{-3}\times(10^{-3}+0.5)]^{1/2}}\times\frac{5\times 3600\,{\rm s}}{2^{1/2}} \\ &\approx 9.3\,{\rm s}. \end{aligned}$$

The maximum transit duration variation is therefore of order 10 seconds.

Exercise 7.11 The Hill radius for the planet is given by Equation 7.24 as

$$\begin{aligned} R_H &= a\left(\frac{M_P}{3M_*}\right)^{1/3} \\ &= 0.05\ \mathrm{AU} \times \left(\frac{10^{-3}\ \mathrm{M_\odot}}{3\ \mathrm{M_\odot}}\right)^{1/3} \\ &= 0.0035\ \mathrm{AU}. \end{aligned}$$

Hence the Hill radius is about 5.2×10^8 m or just over $7\ \mathrm{R_J}$.

Exercise 7.12 From Equation 7.16, the Roche limit is

$$\begin{aligned} d_R &= R_M\left(\frac{2M_P}{M_M}\right)^{1/3} \\ &= 6.4 \times 10^6\ \mathrm{m} \times \left(\frac{2 \times 10^{-3}}{3 \times 10^{-6}}\right)^{1/3} \\ &= 5.6 \times 10^7\ \mathrm{m}. \end{aligned}$$

The Roche limit for this planet and moon is therefore about 56 thousand kilometres.

Exercise 7.13 We first note that the reduced mass is

$$\mu_T \approx \frac{M_T}{M_P} \approx \frac{3 \times 10^{-6}\ \mathrm{M_\odot}}{10^{-3}\ \mathrm{M_\odot}} \approx 3 \times 10^{-3}.$$

The maximum TTV is given by Equation 7.34 as

$$\delta t \approx \frac{3 \times 10^{-3} \times 4\ \mathrm{days} \times 25^\circ}{360^\circ} \approx 8.3 \times 10^{-4}\ \mathrm{days} = 72\ \mathrm{seconds}.$$

Exercise 8.1 We need to apply

$$\text{geometric transit probability} = \frac{R_* + R_P}{a} \approx \frac{R_*}{a} \qquad \text{(Eqn 1.21)}$$

and substitute in the appropriate solar and terrestrial values. Thus

$$\begin{aligned} \text{geometric transit probability} &\approx \frac{\mathrm{R_\odot}}{1\ \mathrm{AU}} \\ &\approx \frac{6.96 \times 10^8\ \mathrm{m}}{1.50 \times 10^{11}\ \mathrm{m}} \\ &\approx 0.0047. \end{aligned}$$

We are told to assume that Kepler will observe 5000 such stars, so the number of transiting exo-Earths detected will be

$$\begin{aligned} \text{total exo-Earths detected} &= \text{geometric transit probability} \times 5000 \\ &= 0.0047 \times 5000 = 23. \end{aligned}$$

Kepler is predicted to find just over 23 exo-Earths, with the (optimistic) assumption that each star hosts a single planet in an orbit like Earth's. The actual anticipated number is less than this, by a factor (of perhaps roughly 10 or so) that depends on the planeticity and the value of $\alpha(a, M_P)$ over the range within which planets would be classed as exo-Earths (cf. Subsection 4.2.2).

Exercise 8.2 (a) The density is simply the mass divided by the volume, so

$$\begin{aligned}\rho &= \frac{3\,M_P}{4\pi\,R_P^3}\\ &= \frac{3}{4\pi}\,\frac{6.55\,M_\oplus}{(2.68\,R_\oplus)^3}.\end{aligned} \qquad \text{(S8.1)}$$

This gives

$$\begin{aligned}\rho &= \frac{0.75 \times 6.55 \times 5.97 \times 10^{24}\,\text{kg}}{\pi \times (2.68 \times 6.37 \times 10^6\,\text{m})^3}\\ &= 1876\,\text{kg m}^{-3}.\end{aligned}$$

In terms of the density of Earth, Equation S8.1 tells us that

$$\rho = \frac{6.55}{19.25}\rho_\oplus = 0.340\rho_\oplus.$$

Thus we see that GJ 1214 b is just over one-third as dense as Earth.

(b) The magnitude of the acceleration due to gravity on the surface of a planet is

$$g_P = \frac{GM_P}{R_P^2}, \qquad \text{(Eqn 4.48)}$$

so on the surface of GJ 1214 b we have

$$\begin{aligned}g_P &= \frac{6.673 \times 10^{-11}\,\text{N m}^2\,\text{kg}^{-2} \times 6.55 \times 5.97 \times 10^{24}\,\text{kg}}{(2.68 \times 6.37 \times 10^6\,\text{m})^2}\\ &= 8.95\,\text{m s}^{-2}.\end{aligned}$$

This is very similar to the value of g on Earth, which is $9.81\,\text{m s}^{-2}$.

(c) Humans would be rather comfortable on a planet with this value of gravity. Golf would be perfectly feasible, and it would appear to be possible to hit a ball somewhat farther than on Earth (assuming comparable values of air resistance) because the downwards acceleration during flight would be slightly less than on Earth, so the ball would travel further before landing.

Exercise 8.3 (a) The appropriate equation to use is

$$\frac{A_S}{A_{RV}} \approx 4.7 \left(\frac{V_S \sin i_S}{5\,\text{km s}^{-1}}\right)\left(\frac{R_P}{R_\oplus}\right)^2\left(\frac{M_P \sin i}{M_\oplus}\right)^{-1}\left(\frac{P}{\text{yr}}\right)^{1/3}\left(\frac{\rho_*}{\rho_\odot}\right)^{2/3}, \qquad \text{(Eqn 5.31)}$$

and everything to the right of the first term in parentheses is 1, because each of the terms is normalized to the Earth or the Sun. We are told that the Sun's spin period is 25 days, so we have

$$\begin{aligned}V_S &= \frac{2\pi\,R_\odot}{P_{\text{rot}}}\\ &= \frac{2\pi \times 6.96 \times 10^8\,\text{m}}{25 \times 24 \times 3600\,\text{s}}\\ &= 2.0 \times 10^3\,\text{m s}^{-1} \quad \text{(to 2 s.f.)}.\end{aligned}$$

The only thing left to evaluate is $\sin i_S$. We are viewing so that the Earth transits with an impact parameter of $b = 0$, so we know $i = 90°$. The ecliptic is the plane of the Earth's orbit, so $i_S = (90 - 7.15)°$. Putting these values into Equation 5.31, we have

$$\begin{aligned}\frac{A_S}{A_{RV}} &\approx 4.7\left(\frac{2.0 \times 10^3\,\mathrm{m\,s^{-1}}\sin 82.85°}{5\,\mathrm{km\,s^{-1}}}\right)\\ &\approx 1.9 \quad \text{(to 2 s.f.)}.\end{aligned}$$

The amplitude of the Rossiter–McLaughlin effect is almost twice the reflex radial velocity amplitude.

(b) With 8 planets, the Sun's reflex radial velocity curve will be a superposition of the effects of all of them. Since the Earth is a low-mass component of the Solar System, it will be difficult to unambiguously isolate the Sun's reflex radial velocity due to the Earth. Measurements would need to be made for longer than the orbital period of Jupiter. The Rossiter–McLaughlin effect could be measured in a few hours, and would verify that a terrestrial-planet-sized body was transiting the host star. The reflex radial velocity amplitude allows the mass of the planet to be deduced; without a measurement of this amplitude, the mass of the planet would remain unknown.

Acknowledgements

Grateful acknowledgement is made to the following sources:

Figures

Cover image: An artist's impression of the 'hot Jupiter' exoplanet HD 189733 b transiting across the face of its star (ESA, C. Carreau);

Figure 1: © Lynette R. Cook;

Figure 1.1: NASA Ames Research Center (NASA-ARC); Figures 1.2 and 1.3: Richard Powell, www.atlasoftheuniverse.com, used under a Creative Commons Attribution-ShareAlike 2.5 License; Figure 1.4; generated by the author from RECONs data; Figure 1.5a: adapted from a European Space Agency image; Figure 1.5b: The International Astronomical Union/Martin Kornmesser; Figure 1.5c: Lunar and Planetary Institute/NASA; Figures 1.6a & b and 1.7: NASA Jet Propulsion Laboratory (NASA-JPL); Figure 1.6c: NASA/Goddard Space Center; Figures 1.8 & 8.12: adapted from Kaltenegger, L. (2008) 'Assembling the Puzzle — Evolution of an Earth-Like Planet', Harvard–Smithsonian Center for Astrophysics; Figure 1.9: ESA; Figure 1.10: NASA/ESA, P. Kalas & J. Graham (UC Berkeley), M. Clampin (NASA/GSFC); Figure 1.11: Marois, C. et al. (2008) 'Direct imaging of multiple planets orbiting the star HR8799', *Science*, **322**, 1348, American Association for the Advancement of Science; Figure 1.14: adapted from Beaugé, C., Ferraz-Mello, S. and Michtchenko, T. A. in *Extrasolar Planets: Formation, Detection and Dynamics*, Dvorak (ed.), Wiley, (2008); Figure 1.15: adapted from Bailey, J. et al. (2009) 'A Jupiter-like planet orbiting the nearby M dwarf GJ 832', *The Astrophysical Journal*, **690**, 743, The American Astronomical Society; Figures 1.16 & 1.17: adapted from Mayor, M. et al. (2004) 'The CORALIE survey for southern-solar planets', *Astronomy and Astrophysics*, **415**, 391, EDP Sciences; Figure 1.18: Photos by David Cortner; Figures 1.19 & 1.20: adapted from Sackett, P. D. (1998) 'Searching for unseen planets via occultation and microlensing', Mariotti, J. and Alloin, D. M. (eds.), 'Planets outside the solar system: theory and observations' (NATO Science series C). Springer; Figure 1.21: Atlas Image obtained as part of the Two Micron All Sky Survey (2MASS), a joint project of the University of Massachusetts and the Infrared Processing and Analysis Center/CalTech, funded by the National Aeronautics and Space Administration and the National Science Foundation, annotations due to K. D. Horne and C. A. Haswell;

Figure 2.1: © University Corporation for Atmospheric Research; Figure 2.2a: Observatoire de Haute Provence; Figure 2.2b: Gdgourou, used under a Creative Commons Attribution-ShareAlike 3.0 Licence; Figure 2.3: adapted from Mazeh, T. et al. (2000), 'The Spectroscopic Orbit of the Planetary Companion transiting HD209458', *The Astrophysical Journal*, **532**, 55, The American Astronomical Society; Figure 2.4: adapted from Charbonneau, D., et al. (2000), 'Detection of planetary transits across a sun-like star', *The Astrophysical Journal*, **529**, 45, The American Astronomical Society; Figure 2.6: © François du Toit; Figure 2.7 adapted from Charbonneau, D., Brown, T. M., Latham, D. W. and Mayor, M. (2000) *The Astrophysical Journal Letters*, **529**, L45; Figure 2.9(a): NASA Jet Propulsion Laboratory (NASA-JPL); Figures 2.10 & 2.11: The SuperWASP Consortium; Figure 2.12(a) adapted from SuperWASP archive by A. J. Norton; Figures 2.12b & 2.20: adapted from Pollacco, D. L., et al. (2006), 'The WASP

Project and the SuperWASP Cameras', *Publications of the Astronomical Society of the Pacific*, **118**, 1407, Astronomical Society of the Pacific; Figure 2.14: adapted from SuperWASP archive by the author; Figure 2.15 adapted from a WASP consortium powerpoint; Figure 2.16: adapted from Collier-Cameron, A., et al. (2007), 'Efficient identification of exoplanetary transit candidates from SuperWASP light curves', *Monthly Notices of the Royal Astronomical Society*, **380**, 1230, The Royal Astronomical Society; Figure 2.17: produced using software written by Rob Hynes; Figure 2.18: adapted from Exoplanet Encyclopedia data by A. J. Norton, A. Dackombe and the author; Figure 2.19: Parley, N. R., (2008) *Serendipitous Asteroid Survey Using SuperWASP*, PhD thesis, The Open University;

Figures 3.1, 3.5 & 3.17: adapted from Brown, T. M. et al. (2001), 'Hubble Space Telescope Time-Series Photometry of the Transiting Planet of HD209458', *The Astrophysical Journal*, **552**, 699, The American Astronomical Society; Figure 3.3: adapted from Sackett, P. D. arXiv:astro-ph/9811269; Figure 3.6: SOHO/ESA/NASA; Figure 3.9a: adapted from Van Hamme, W. (1993), 'New Limb-Darkening Coefficients for modelling Binary Star Light Curves', *The Astronomical Journal*, **106**, Number 5, 2096, The American Astronomical Society; Figure 3.9b: adapted from Morrill J. S. and Korendyke C. M., (2008), 'High-Resolution Center-to-limb variation of the quiet solar spectrum near MgII', *The Astrophysical Journal*, **687**, 646, The American Astronomical Society; Figures 3.10 & 3.11: adapted from Knutson H. A. et al. (2007), 'Using Stellar Limb-Darkening to Refine the Properties of HD 209458 b', *The Astrophysical Journal*, **655**, 564, The American Astronomical Society; Figures 3.16 & 3.18: adapted from Charbonneau, D. et al. (2000), 'Detection of Planetary Transits across a Sun-like Star', *The Astrophysical Journal*, **529**, 45, The American Astronomical Society;

Figures 4.1 & 4.10: adapted from Udry, S. (2008) Extrasolar planets: XVI Canary Islands Winter School of Astrophysics, CUP; Figure 4.2: CNES; Figures 4.3, 4.4, 4.5, 4.6, 4.7 & 4.9: adapted from Exoplanet Encyclopedia data by A. J. Norton, A. Dackombe and the author; Figures 4.8b & c: adapted from Johnson, J. A. (2009), Draft version of submission *International Year of Astronomy Invited Review on Exoplanets*, © John Asher Johnson; Figure 4.11: adapted from Leger, A. et al. (2009), 'Transiting exoplanets from the CoRoT space mission: VIII. CoRoT-7b: the first super-Earth with measured radius', *Astronomy and Astrophysics*, **506**, 287, ESO; Figures 4.12, 4.13 & 4.15: adapted from Guillot, T. (2005) 'The Interiors of Giant Planets: Models and Outstanding Questions', *Annual Review of Earth and Planetary Sciences*, **33**, 493, © Annual Reviews; Figures 4.14 & 4.17: adapted from Liu, X., Burrows, A. and Igbui, L. (2008) 'Theoretical Radii of Extrasolar Giant Planets: the cases of TrES-4, XO-3b and HAT-P-1b', *The Astrophysical Journal*, **687**, 1191, The American Astronomical Society; Figure 4.16: adapted from Arras, P. and Bildsten, L., (2006), 'Thermal Structure and Radius Evolution of Irradiated Gas Giant Planets', *The Astrophysical Journal*, **650**, 394, The American Astronomical Society; Figure 4.19: adapted from Sasselov, D. (2008) 'Astronomy: Extrasolar Planets', *Nature*, **451**, 29; Figure 4.20: adapted from Charbonneau, D. et al. (2006), 'When Extrasolar Planets Transit Their Parent Stars' review chapter in *Protostars and Planets V* (Space Science Series), Jewitt, D. and Reipurth, B. (eds.) University of Arizona Press; Figure 4.21: adapted from Southworth, J. (2008), 'Homogeneous studies of transiting extrasolar planets - I. Light-curve analyses', Monthly Notices of the Royal Astronomical Society, **386**, 1644, The Royal Astronomical Society;

Figure 5.1: adapted from a drawing of Frances Bagenal, University of Colorado; Figure 5.2: adapted from Brown, T. M. (2001) 'Transmission spectra as diagnostics of extrasolar giant planet atmospheres', *The Astrophysical Journal*, **553**, 1006, The American Astronomical Society; Figure 5.4: adapted from Marley, M. S. (2009) 'The atmospheres of extrasolar planets', *Physics and Astrophysics of Planetary Systems*, EDP Sciences; Figure 5.5a: © 2009 Astrobio.net; Figures 5.5b & c and 5.6: adapted from Charbonneau, D. et al. (2002) 'Detection of an extrasolar planet atmosphere', *The Astrophysical Journal*, **568**, 377, The American Astronomical Society; Figure 5.7: adapted from Vidal-Madjar, A. et al. (2003) 'An extended upper atmosphere around the extrasolar planet HD209458b', *Nature*, **422**, 143, Nature Publishing Group; Figure 5.8: adapted from Charbonneau, D. (2003) 'Atmosphere out of that world', *Nature*, **422**, 124, Nature Publishing Group; Figures 5.9 & 5.10a: adapted from Vidal-Madjar, A. et al. (2004) 'Detection of oxygen and carbon in the hydrodynamically escaping atmosphere of the extrasolar planet HD 209458 B', *The Astrophysical Journal*, **604**, 69, The American Astronomical Society; Figure 5.10b: NASA/MSSTA; Figure 5.11: adapted from Sing, D. K. et al. (2009) 'Transit spectrophotometry of the exoplanet HD189733 b. I. Searching for water but finding haze with HST NICMOS', *Astronomy & Astrophysics*, **505**, 819, European Southern Observatory; Figure 5.12: adapted from Désert, J. et al. (2009) 'Search for carbon monoxide in the atmosphere of the transiting exoplanet HD 189733b', *The Astrophysical Journal*, **699**, 478, The American Astronomical Society; Figure 5.13: adapted from Redfield, S. et al. (2008) 'Sodium Absorption from the exoplanetary atmosphere of HD 189733 B detected in the optical transmission spectrum', *The Astrophysical Journal*, **673**, 87, The American Astronomical Society; Figure 5.14: adapted from Scott Gaudi, B. and Winn, J. N. (2007) 'Prospects for the characterization and confirmation of transiting exoplanets via the Rossiter–McLaughlin effect', *The Astrophysical Journal*, **655**, 550, American Astronomical Society; Figures 5.15 & 5.16: adapted from Winn, J. N. (2007) 'Exoplanets and the Rossiter–McLaughlin Effect', *Transiting Extrasolar Planets Workshop*, ASP Conference Series, **336**, 170, Astronomical Society of the Pacific; Figures 5.19a & b: adapted from Triaud, A. H. M. J. et al. (2009) 'The Rossiter–McLaughlin Effect of CoRoT-3b & HD 189733b', *Astronomy and Astrophysics*, **506**, 377, European Space Operations Centre ESOC; Figure 5.19c: adapted from Bouchy, F. et al. (2008) 'Transiting exoplanets from the CoRoT space mission III. The spectroscopic transit of CoRoT-Exo-2b', *Astronomy and Astrophysics*, **482**, 25, European Space Operations Centre; Figures 5.19d & e: adapted from Narita, N. et al. (2009) 'Improved Measurement of the Rossiter–McLaughlin effect in the Exoplanetary System HD 17156', *Publications of the Astronomical Society of Japan*, **61**, 991, Astronomical Society of Japan; Figure 5.19f: adapted from Hébrard, G. et al. (2009) 'Misaligned spin-orbit in the XO-3 planetary system?', *Astronomy and Astrophysics*, **488**, 763, European Southern Observatory; Figure 5.20: adapted from Fabrycky, D. C. and Winn, J. N. (2009) 'Exoplanetary Spin-Orbit Alignment: results from the ensemble of Rossiter–McLaughlin observations', *The Astrophysical Journal*, **696**, 1230, The American Astronomical Society;

Figure 6.1 adapted from Seager, S., Deming, D. and Valenti, J. A. (2008) in *Astrophysics in the Next Decade: The James Webb Space Telescope and Concurrent Facilities*, Thronson, H. A., Stiavelli, M., Tielens, A. (eds.); Figure 6.3: NASA/JPL-Caltech/R. Hurt (SSC); Figure 6.4: adapted from NASA/JPL-Caltech/D. Charbonneau (Harvard–Smithsonian CfA) and NASA/JPL-Caltech/D. Deming (Goddard Space Flight Center); Figure 6.5: adapted from Deming, D. et al. (2006) 'Strong infrared emission from the extrasolar planet HD 189733b', *The Astrophysical Journal*, **644**, 560, The American Astronomical Society; Figure 6.6: adapted from Borucki, W. J. et al. (2009) 'Kepler's optical phase curve of the exoplanet HAT-P-7b', *Science*, **325**, 709, American Association for the Advancement of Science; Figure 6.7: adapted from Charbonneau, D. et al. (2008) 'The broadband infrared emission spectrum of the exoplanet HD 189733b', *The Astrophysical Journal*, **686**, 1341, American Association for the Advancement of Science; Figure 6.8a: adapted from Showman, A. P. et al. (2009) 'Atmospheric circulation of hot jupiters: coupled radiative-dynamical general circulation model simulations of HD189733b and HD209458b', *The Astrophysical Journal*, **699**, 564, The American Astronomical Society; Figure 6.8b: adapted from Knutson, H. A. et al. (2008) 'The 3.6–8.0 μm Broadband Emission Spectrum of HD 209458 b: Evidence for an Atmospheric Temperature Inversion' *The Astrophysical Journal*, **673**, 526, IOP Publishing; Figure 6.8c: adapted from Machalek, P. et al. (2008) 'Thermal emission of exoplanet XO-1b', *The Astrophysical Journal*, **684**, 1427, The American Astronomical Society; Figure 6.8d: adapted from Knutson, H. A. et al. (2009) 'Detection of a temperature inversion in the broadband infrared emission spectrum of TrES-4', *The Astrophysical Journal*, **691**, 866, The American Astronomical Society; Figures 6.8e & 6.9: adapted from Machalek, P. et al. (2008) 'Detection of thermal emission of XO-2b: evidence for a weak temperature inversion', *The Astrophysical Journal*, **701**, 514, The American Astronomical Society; Figure 6.10: adapted from Deming, D. (2008) 'Emergent exoplanet flux: review of the Spitzer results', *Transiting Planets — Proceedings of the IAU Symposium*, **253**, 197, International Astronomical Union; Figures 6.11 & 6.12: adapted from Harrington, J. et al. (2006) 'The phase-dependent infrared brightness of the extrasolar planet υ Andromedae b', *Science*, **314**, 623, American Association for the Advancement of Science; Figures 6.13 & 6.14: adapted from Knutson, H. A. et al. (2007) 'A map of the day-night contrast of the extrasolar planet HD 189733b', *Nature*, **447**, 183, Nature Publishing Group;

Figure 7.1: adapted by A. J. Norton from Exoplanet Encyclopaedia; Figure 7.3: adapted from Jones, B. W., Sleep, P. N. and Underwood, D. R. (2006) 'Which exoplanetary systems could harbour habitable planets?', *International Journal of Astrobiology*, **5**(3), 251, CUP; Figure 7.5: NASA/JPL-Caltech/T. Pyle (SSC); Figure 7.6: adapted from Agol, E. et al. (2005) 'On detecting terrestrial planets with timing of giant planet transits', *Monthly Notices of the Royal Astronomical Society*, **359**, 567, The Royal Astronomical Society;

Figure 8.2: adapted from ESA/SRE (2009) PLATO-Next Generation Planet Finder-Assessment Study Report 4, December 2009. ESA; Figure 8.3: MEarth Project; Figure 8.4: adapted from Palle E. et al. (2009) 'Earth's Transmission spectrum from lunar eclipse observations', *Nature*, **459**, 814, Nature Publishing Group; Figures 8.5 & 8.6: adapted from Miller-Ricci, E. and Fortney, J. J. (2010) 'The Nature of the Atmosphere of the Transiting Super-Earth GJ 1214b', *The*

Astrophysical Journal Letters, **716**, L74, The American Astronomical Society; Figure 8.7: adapted from Seager, S., Deming, D. and Valenti, J. A. (2008) 'Transiting exoplanets with JWST', Thronson, H. A., Tielens, A., Stiavelli, M., (eds.) *Astrophysics in the Next Decade: JWST and Concurrent Facilities*, 123, Astrophysics & Space Science Library. Springer; Figure 8.8: Lahav, N. (1999) *Biogenesis*, courtesy of Noam Lahav; Figure 8.10: Woods Hole Oceanographic Institute, Deep Submergence Operations Group, Dan Fornari.

Every effort has been made to contact copyright holders. If any have been inadvertently overlooked the publishers will be pleased to make the necessary arrangements at the first opportunity.

Index

Items that appear in the Glossary have page numbers in **bold type**. Ordinary index items have page numbers in Roman type.